Statistical Data Analysis Explained

Statistical Data Analysis Explained

Applied Environmental Statistics with R

Clemens Reimann
Geological Survey of Norway

Peter Filzmoser
Vienna University of Technology

Robert G. Garrett
Geological Survey of Canada

Rudolf Dutter
Vienna University of Technology

John Wiley & Sons, Ltd

FSC
Mixed Sources
Product group from well-managed
forests and other controlled sources
Cert no. SGS-COC-2953
www.fsc.org
© 1996 Forest Stewardship Council

Contents

Preface

Although several books already exist on statistical data analysis in the natural sciences, there are few books written at a level that a non-statistician will easily understand. In our experience many colleagues in earth and environmental sciences are not sufficiently trained in mathematics or statistics to easily comprehend the necessary formalism. This is a book written in colloquial language, avoiding mathematical formulae as much as possible (some may argue too much) trying to explain the methods using examples and graphics instead. To use the book efficiently, readers should have some computer experience and some basic understanding of statistical methods. We start with the simplest of statistical concepts and carry readers forward to a deeper and more extensive understanding of the use of statistics in the natural sciences. Importantly, users of the book, rather than readers, will require a sound knowledge of their own branch of natural science.

In the book we try to demonstrate, based on practical examples, how data analysis in environmental sciences should be approached, outline advantages and disadvantages of methods and show and discuss the do's and don'ts. We do not use "simple toy examples" to demonstrate how well certain statistical techniques function. The book rather uses a single, large, real world example data set, which is investigated in more and more depth throughout the book. We feel that this makes it an interesting read from beginning to end, without preventing the use of single chapters as a reference for certain statistical techniques. This approach also clearly demonstrates the limits of classical statistical data analysis with environmental (geochemical) data. The special properties of environmental data (e.g., spatial dependencies, outliers, skewed distributions, closure) do not agree well with the assumptions of "classical" (Gaussian) statistics. These are, however, the statistical methods taught in all basic statistics courses at universities because they are the most fundamental statistical methods. As a consequence, up to this day, techniques that are far from ideal for the data at hand are widely applied by earth and environmental scientists in data analysis. Applied earth science data call for the use of robust and non-parametric statistical methods. These techniques are extensively used and demonstrated in the book. The focus of the book is on the exploratory use of statistical methods extensively applying graphical data analysis techniques.

The book concerns the application of statistical and other computer methods to the management, analysis and display of spatial data. These data are characterised by including locations (geographic coordinates), which leads to the necessity of using maps to display the data and the results of the statistical methods. Although the book uses examples from applied geochemistry, the principles and ideas equally well apply to other natural sciences, e.g., environmental sciences, pedology, hydrology, geography, forestry, ecology, and health sciences/epidemiology. That is, to anybody using spatially dependent data. The book will be useful to postgraduate students, possibly final year students with dissertation projects, students and others interested in the application of modern statistical methods (and not so much in theory), and natural scientists and other applied statistical professionals. The book can be used as a textbook, full of practical examples or in a basic university course on exploratory data analysis for spatial data. The book can also serve as a manual to many statistical methods and will help the reader to better understand how different methods can be applied to their data – and what should not be done with the data.

The book is unique because it supplies direct access to software solutions (based on R, the Open Source version of the S-language for statistics) for applied environmental statistics. For all graphics and tables presented in the book, the R-codes are provided in the form of executable R-scripts. In addition, a graphical user interface for R, called DAS+R, was developed by the last author for convenient, fast and interactive data analysis. Providing powerful software for the combination of statistical data analysis and mapping is one of the highlights of the software tools. This software may be used with the example data as a teaching/learning tool, or with the reader's own data for research.

Clemens Reimann   **Peter Filzmoser**   **Robert G. Garrett**   **Rudolf Dutter**
Geochemist   Statistician   Geochemist   Statistician

Trondheim, Vienna, Ottawa
September 1, 2007.

Acknowledgements

This book is the result of a fruitful cooperation between statisticians and geochemists that has spanned many years. We thank our institutions (the Geological Surveys of Norway (NGU) and Canada (GSC) and Vienna University of Technology (VUT)) for providing us with the time and opportunity to write the book. The Department for International Relations of VUT and NGU supported some meetings of the authors.

We thank the Wiley staff for their very professional support and discussions.

Toril Haugland and Herbert Weilguni were our test readers, they critically read the whole manuscript, made many corrections and valuable comments.

Many external reviewers read single chapters of the book and suggested important changes.

The software accompanying the book was developed with the help of many VUT students, including Andreas Alfons, Moritz Gschwandner, Alexander Juschitz, Alexander Kowarik, Johannes Löffler, Martin Riedler, Michael Schauerhuber, Stefan Schnabl, Christian Schwind, Barbara Steiger, Stefan Wohlmuth and Andreas Zainzinger, together with the authors.

Friedrich Leisch of the R core team and John Fox and Matthias Templ were always available for help with R and good advice concerning R-commander.

Friedrich Koller supplied lodging, many meals and stimulating discussions for Clemens Reimann when working in Vienna. Similarly, the Filzmoser family generously hosted Robert G. Garrett during a working visit to Austria.

NGU allowed us to use the Kola Project data; the whole Kola Project team is thanked for many important discussions about the interpretation of the results through many years.

Arne Bjørlykke, Morten Smelror, and Rolf Tore Ottesen wholeheartedly backed the project over several years.

Heidrun Filzmoser is thanked for translating the manuscript from Word into Latex. The families of the authors are thanked for their continued support, patience with us and understanding.

Many others that are not named above contributed to the outcome, we wish to express our gratitude to all of them.

About the authors

Clemens REIMANN

Clemens Reimann (born 1952) holds an M.Sc. in Mineralogy and Petrology from the University of Hamburg (Germany), a Ph.D. in Geosciences from Leoben Mining University, Austria, and a D.Sc. in Applied Geochemistry from the same university. He has worked as a lecturer in Mineralogy and Petrology and Environmental Sciences at Leoben Mining University, as an exploration geochemist in eastern Canada, in contract research in environmental sciences in Austria and managed the laboratory of an Austrian cement company before joining the Geological Survey of Norway in 1991 as a senior geochemist. From March to October 2004 he was director and professor at the German Federal Environment Agency (Umweltbundesamt, UBA), responsible for the Division II, Environmental Health and Protection of Ecosystems. At present he is chairman of the EuroGeoSurveys geochemistry expert group, acting vice president of the International Association of GeoChemistry (IAGC), and associate editor of both Applied Geochemistry and Geochemistry: Exploration, Environment, Analysis.

Peter FILZMOSER

Peter Filzmoser (born 1968) studied Applied Mathematics at the Vienna University of Technology, Austria, where he also wrote his doctoral thesis and habilitation devoted to the field of multivariate statistics. His research led him to the area of robust statistics, resulting in many international collaborations and various scientific papers in this area. His interest in applications of robust methods resulted in the development of R software packages. He was and is involved in the organisation of several scientific events devoted to robust statistics. Since 2001 he has been dozent at the Statistics Department at Vienna University of Technology. He was visiting professor at the Universities of Vienna, Toulouse and Minsk.

Robert G. GARRETT

Bob Garrett studied Mining Geology and Applied Geochemistry at Imperial College, London, and joined the Geological Survey of Canada (GSC) in 1967 following post-doctoral studies at Northwestern University, Evanston. For the next 25 years his activities focussed on regional geochemical mapping in Canada, and overseas for the Canadian International Development Agency, to support mineral exploration and resource appraisal. Throughout his work there has been a use of computers and statistics to manage data, assess their quality, and maximise the knowledge extracted from them. In the 1990s he commenced collaborations with soil and agricultural scientists in Canada and the US concerning trace elements in crops. Since then he has been involved in various Canadian Federal and university-based research initiatives aimed at providing sound science to support Canadian regulatory and international policy activities concerning risk assessments and risk management for metals. He retired in March 2005 but remains active as an Emeritus Scientist.

Rudolf DUTTER

Rudolf Dutter is senior statistician and full professor at Vienna University of Technology, Austria. He studied Applied Mathematics in Vienna (M.Sc.) and Statistics at Université de Montréal, Canada (Ph.D.). He spent three years as a post-doctoral fellow at ETH, Zurich, working on computational robust statistics. Research and teaching activities followed at the Graz University of Technology, and as a full professor of statistics at Vienna University of Technology, both in Austria. He also taught and consulted at Leoben Mining University, Austria; currently he consults in many fields of applied statistics with main interests in computational and robust statistics, development of statistical software, and geostatistics. He is author and coauthor of many publications and several books, e.g., an early booklet in German on geostatistics.

1

Introduction

Statistical data analysis is about studying data – graphically or via more formal methods. Exploratory Data Analysis (EDA) techniques (Tukey, 1977) provide many tools that transfer large and cumbersome data tabulations into easy to grasp graphical displays which are widely independent of assumptions about the data. They are used to "visualise" the data. Graphical data analysis is often criticised as non-scientific because of its apparent ease. This critique probably stems from many scientists trained in formal statistics not being aware of the power of graphical data analysis.

Occasionally, even in graphical data analysis mathematical data transformations are useful to improve the visibility of certain parts of the data. A logarithmic transformation would be a typical example of a transformation that is used to reduce the influence of unusually high values that are far removed from the main body of data.

Graphical data analysis is a creative process, it is far from simple to produce informative graphics. Among others, choice of graphic, symbols, and data subsets are crucial ingredients for gaining an understanding of the data. It is about iterative learning, from one graphic to the next until an informative presentation is found, or as Tukey (1977) said "It is important to understand what you can do before you learn to measure how well you seem to have done it".

However, for a number of purposes graphics are not sufficient to describe a given data set. Here the realms of descriptive statistics are entered. Descriptive statistics are based on model assumptions about the data and thus more restrictive than EDA. A typical model assumption used in descriptive statistics would be that the data follow a normal distribution. The normal distribution is characterised by a typical bell shape (see Figure 4.1 upper left) and depends on two parameters, mean and variance (Gauss, 1809). Many natural phenomena are described by a normal distribution. Thus this distribution is often used as the basic assumption for statistical methods and estimators. Statisticians commonly assume that the data under investigation are a random selection of many more possible observations that altogether follow a normal distribution. Many formulae for statistical calculations, e.g., for mean, standard deviation and correlation are based on a model. It is always possible to use the empirical data at hand and the given statistical formula to calculate "values", but only if the data follow the model will the values be representative, even if another random sample is taken. If the distribution of the samples deviates from the shape of the model distribution, e.g., the bell shape of the normal distribution, statisticians will often try to use transformations that force the data to approach a normal

Statistical Data Analysis Explained   Clemens Reimann, Peter Filzmoser, Robert G. Garrett, Rudolf Dutter
© 2008 John Wiley & Sons, Ltd.

distribution. For environmental data a simple log-transformation of the data will often suffice to approach a normal distribution. In such a case it is said that the data come from a lognormal distribution.

Environmental data are frequently characterised by exceptionally high values that deviate widely from the main body of data. In such a case even a data transformation will not help to approach a normal distribution. Here other statistical methods are needed, that will still provide reliable results. Robust statistical procedures have been developed for such data and are often used throughout this book.

Inductive statistics is used to test hypotheses that are formulated by the investigator. Most methods rely heavily on the normal distribution model. Other methods exist that are not based on these model assumptions (non-parametric statistical tests) and these are often preferable for environmental data.

Most data sets in applied earth sciences differ from data collected by other scientists (e.g., physicists) because they have a spatial component. They present data for individual specimens, termed as samples by earth scientists, that were taken somewhere on Earth. Thus, in addition to the measured values, for example geochemical analyses, the samples have spatial coordinates. During data analysis of environmental and earth science research results this spatial component is often neglected, to the detriment of the investigation. At present there exist many computer program systems either for data analysis, often based on classical statistics that were developed for physical measurements, or for "mapping" of spatial data, e.g., geographical information systems (GIS). For applied earth sciences a data analysis package that takes into account the special properties of spatial data and permits the inclusion of "space" in data analysis is needed. Due to their spatial component, earth science and environmental data have special properties that need to be identified and understood prior to statistical data analysis in order to select the "right" data analysis techniques. These properties include:

- The data are spatially dependent (the closer two sample sites are the higher the probability that the samples show comparable analytical results) – all classical statistical tests assume independence of individuals.
- At each sample site a multitude of different processes can have had an influence on the measured analytical value (e.g., for soil samples these include: parent material, topography, vegetation, climate, Fe/Mn-oxyhydroxides, content of organic material, grain size distribution, pH, mineralogy, presence of mineralisation or contamination). For most statistical tests, however, it is necessary that the samples come from the same distribution – this is not possible if different processes influence different samples in different proportions. A mixture of results caused by many different underlying processes may mimic a lognormal data distribution – but the underlying "truth" is that the data originate from multiple distributions and should not be treated as if they were drawn from a single normal distribution.
- Like much scientific data (consider, for example, data from psychological investigations), applied earth sciences data are imprecise. They contain uncertainty. Uncertainty is unavoidably introduced at the time of sampling, sample preparation and analysis (in psychological investigations some people may simply lie). Classical statistical methods call for "precise" data. They will often fail, or provide wrong answers and "certainties", when applied to imprecise data. In applied earth sciences the task is commonly to optimally "visualise" the results. This book is all about "visualisation" of data behaviour.
- Last but not least environmental data are most often compositional data. The individual variables are not independent of each other but are related by, for example, being expressed as a

percentage (or parts per million – ppm (mg/kg)). They sum up to a constant, e.g., 100 percent or 1. To understand the problem of closed data, it just has to be remembered how percentages are calculated. They are ratios that contain all variables that are investigated in their denominator. Thus, single variables of percentage data are not free to vary independently. This has serious consequences for data analysis. Possibilities of how to deal with compositional data and the effect of data closure are discussed in Chapter 10 (for an in depth discussion see Aitchison, 1986, 2003; Buccianti *et al.*, 2006).

The properties described above do not agree well with the assumptions of "classical" (Gaussian) statistics. These are, however, the statistical methods taught in all basic statistics courses at universities because they are the most fundamental statistical methods. As a consequence, up to this day techniques that are far from ideal for the data at hand are widely applied by earth scientists in data analysis. Rather, applied earth science data call for the use of robust and non-parametric statistical methods. Instead of "precise" statistics so called "simple" exploratory data analysis methods, as introduced by Tukey in 1977, should always be the first choice. To overcome the problem of "closed data" the data array may have to be opened (see Section 10.5) prior to any data analysis (note that even graphics can be misleading when working with closed data). However, working with the resulting ratios has other severe shortcomings and at present there is no ideal solution to the problems posed by compositional data. All results based on correlations should routinely be counterchecked with opened data. Closure cannot be overcome by not analysing the major components in a sample or by not using some elements during data analysis (e.g. by focussing on trace elements rather than using the major elements). Even plotting a scatterplot of only two variables can be severely misleading with compositional data (compare Figures 10.7 and 10.8).

Graphical data analyses were largely manual and labour intensive 30 or 40 years ago, another reason why classical statistical methods were widely used. Interactive graphical data analysis has become widely available in these days of the personal computer – provided the software exists to easily prepare, modify and finally store the graphics. To make full use of exploratory data analysis, the software should be so powerful and convenient that it becomes fun to "play" with the data, to look at the data in all kinds of different graphics until one intuitively starts to understand their message. This leads to a better understanding about what kind of advanced statistical methods can and should (or rather should not) be applied to the data at hand. This books aims at providing such a package, where it becomes easy and fun to "play" with spatial data and look at them in many different ways in a truly exploratory data analysis sense before attempting more formal statistical analyses of the data. Throughout the book it is demonstrated how more and more information is extracted from spatial data using "simple" graphical techniques instead of advanced statistical calculations and how the "spatial" nature of the data can and should be included in data analysis.

Such a data analysis and mapping package for spatial data was developed 20 years ago under the name "DAS" (Data Analysis System – Dutter *et al.*, 1990). A quite similar system, called IDEAS, was used at the Geological Survey of Canada (GSC) (Garrett, 1988). DAS was only available for the DOS environment. This made it more and more difficult to run for people used to the Microsoft Windows operating system. During recent years, R has developed into a powerful and much used open source tool (see: http://www.r-project.org/) for advanced statistical data analysis in the statistical community. R could actually be directly used to produce all the tables and graphics shown in this book. However, R is a command-line language, and as such it requires more training and experience than the average non-statistician will usually have, or be willing to invest in gaining. The R-scripts for all the graphics and tables shown in this book are provided and can be used to learn R and to produce these outputs with the reader's own data. The program package accompanying this book provides the link between DAS and R and is called DAS+R. It uses the power of R and the experience of the authors in practical data analysis of applied geoscience data. DAS+R allows the easy and fast production of tables, graphics and maps, and the creation and storage of data subsets, through a graphical user interface that most scientists will find intuitive and be able to use with very little training. R provides the tools for producing and changing tables, graphics and maps and, if needed, the link to some of the most modern developments in advanced statistical data analysis techniques.

To demonstrate the methods, and to teach a user not trained to think "graphically" in data analysis, an existing multidimensional data set from a large environmental geochemical mapping project in the European Arctic, the Kola Ecogeochemistry Project (Reimann *et al.*, 1998a), is used as an example. The Kola Project data include many different sample materials, more than 60 chemical elements were determined, often by several different analytical techniques. The book provides access to the data set and to the computer scripts used to produce the example statistical tables, graphics and maps. It is assumed that the reader has a spreadsheet package like Microsoft Excel^{TM} or Quattro Pro^{TM} installed and has the basic knowledge to effectively use a spreadsheet program.

The Kola Project data set is used to demonstrate step by step how to use exploratory data analysis to extract more and more information from the data. Advantages and disadvantages of certain graphics and techniques are discussed. It is demonstrated which techniques may be used to good effect, and which should better be avoided, when dealing with spatial data. The book and the software can be used as a textbook for teaching exploratory data analysis (and many aspects of applied geochemistry) or as a reference guide to certain techniques. The program system can be used with the reader's own data, and the book can then be used as a "handbook" for graphical data analysis. Because the original R scripts are provided for all tables, graphics, maps, statistical tests, and more advanced statistical procedures, the book can also be used to become familiar with R programming procedures.

Because many readers will likely use the provided software to look at their own data rather than to study the Kola Project data, the book starts out with a description of the file structure that is needed for entering new data into R and DAS+R (Chapter 2). Some common problems encountered when editing a spreadsheet file as received from a laboratory to the DAS+R (or R) format are discussed in the same chapter. A selection of graphics for displaying data distributions are introduced next (Chapter 3), before the more classical distribution measures are introduced and discussed (Chapter 4). The spatial structure of applied earth science data should be an integral part of data analysis and thus spatial display, mapping, of the data is discussed next (Chapter 5) and before further graphics used in exploratory data analysis are introduced (Chapter 6). A "classical" task in the analysis of applied geochemical data is the definition of background and threshold, coupled with the identification of "outliers" and of element sources – techniques used up to this time are discussed in their own chapter (7). A key component of exploratory data analysis lies in comparing data. The use of data subsets is an especially powerful tool. The definition of subsets of data and using a variety of graphics for comparing them are introduced in Chapter 8, while Chapter 9 covers the more formal statistical tests. Statistical tests often require that the data are drawn from a normal distribution. Many problems can arise when using formal statistics with applied earth science data that are not normally distributed or drawn from multiple statistical populations. Chapter 10 covers techniques that may be used to improve the data behaviour for statistical analysis as a preparation for entering the realms of multivariate data analysis. The following chapters cover some of the widely used multivariate techniques such as correlation analysis (Chapter 11), multivariate graphics (Chapter 12), multivariate outlier detection (Chapter 13), principal component and factor analysis (Chapter 14), cluster analysis (Chapter 15), regression analysis (Chapter 16) and discriminant analysis (Chapter 17). In all chapters the advantages and disadvantages of the methods as well as the data requirements are discussed in depth. Chapter 18 covers different aspects of an integral part of collecting data in applied earth sciences: quality control. One could argue that Chapter 18 should be at the front of this book, due to its importance. However, quality control is based on graphics and statistics that needed to be introduced first and thus it is treated in Chapter 18, notwithstanding the fact that it should be a very early consideration when designing a new project. Chapter 19 provides an introduction to R and the R-scripts used to produce all diagrams and tables in this book. The program system and graphical user interface under development to make R easily accessible to the non-statistician is also explained.

The following books can be suggested for further reading. Some cover "graphical statistics", some the interpretation of geoscience and environmental data, some give an introduction to the computer language S (the base of R) and the commercial software package S-Plus (comparable to R), and several provide the mathematical formulae that were consciously avoided in this book. Davis (1973, 2002) is still one of the "classic" textbooks about computerised data analysis in the geosciences. Tukey (1977) coined the term "Exploratory Data Analysis" (EDA) and introduced a number of powerful graphics to visualise data (e.g., the boxplot). Velleman and Hoaglin (1981) provide an introduction to EDA including the computer codes for standard EDA methods (Fortran programs). Chambers *et al.* (1983) contains a comprehensive survey of graphical methods for data analysis. Rollinson (1993) is a classical textbook on data analysis in geochemistry, the focus is on interpretation rather than on statistics. Rock (1988), and Helsel and Hirsch (1992) provide an excellent compact overview of many statistical techniques used in the earth sciences with an early focus on robust statistical methods. Cleveland's papers (1993, 1994) are general references for visualising and graphing data. Millard and Neerchal (2001) provide an extensive introduction to environmental statistics using S-Plus. Venables and Ripley (2002) explain a multitude of statistical methods and their application using S. Murrell (2006) provides an excellent and easy to read description of the graphics system in R.

1.1 The Kola Ecogeochemistry Project

The Kola Ecogeochemistry Project (web site http://www.ngu.no/Kola) gathered chemical data for up to more than fifty chemical elements from four different primary sample materials (terrestrial moss, and the O-, B-, and C-horizon of podzolic soils) in parts of northern Finland, Norway and Russia. Two additional materials were collected for special purposes (Topsoil: 0–5 cm, and lake water – the latter in the Russian survey area only). The size of the survey area in the European Arctic (Figure 1.1) was $188\,000\,km^2$. The four primary materials were collected because in combination they can reflect atmospheric input (moss), interactions of the biosphere with element cycles (moss and O-horizon), the atmosphere–biosphere–lithosphere interplay (O-horizon), the influence of soil-forming processes (B-horizon), and the regional geogenic background distribution (the lithosphere) (C-horizon) for the elements investigated. Topsoil was primarily collected for the determination of radionuclides, but later a number of additional parameters were determined. Lake water reflects the hydrosphere, samples were

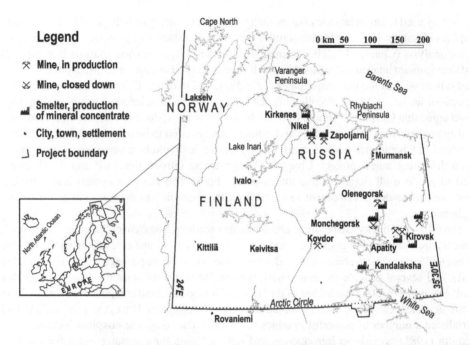

Figure 1.1 Location of the Kola Ecogeochemistry Project survey area

collected in Russia only because the 1000 lakes project (Henriksen *et al.*, 1996, 1997) collected lake water samples over all of Scandinavia at the same time (1995). All results for the four primary sample materials and topsoil are documented in the form of a geochemical atlas (Reimann *et al.*, 1998a). Lake water geochemistry is documented in a number of publications in international journals (see, e.g., Reimann *et al.*, 1999a, 2000a).

The main aim of the project was the documentation of the impact of the Russian nickel industry on the vulnerable Arctic environment. The result was a database of the concentration of more than 50 chemical elements in the above sample materials, reflecting different compartments of the ecosystem, in the year 1995. Each material can be studied for itself, the main power of the project design, however, lies in the possibility to directly compare results from all the different sample materials at the same sites.

This book provides access to all the regional Kola data, including topsoil and lake waters. Throughout the book examples were prepared using these data and the reader is thus able to reproduce all these diagrams (and many others) by using DAS+R and the Kola data. There are certainly many features hidden in the data sets that have not yet been covered by publications. Feel free to use the data for your own publications, but if a fellow scientist wants to use these data for publications, due reference should be given to the original source of the data (Reimann *et al.*, 1998a).

1.1.1 Short description of the Kola Project survey area

The survey area is described in detail in the geochemical atlas (Reimann *et al.*, 1998a). This book also provides a multitude of maps, which can be helpful when interpreting the data.

The project covered the entire area north of the Arctic Circle between longitudes 24° and 35.5° east and thence north to the Barents Sea (Figure 1.1). Relative to most of Europe, the Finnish and Norwegian parts of the area are still almost pristine. Human activities are mostly limited to fishery, reindeer-herding and forestry (in the southern part of the project area). Exceptions are a large iron ore mine and mill at Kirkenes, N-Norway; a small, brown coal-fired power station near Rovaniemi at the southern project border in Finland and some small mines. Population density increases gradually from north to south. In contrast, the Russian part of the project area is heavily industrialised with the nickel refinery at Monchegorsk, the nickel smelter at Nikel and the Cu/Ni-ore roasting plant at Zapoljarnij, which are three of the world's largest point-source emitters of SO_2 and Cu, Ni and Co and a number of other metals. These three sources together accounted for emissions of 300 000 t SO_2, 1900 t Ni and 1100 t Cu in 1994 (Reimann *et al.*, 1997c). Apatite ore is mined and processed near Apatity, iron ore at Olenegorsk and Kovdor, Cu-Ni-ore near Zapoljarnij. An aluminium smelter is located near Kandalaksha. The major towns of Murmansk and Apatity have large oil- and coal-fired thermal heating and power plants.

Topographically, large parts of the area can be characterised as highlands. In Norway, the general landscape in the coastal areas is quite rugged, and the mountains reach elevations of 700 m above sea level (a.s.l.). In Russia, in the south-western part of the Kola Peninsula, there are mountains reaching 200–500 m a.s.l. Near Monchegorsk and Apatity and near the coast of the White Sea there are some higher mountains (over 1000 m a.s.l.).

The geology of the area is complex and includes a multitude of different bedrock types (Figure 1.2). Some of the rock types occurring in the area are rare on a global scale and have unusual geochemical compositions. The alkaline intrusions that host the famous apatite deposits near Apatity are an example. The main rock types in the area that are occasionally mentioned in later chapters due to their special geochemical signatures, are:

- Sedimentary rocks of Caledonian (c. 1600–400 million years (Ma) old) and Neoproterozoic (1000–542 Ma) age that occur along the Norwegian coast and on the Rhybachi Peninsula in Russia (lithologies 9 and 10 in the data files).
- The rocks of the granulite belt that runs from Norway through northern Finland into Russia. These rocks are of Archean age (2300–1900 Ma) and "foreign" to the area (see Reimann and Melezhik, 2001). They are subdivided into "felsic" and "mafic" granulites (lithologies 31 and 32 in the data files).
- Diverse greenstone belts, which occur throughout the area. These are Palaeoproterozoic (2400–1950 Ma) rocks of volcanic origin (lithologies 51 and 52 in the data files). These rocks host many of the ore occurrences in the area, e.g., the famous Cu-Ni-deposits near Zapoljarnij in Russia.
- Alkaline and ultramafic alkaline intrusions of Palaeoproterozoic to Palaeozoic age (1900–470 Ma) (see above). These host the important phosphate deposits near Apatity (lithologies 81, 82 and 83 in the data files).
- Granitic intrusions of Palaeoproterozoic age (1960–1650 Ma), occurring in the south-western corner of the survey area in Finland and as small bodies throughout the survey area (lithology 7 in the data files).
- Large gneiss masses of Archean (3200–2500 Ma) and uncertain age (lithologies 1, 4 and 20 in the data files) that do not show any geochemical peculiarities.

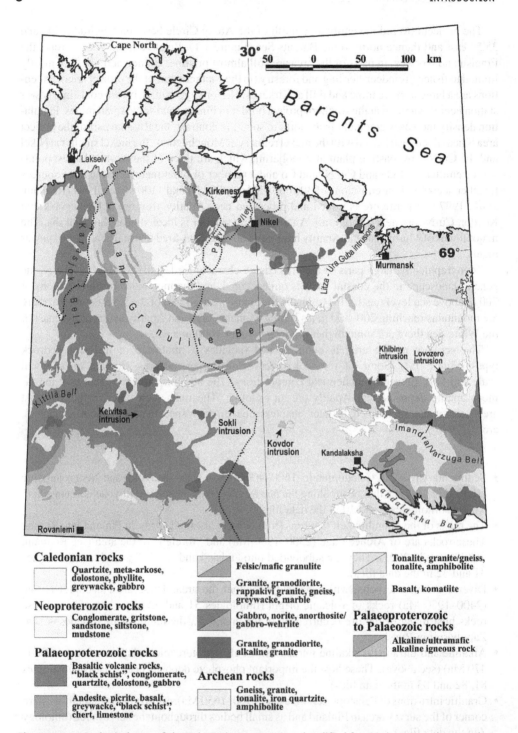

Figure 1.2   Geological map of the Kola Project survey area (modified from Reimann et al., 1998a). A colour reproduction of this figure can be seen in the colour section, positioned towards the centre of the book

The study area is part of the glaciated terrain of Northern Europe. The main Quaternary deposits are till and peat. There are also large areas without any surficial cover, dominated by outcrops and boulder fields (Niemelä *et al.*, 1993).

The north–south extent of the survey area is about 500 km. Within this distance, three vegetation zones gradually replace each other (for a vegetation map in colour see Kashulina *et al.*, 1997; Reimann *et al.*, 1998a). The southern and central parts of the area fall into the northern boreal coniferous zone. Towards the north, this zone gradually gives way to subarctic birch forest, followed by the subarctic tundra zone close to the coast of the Barents Sea. These changes in vegetation zones can also occur with altitude. Major characteristics of the forest ecosystems in this area are the sparseness of the tree layer, a slow growth rate, and the large proportion of ground vegetation in the total biomass production.

The dominant soil-forming process in forested and treeless regions of Northern Europe is podzolisation of mineral soils (Borggaard, 1997). Podzols are thus the most important soil type present throughout the survey area. Soils in the area are young. Their age ranges between 5000 and 8000 years, having formed since the retreat of the continental ice sheet in Northern Europe. A typical podzol profile consists of five main layers, the O (organic), E (eluvial), B (illuvial), BC (transitional) and C (weathered parent)-horizon. Colour photographs of typical podzol profiles from the area can be found in Reimann *et al.* (1998a) or on the Kola Project web site: `http://www.ngu.no/Kola/podzol.html`. The O-horizon of podzol is characterised by a low pH-value (MEDIAN in the survey area 3.85 – Reimann *et al.*, 1998a). pH increases systematically with depth. In the C-horizon it reaches a MEDIAN value of 5.8. The O-horizon varies in thickness from less than 0.5 cm to more than 20 cm, the MEDIAN for the survey area being 2.5 cm. The thickness of the soil profiles can vary considerably – the depth to the top of the C-horizon varies between 10 and 123 cm, the MEDIAN being 35 cm (maps for all these parameters are presented in Reimann *et al.*, 1998a).

Major climatic differences occur within the survey area. In the northwest, summers are cool and winters mild, with most precipitation between September and January (coastal climate). In the central part, warm summers and relatively cold winters are typical; the main precipitation takes the form of rain showers in July and August (continental climate). The easternmost part is similar to the central part, but with colder winters. Precipitation is in general low. The average for the whole area is <500 mm/year. The yearly average temperature is $-1°C$. Maps showing average rainfall and temperature for summer and winter can be found in Reimann *et al.* (1998a).

1.1.2 Sampling and characteristics of the different sample materials

Sampling of the 188 000 km^2 area took place from July to September 1995. The average sample density was 1 site per 300 km^2. Samples were collected from 617 sites. Depending on availability the total number of samples, *n*, ranges for any one sample material from 594 (moss) to 617 (O-horizon). Sample site selection for the project was such that only locations where podzol had developed on glacial drift were visited, giving genetically comparable samples over the whole area. At each site the following samples were collected (for more detailed information see Äyräs and Reimann, 1995; Reimann *et al.*, 1998a):

Moss (*n* = 594): Terrestrial moss, preferably the species *Hylocomium splendens*, and if it was absent *Pleurozium schreberi*, was collected. Only shoots representing the previous three years' growth were taken.

Terrestrial moss (*Hylocomium splendens* and *Pleurozium schreberi*) receives most of its nutrients from the atmosphere. In Scandinavia it has been used to monitor the atmospheric deposition of chemical elements for more than 30 years (see, e.g., Rühling and Tyler, 1968, 1973; Tyler, 1970). It was thus included in the Kola Project to reflect the input of elements from the atmosphere (wet and dry deposition, including both geogenic and anthropogenic dust) over the previous 2–3 years. At the same time the moss reflects element concentrations in an important component of the arctic ecosystem in terms of total biomass production and ecological function. It is an important supplier of litter for the formation of the organic horizon of the soils (Kashulina *et al.*, 1997). Results of an interspecies comparison between *Hylocomium splendens* and *Pleurozium schreberi* are presented in Halleraker *et al.* (1998). A review of the moss technique and a discussion of possible problems related to it as well as a comparison with the chemical composition of lichen and crowberry from the Kola Project area are given in Reimann *et al.* (1999b, 2001c). Results of regional mapping are also discussed in Äyräs *et al.* (1997a); de Caritat *et al.* (2001); Reimann *et al.* (1997b).

O-horizon ($n = 617$): This was collected with a custom-built tool to facilitate and systematise sampling, and to allow easy measurement of the volume of each sample (Äyräs and Reimann, 1995). For the O-horizon samples only the top 3 cm of the organic layer were taken. If the total thickness of the O-horizon was less than 3 cm, only the organic layer was sampled, and the thickness was recorded on the field notes. From seven to ten sub-samples were collected at each site to give a composite sample with a minimum volume of one litre.

The O-horizon consists mostly of plant residues in differing stages of decay and humus, almost inevitably mixed with some minerogenic particles. Due to its location and genesis, the organic horizon reflects the complex interplay between the lithosphere, the biosphere and the atmosphere. The O-horizon is a major sink for plant nutrients in northern ecosystems. It can accumulate and enrich many chemical elements, e.g., via organic complexing and the formation of metallo-organic compounds. For many elements, both from natural and anthropogenic sources, it thus acts as a very effective "geochemical barrier" as defined first by Goldschmidt (1937). A separate interpretation of the regional distribution patterns found in the O-horizon from the survey area is given in Äyräs and Kashulina (2000). Other publications discussing special properties of the O-horizon include Reimann *et al.* (1998c, 2000b, 2001a).

Podzol profiles (B-horizon, $n = 609$**; C-horizon,** $n = 605$**):** Before sampling a podzol profile, homogeneity of the soil cover was checked over an area of 10×10 m. The exact location of the profile was chosen so that both ground vegetation and micro-topography were representative. The sampling pits were dug by spade to the C-horizon. Samples of the O-, E-, B-, BC- and C-horizons were collected, starting from the bottom to avoid contamination and mixing of the horizons. With the exception of the C-horizon, each layer was sampled over its complete thickness. If there were distinguishable layers within the B-horizon, these were collected in the same ratio as present in the profile. Each sample weighed about 1–1.5 kg, depending on parameters like grain size, mineralogy, and water content. At present only the samples of the B- and C-horizon have been studied, all other samples have been archived in air dried condition for future reference.

The B-horizon is clearly affected by soil-forming processes. Compared to the C-horizon it is relatively enriched in clay minerals, organic matter, and free and organically-bound amorphous Fe- and Al-oxides and -hydroxides, which have been leached from the upper soil horizons. It is less active than the O-horizon, but can still act as a second "geochemical barrier" for many elements (independent of origin) within the soil profile, e.g., via co-precipitation with the Fe-oxides/-hydroxides. Some soil-forming processes within podzol profiles from the Kola Ecogeochemistry Project area are discussed in Räisänen *et al.* (1997) and Kashulina *et al.* (1998a).

The deepest soil horizon of podzols, the C-horizon, is only slightly influenced by soil-forming processes and sometimes by anthropogenic contamination, and thus mostly reflects the natural, geogenic, element pool and regional variations therein. A geological interpretation of the C-horizon results is given in Reimann and Melezhik (2001b).

Topsoil (*n* = 607): Samples were taken for the analysis of radionuclides. Topsoil samples (0–5 cm) were collected at all sample sites using the same procedure as for the O-horizon. These samples were also analysed by Instrumental Neutron Activation Analysis (INAA) for about 30 elements. These results are not used within this book but provided on the web page http://www.statistik.tuwien.ac.at/StatDA/R-scripts/data/.

For reasons of completeness, analytical results of lake water samples collected in the Russian survey area are also included on the web page of the book. These are not used within this book, interested readers can find a description of the sampling procedures and some first interpretations of the results in Reimann *et al.* (1999a).

In addition to the sample media documented in the atlas and provided with this book, snow, rain water, stream water, lake water, ground water, bedrock, organic stream sediments, topsoil 0–5 cm, and overbank sediments were studied during different stages of the project (Äyräs and Reimann, 1995; Äyräs *et al.*, 1997a,b; Boyd *et al.*, 1997; de Caritat *et al.*, 1996a,b, 1997a,b, 1998a,b; Chekushin *et al.*, 1998; Gregurek *et al.*, 1998a,b, 1999a,b; Halleraker *et al.*, 1998; Kashulina *et al.*, 1997, 1998a,b; Niskavaara *et al.*, 1996, 1997; Reimann *et al.*, 1996, 1997a,b,c, 1998c, 1999a,b, 2000a,b; Volden *et al.*, 1997), often on a spatially detailed scale. These data can be used to assist in the interpretation of the observed regional features. They are not provided here but are available from the Norwegian Geological Survey (NGU) upon request. A complete and regularly updated list of publications with data from the Project can be found at the following project website: http://www.ngu.no/Kola/publist.html.

1.1.3 Sample preparation and chemical analysis

Detailed descriptions of sample preparation and analysis are provided in Reimann *et al.* (1998a). In general, selection of elements, extractions and detection limits was based on a conscious decision to make all results as directly comparable as possible.

Both the Geological Survey of Finland (GTK) and NGU chemical laboratories and the methods described above for analysing the Kola samples are accredited according to EN 45001 and ISO Guide 25. All methods have been thoroughly validated and trueness, accuracy and precision are monitored continuously. Some of the quality control results are presented in Reimann *et al.* (1998a) – and in the last chapter of this book. The data files include information about the analytical method used and the detection limits reached.

2

Preparing the Data for Use in R and DAS+R

Finally – the long-expected analytical results have arrived from the laboratory. With print-out in hand and the spreadsheet-file with the results on the computer hard-disk the scientist burns to look at the data. Are there any interesting patterns and information hidden in the data, and if yes, what are the locations of the respective individuals/samples?

First things first. It is prudent, some would argue essential, to make a back-up copy of the received data, giving the file a name that clearly identifies it as the original received data, ideally making it "read-only", and placing a copy in a project archive. Depending on the importance of the project, copies of the project archive, which will grow as other critical files and output are generated, may be stored on an institutional server, or even on some media, e.g., a CD, off-site. The original received data should never be modified. If this procedure is not followed, it is almost guaranteed that serious problems will be encountered when returning to a project after some months or years of working with other data. It is important that the data, both field and laboratory, and all related information are captured as soon after they are created as possible. Otherwise they may get lost over time as scientists move on to other projects, switch computers or even jobs. In an ideal world the scientist's institution will have a geochemical database where analytical results, together with other vital project information, are stored for easy retrieval.

In the less than ideal, but often real, world the scientist will work in an institution without an established geochemical database and without an Information Technology (IT) support group. Most likely the scientist has learned to use a spreadsheet like Microsoft Excel™ or Quattro Pro™ for daily work and would like to keep the data in this format (note that the input (and output) tables of a database are commonly such spreadsheet files). Although we do not want to advocate the use of spreadsheets exclusively, we will discuss how to work with spreadsheets and the pitfalls that exist when working with them. For speed of graphical data analysis, DAS+R has been constructed in a way that it can handle many situations that would usually require an additional database program. It is for example possible to keep auxiliary information like method of analysis, detection limits of the analytical method and comments concerning the data set, as well as any variable. This auxiliary information, if provided in the data file, can be read into DAS+R together with the data. It is also possible to link and/or append data files, to create and store data subsets, and to manipulate variables using different mathematical operators. Linking and/or appending data in R has the advantage that almost no

Statistical Data Analysis Explained Clemens Reimann, Peter Filzmoser, Robert G. Garrett, Rudolf Dutter
© 2008 John Wiley & Sons, Ltd.

limits exist as to the final length or width of the data file. Today's spreadsheet software is limited as to the allowed width of a file, e.g., many do not allow for more than 256 variables in one file.

2.1   Required data format for import into R and DAS+R

The question could be asked, "what is the problem with using the original data file as received from the laboratory for data analysis?" Firstly, the laboratory's result file will likely not contain the geographic coordinates of the samples, field information, and other ancillary data that may be required in the following data analyses, nor the "keys" identifying field duplicates, analytical duplicates, the project standards (control reference materials), or even the original sample site number. The data file must thus be linked with a "key-file" containing all this information. Secondly, data will often be received from several different laboratories, and the different files must be linked somehow. Thirdly, the received files most likely contain a lot of information that are necessary to know and may be useful to view in a table, but are of no use in statistical analysis. Tables prepared for "nice looks" are actually often especially unsuited as input for data analysis software and may require much editing. Thus it is sensible to not spend time on "table design" at this stage.

However, during data analysis it is often advantageous to have all variables in the same unit in the data file (for example for direct 1:1 comparisons or if a centred logratio transformation (see Section 10.5.2) is necessary). For the Kola data mg/kg was used as it was the most frequent laboratory reporting unit. For example, results of XRF analysis reported as oxides in wt%, e.g., Al_2O_3, were re-calculated as element concentrations in mg/kg resulting in the new variable Al_XRF (Figure 2.1). A data file for import into data analysis software packages should not contain any special signs such as "<", ">", "!", "?", "%" or characters from a language other than English, e.g., "å", "ø", "ü", or "μ" (μg/kg). Sometimes these can cause serious problems during data import or cause the software to crash later on. All variables need to have different and unique names. Missing values (see Section 2.3) need to be marked by an empty field, or some coded value, e.g., -9999, which can be used to tell the program that this is a missing value. Data below (or above) the analytical method detection limit must be treated in a consistent manner (see Section 2.2) – in the case of the Kola data sets all values below the detection limit (there were no data above an upper detection limit) were set to a value of half of the detection limit. Thus there usually is a certain amount of editing necessary to make the data file software-compatible.

Most statistical software, including R, require as input a rather simple data format, where the first row identifies the variables and all further rows contain the results to be used during data analysis. Some further requirements will be discussed below. DAS+R will read such files, but accepts data files with additional auxiliary information like project name, a description of sampling and analytical methods used, the lower and upper detection limits, and the units in which the data of each variable are reported; these can be directly stored (and thus retrieved during data analysis) with the data. This special DAS+R data file format is constructed such that it can be turned into a "normal" R file with no more than two commands.

The most simple file format that can be read by almost all statistical data analysis packages consists of a header row, identifying the variables, and the results following row by row underneath (Figure 2.1). Such a file should be stored in a simple "csv" (variables and results separated by a comma: csv = comma separated values) or "txt" format.

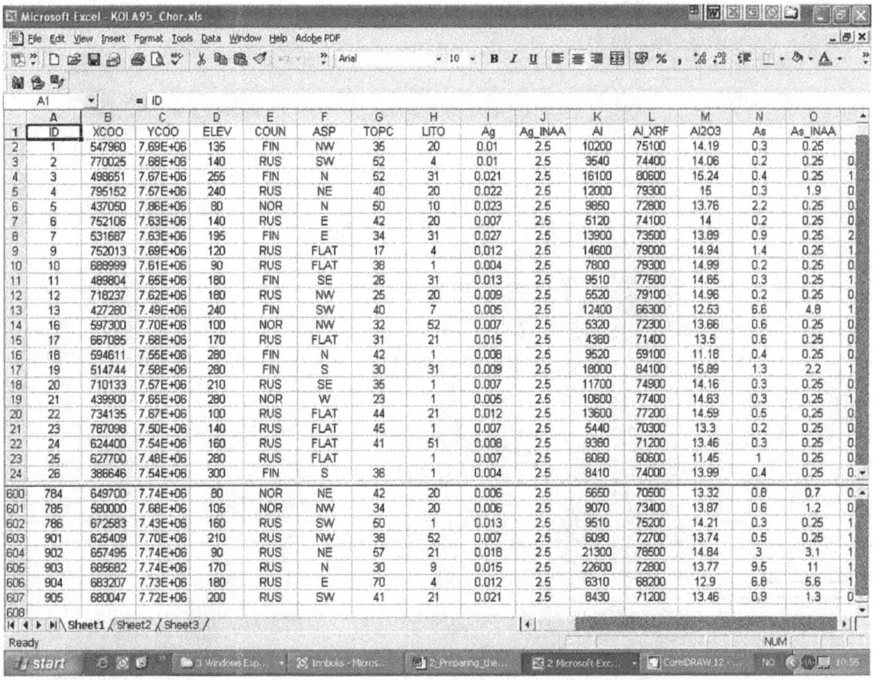

ID	XCOO	YCOO	ELEV	COUN	ASP	TOPC	LITO	Ag	Ag_INAA	Al	Al_XRF	Al2O3	As	As_INAA
1	547960	7.69E+06	135	FIN	NW	35	20	0.01	2.5	10200	75100	14.19	0.3	0.25	
2	770025	7.68E+06	140	RUS	SW	52	4	0.01	2.5	3540	74400	14.06	0.2	0.25	0
3	498651	7.67E+06	255	FIN	N	52	31	0.021	2.5	16100	80600	15.24	0.4	0.25	1
4	795152	7.67E+06	240	RUS	NE	40	20	0.022	2.5	12000	79300	15	0.3	1.9	0
5	437050	7.86E+06	80	NOR	N	50	10	0.023	2.5	9850	72800	13.76	2.2	0.25	0
6	752106	7.63E+06	140	RUS	E	42	20	0.007	2.5	5120	74100	14	0.2	0.25	0
7	531687	7.63E+06	195	FIN	E	34	31	0.027	2.5	13900	73500	13.89	0.9	0.25	2
8	752013	7.69E+06	120	RUS	FLAT	17	4	0.012	2.5	14600	79000	14.94	1.4	0.25	1
9	688999	7.61E+06	90	RUS	FLAT	38	1	0.004	2.5	7800	79300	14.99	0.2	0.25	0
10	489804	7.65E+06	180	FIN	SE	26	31	0.013	2.5	9510	77500	14.65	0.3	0.25	1
11	718237	7.63E+06	180	RUS	NW	25	20	0.009	2.5	5620	79100	14.96	0.2	0.25	0
12	427280	7.49E+06	240	FIN	SW	40	7	0.005	2.5	12400	66300	12.53	6.6	4.8	1
13	597300	7.70E+06	100	NOR	NW	32	52	0.007	2.5	5320	72300	13.66	0.6	0.25	0
16	667085	7.68E+06	170	RUS	FLAT	31	21	0.015	2.5	4360	71400	13.5	0.6	0.25	0
18	594611	7.55E+06	280	FIN	N	42	1	0.008	2.5	9520	59100	11.18	0.4	0.25	0
19	514744	7.58E+06	280	FIN	S	30	31	0.009	2.5	18000	84100	15.89	1.3	2.2	1
20	710133	7.57E+06	210	RUS	SE	35	1	0.007	2.5	11700	74900	14.16	0.3	0.25	0
21	439900	7.65E+06	280	NOR	W	23	1	0.005	2.5	10800	77400	14.63	0.3	0.25	1
22	734135	7.67E+06	100	RUS	FLAT	44	21	0.012	2.5	13600	77200	14.59	0.5	0.25	0
23	787098	7.50E+06	140	RUS	FLAT	45	1	0.007	2.5	5440	70300	13.3	0.2	0.25	0
24	624400	7.54E+06	160	RUS	FLAT	41	51	0.008	2.5	9380	71200	13.46	0.3	0.25	0
25	627700	7.48E+06	280	RUS	FLAT		1	0.007	2.5	6060	60600	11.45	1	0.25	0
26	386646	7.54E+06	300	FIN	S	36	1	0.004	2.5	8410	74000	13.99	0.4	0.25	0
784	649700	7.74E+06	80	NOR	NE	42	20	0.006	2.5	5650	70500	13.32	0.8	0.7	0
785	580000	7.68E+06	105	NOR	NW	34	20	0.006	2.5	9070	73400	13.87	0.6	1.2	0
786	672583	7.43E+06	160	RUS	SW	50	1	0.013	2.5	9510	75200	14.21	0.3	0.25	1
901	625409	7.70E+06	210	RUS	NW	38	52	0.007	2.5	6090	72700	13.74	0.5	0.25	1
902	657495	7.74E+06	90	RUS	NE	57	21	0.018	2.5	21300	78500	14.84	3	3.1	1
903	685682	7.74E+06	170	RUS	N	30	9	0.015	2.5	22600	72800	13.77	9.5	11	1
904	683207	7.73E+06	180	RUS	E	70	4	0.012	2.5	6310	68200	12.9	6.8	5.6	1
905	680047	7.72E+06	200	RUS	SW	41	21	0.021	2.5	8430	71200	13.46	0.9	1.3	0

Figure 2.1 Screen snapshot of a simple Microsoft Excel™-file of Kola Project C-horizon results. These are ready for import by most statistical data analysis packages, including R, once stored in a simple format, e.g., the "csv" format

The special DAS+R file format, fine-tuned for use in applied geochemistry, allows the storing of much more information about the data and about each single variable, so it can be retrieved and used during data analysis. This format requires an empty first column, providing a number of predetermined key words (up to but not more than eleven) telling the software about the auxiliary information that it is expected to handle (see Figure 2.2). The sequence in which the keywords are provided is flexible, only HEADER and VARIABLE need to be specified, none of the other keywords need to be used, the software checks for the keywords and stores the information if it detects one or several of the keywords. Of course other variable-relevant information than "EXTRACTION" and "METHOD" could be stored using these fields. The keywords are:

HEADER: holding information on the data file, e.g., "Kola Project, regional mapping 1995, C-horizon".

COMMENT DATASET: this row can contain free text with additional comments that will be kept with the data and are valid for the whole file or large parts thereof, e.g., "<2 mm fraction, air dried, laboratories: Geological Survey of Finland (all ICP and AAS results); Geological Survey of Norway (all XRF results); ACTLABS (all INAA results)".

These two keywords are not linked to any specific variable but to the data file as a whole.

Microsoft Excel - KOLA95_MOSS.csv

File Edit View Insert Format Tools Data Window Help Adobe PDF

	A	B	C	D		U	V	W	X	Y	Z	AA	AB	AC
1	HEADER					KOLA PROJECT, regional sampling 1995 (Finland (FIN), Norway (NOR) and Russia (RUS)), all analytic								
2	COMMENT DATASET					Terrestrial moss (Hylocomium splendens OR Pleurozium schreberi - see SPECIES), air dried, milled (								
3	SAMPLE IDENTIFIER	ID												
4	COORDINATES		XCOO	YCOO										
5	COMMENT VARIABLES									analysed 2000; see Niskavaara et al. 2004				
6	EXTRACTION					conc.HNO	conc.HNO	conc.HNO	ashed, HCl	conc.HNO	conc.HNO	conc.HNO	conc.HNO	conc.HNO	co
7	METHOD					ICP-MS	ICP-MS	ICP-MS	GF-AAS	ICP-MS	ICP-MS	ICP-AES	ICP-MS	ICP-AES	ICF
8	UDL													
9	LDL					0.01	10	0.05	0.05	0.5	0.5	0.03	0.004	6
10	UNIT		m east	m north		mg/kg	mg/kg	mg/kg	micro g/kg	mg/kg	mg/kg	mg/kg	mg/kg	mg/kg	mg
11	VARIABLE	ID	XCOO	YCOO		Ag	Al	As	Au	B	Ba	Be	Bi	Ca	Cd
12		1	547960.4	7893790		0.016	71.2	0.123		1.74	14	0.015	0.002	2310
13		2	770024.8	7679167		0.073	245	0.299		2.77	17.4	0.015	0.039	2460
14		3	498650.6	7668151		0.032	103	0.176		1.89	20.9	0.015	0.012	3430
15		4	795151.9	7569386		0.118	307	0.423		2.3	21.8	0.015	0.033	2860
16		5	437050	7855900		0.038	253	0.119		4.65	31.1	0.015	0.002	3190
17		6	752105.9	7626023		0.106	238	0.342		2.18	19.9	0.015	0.052	2660
18		7	531687.1	7630816		0.036	257	0.259		1.55	24.9	0.015	0.02	2750
19		8	537800	7882500		0.018	177	0.167		7.08	15.3	0.015	0.009	4120
20		9	752012.5	7686449		0.089	531	0.272		1.74	31.7	0.04	0.027	2690
21		10	688999.4	7610298		0.036	146	0.172		1.41	15.5	0.015	0.014	2940
22		11	489804.2	7652008		0.021	77.5	0.111		1.41	13	0.015	0.002	2450
23		12	718237.1	7623814		0.068	647	0.316		1.49	15.1	0.015	0.018	2070
24		13	427279.8	7487776		0.023	154	0.121		0.91	20.1	0.015	0.008	2440
25		16	597300	7699300		0.038	184	0.336		0.25	9.4	0.015	0.017	1800
26		17	667085.1	7883133		0.047	147	0.41		2.14	19.5	0.015	0.011	2610
27		18	594611.3	7550950		0.034	116	0.149		0.81	22.9	0.015	0.01	2820
602		903	685681.5	7741352		0.037	267	0.296	0.14	3.26	15.1	0.015	0.021	2190
603		904	683207.4	7730755		0.039	283	0.311	0.2	5.2	91.2	0.12	0.02	3760
604		905	680047.2	7720895		0.05	216	0.314	0.14	4.3	12.9	0.015	0.031	2140
605		906	749686.8	7557109		0.398	1190	1.4		4.31	85.7	0.04	0.276	6370
606														
607														

KOLA95_MOSS

Ready

Figure 2.2 Kola Project Moss data file prepared with auxiliary information (project name, sample type and preparation, analytical method, detection limit, unit) that may be needed during data analysis. This format will be accepted by DAS+R. If the data file is to be used outside of DAS+R, for example in R, the first column and the uppermost ten rows should be deleted or commented out

SAMPLE IDENTIFIER: this keyword is used to identify the table column (variable) that contains the sample or site number (or code) via entering "ID" in the column that contains this information in the VARIABLE record. This can also be done later after the data are imported into DAS+R if necessary. If no sample identifier is provided, the samples are numbered from 1 to n, the number of observations (samples), and identified using this number in the graphics and tables where samples are identified.

COORDINATES: used to identify the two columns holding the information about the geographical coordinates via entering "XCOO" (east) and "YCOO" (north) in the columns containing this information. If the two variables containing the coordinates are not identified, mapping and spatial functions within the software will not be available during data analysis.

COMMENT VARIABLES: can hold a free comment that is linked to each single variable, e.g., a remark on data quality or number (or percentage) of samples below detection.

EXTRACTION: holds a second comment linked to each variable separately like "aqua regia" or "total" (or any other variable-related comment).

METHOD: holds a third comment, the method of determination for each variable like "ICP-AES" or "XRF" (or any other variable-related comment).

UDL: the value of the upper limit of detection for each variable (if not applicable for the data at hand, it is simplest to not provide this keyword and row), e.g., "10000".

LDL: the value of the lower limit of detection for each variable, e.g., "0.01".

UNIT: the unit of the measurement of each variable, e.g., "mg/kg", "wt-percent", "micro_g/kg"
   (it is generally wise to try to avoid "special" signs like "%" or "μ" – see above), and finally
VARIABLE: the line usually starting with the sample identifier and providing the coordinate
   and variable names. *Attention:* a "unique" name must be used for each variable. If the same
   element has been analysed by different methods a simple text or numeric extension can be
   used. Note that variable names should not contain blanks.

The data array can contain empty cells (missing values, see above, R replaces empty cells by
the code NA – not available) but should normally not contain any special signs like "<", ">" or
"!" if the variable is to be used for statistical analyses. Text variables or variables consisting of
a mixture of text and numbers are allowed and will be automatically recognised as such. They
can for example contain important information that is linked to each sample and can be used
to create data subsets or groups (see Chapter 8). Figure 2.2 shows the above example file in
the special DAS+R format with all auxiliary information (except the upper limit of detection)
provided.

Note that different types of variables exist and can be used in DAS+R. When importing a
data file, DAS+R will try to allocate each variable automatically to a certain data type. These
types are later displayed in the software and can be changed. It is important to assign the
"correct" data type to each variable. It determines what can be done with this variable during
data analysis. If the automatic assignment is wrong, this should be a clear indication to check
the data for this variable for inconsistencies before continuing with data analysis.

Logical: TRUE/FALSE or T/F, e.g., used to identify samples that belong to a certain subset.
Integer: a number without decimals.
Double: any real numerical value, i.e. a number with decimals.
Factor: a variable having very few different levels, which can be a character string (text) or
   a number or a combination thereof.
Character: any text; the sample identifier (sample number) is a typical "character" variable
   because it is the unique name of each sample. It can consist of a number, a text string, or
   combination of text and numbers.

2.2   The detection limit problem

A common problem in the analysis of applied geochemical and environmental data is that a
data set as received from the laboratory will often contain a number of variables where a certain
percentage of the samples return values below the detection limit (DL) (see, e.g., Levinson,
1974; AMC, 2001). There are both lower and an upper DLs for any analytical method – in
practice it is the lower DL that will most often be of concern in applied geochemistry and
environmental projects (when analysing ore samples or samples from an extremely contaminated site, the upper DL of the method may also need consideration). In a spreadsheet or file
as received from the laboratory these samples will most often be marked as "<DL" (or "<rl",
"<reporting limit"), where DL is the actual value of the detection limit (e.g., 2 mg/kg). While
values marked "<" are desirable in a printed data table, they cannot be used for any numerical
processing or statistical studies. Means and standard deviations cannot be computed for samples marked "<" – and graphics or maps cannot be easily prepared. Because most statistical
tests are based on estimates of central tendency (e.g., MEAN) and spread (e.g., variance or
standard deviation), such tests cannot be carried out.

Thus selecting analytical procedures that provide adequate DLs is one of the most important considerations in planning any applied geochemistry project. The (lower) DL is commonly understood to be the smallest concentration that can be measured reliably with a particular technique. Different analytical techniques will have different DLs for the same element, and different laboratories may well quote different DLs for the same element using the same analytical technique. In addition the DL will change with the model of the instrument, the technician performing the analysis, and the solid-to-liquid ratio in any procedure requiring sample dissolution.

The choice of the analytical method for a given project is often a compromise between cost, a careful evaluation of expected concentrations, and the need to have a complete data set for the elements required (e.g., for geochemical mapping). In many instances the practitioner will end with a data set containing a substantial number of "<" signs. As mentioned above, these create problems in data analysis. There are several approaches to dealing with variables where some analytical results are "<DL". Options are:

- Delete the whole variable or all samples with values "<DL" from data analysis;
- Mark all observations "<DL" as missing;
- Model a distribution in the interval [0, DL], and assign an arbitrarily chosen value from this distribution to each sample <DL;
- Try to predict a value for this variable in each sample via multiple regression (imputation) techniques using all other analytical results; or
- Set all values marked "<DL" to an arbitrarily chosen low number, e.g., half the DL.

None of these solutions is ideal. To delete samples from data analysis is not acceptable, it will shift all statistical estimates towards the "high" end (or "low" end – in the rare case of observations above the upper limit of detection), although there is information that the concentration in a considerable number of samples is low (high). The same happens if the values are marked as "missing".

Modelling a distribution based on the existing data and an assumption about the shape of the distribution (e.g., lognormal) will give the statistically most satisfying result. On the other hand it is then no longer easy to differentiate between true (measured) values and "modelled" values that have no geochemical legitimacy beyond the model, especially in the multivariate context. This is again an undesirable situation.

To predict the values <DL, for example via regression techniques using all other analytical results or kriging (for spatial data – Chapter 5), is only sensible when the value is really needed, e.g., for mapping. The problem of differentiating between true (measured) values and "modelled" values that have no geochemical legitimacy beyond the model, especially in the multivariate context, remains.

Thus the most widely applied solution to the "<DL" problem is to set the value to an arbitrarily chosen value of half the DL, some practitioners use the value of the DL (but note that there may be "true" measurements at this value, thus 1/2 of the DL is the better choice). This is easy to do in a spreadsheet and works well as long as the number of values <DL is relatively low (say less than 10–15 percent of all samples). A more realistic value than half the DL can often be estimated from an ECDF- or CP-plot (see Chapter 3). This method is acceptable for constructing statistical graphics or maps – the value of half the DL is used as a "place holder" and marks "some low value" in the graphics. Often it is easy to recognise <DL situations in a graphic, e.g., a scatterplot (e.g., Figure 18.8) or an ECDF- or CP-plot

by a straight line of values at the lower end of the data (e.g., Figure 18.7). An element that contains values below the detection limit is called "censored". In case of data below the limit of detection, they are stated to be "left censored"; conversely, in case of data above the upper limit of detection, they are stated to be "right censored".

To just replace the values <DL by an arbitrary low value is not, however, acceptable if we need to calculate a "reliable" estimate (e.g., for complying with some environmental regulation) of central tendency and spread (see Chapter 4), as needed for most statistical tests. Helsel and Hirsch (1992) and Helsel (2005) provide a detailed description of these problems and possible solutions. A first and easy solution is to use MEDIAN (Section 4.1.4) and MAD (median absolute deviation, see Section 4.2.4) as estimators of central value and spread instead of MEAN (Section 4.1.1) and SD (standard deviation, Section 4.2.3) (but remember that the values representing "<DL" need to be kept in the data set as the lowest (highest) values for a correct ranking). MEDIAN and MAD can be used as long as no more than 50 per cent of the data are below the limit of detection.

To further complicate the issue, analytical methods exist where the laboratory will report different DLs for different samples depending on the matrix of the actual sample (typical for Instrumental Neutron Activation Analysis, INAA). It is also not unusual to find different detection limits over time because a laboratory has improved its methodology or instrumentation, or when employing different laboratories reporting with different DLs. One possible solution in this situation is to use the value of half the lowest DL reported as "place holder" for all samples <DL.

To solve these problems, the reporting of values below detection limit as one value has been officially discouraged in favour or reporting all measurements with their individual uncertainty value (AMC, 2001). However, at the time of writing, most laboratories are still unwilling to deliver instrument readings for those results that they consider as "below detection" to their clients and the replacement of values that are marked "<DL" by a value of half the DL prior to data analysis is still a practical, if not perfect, solution for many applications.

In general a conscious decision is required as to whether a variable with censored data is to be included in a multivariate data analysis. A variable can, for example, contain 10 per cent of censored data, an amount that could probably be handled by the selected multivariate method. However, a second variable can also contain 10 per cent of censored observations, but the samples with censored values do not need to be the same as those with censored values in the first variable. In the worst case all censored values occur in different samples and this will accumulate to 20 per cent of censored observations for the multivariate data set. Before entering multivariate analysis, it is thus important to check how many observations (samples) are plagued with missing values for any variable. The proportion of such samples should be as low as possible. If standard multivariate methods are to be used, the total proportion of samples with censored data should probably not exceed 10 per cent. Even when planning to use robust methods (which could in theory handle up to 50 per cent of "outliers"), care is necessary because further data quality issues or missing values in the remainder of the data set may exist.

The problem of censored data can create a serious dilemma when dealing with several sample materials that are to be compared. For example, when excluding variables with more than 5 per cent of censored data from all four sample materials collected during the Kola Project (moss, O-, B-, and C-horizon), only 24 (out of more than 40) variables remain in common for direct comparison.

To avoid as many "DL-problems" as possible, one of the most important first steps in the design of the Kola Project was to construct a list of all the chemical elements for determination, setting a "priority" for each element (how important is the element for project success), and identifying the detection limit that was needed to obtain a complete data set for this element for all of the sample materials. This list can then be compared against the analytical packages offered by different laboratories together with their respective detection limits. The list can then be used to discuss detection limits, costs (if appropriate), and terms with the most suitable laboratories. This list is of course an "ideal" that cannot be achieved for all elements. For some elements no analytical technique may exist that provides a low enough DL for obtaining a "complete" data set (no sample <DL) at the time of the project. For others, procedures may exist but be so expensive that it is not possible to use them within a given budget. Thus in the end the final choice of the elements to be determined, the detection limits, and the choice of the laboratory that undertakes the work will be a compromise. It is always good practice to archive a complete sample set to be recoverable later when analytical techniques are improved (or cheaper) or interest in some element(s) arises that was of no interest at the time of the original project.

2.3 Missing values

In any large data set some values may be missing due to a variety of reasons. It may be that there was not enough sample material to carry out a certain determination; it may be that a single number got lost somewhere in the data processing. Such situations lead to values that are missing at random. A different situation arises if there are missing samples due to some particular event. In that case the missing values could be concentrated in particular parts of the data distribution or survey area.

If any variable or sample has many missing values, this will again create problems for statistical data analyses. In contrast to censored data having some information value (e.g., that the concentration in the sample is very low, lower than the "limit of detection"), missing data are "no information" and are not limited to low (or very high – upper limit of detection) concentrations. Several possibilities to treat missing values exist:

- Delete the whole variable or all samples with missing values from data analysis;
- Set the missing result to an arbitrarily chosen value; or
- Try to predict the missing values, for example via multiple regression (imputation) techniques using all other analytical data (Chapter 16).

None of these suggestions is an optimal solution. The most usual approach to the missing value problem is to just ignore them. This is a possibility as long as the number of missing values is low. When a data set contains missing values, it is generally a good idea to only use robust statistical methods.

To predict a missing value, for example via regression techniques (imputation – Chapter 16), using all other analytical results or kriging (for spatial data – Chapter 5), is only sensible when the value is really needed, e.g., for mapping.

If the missing values do not occur at random, this will result in biased statistical estimates, independently of the selected approach for accommodating the missing values.

For multivariate data analysis the same problem of "accumulating" missing values occurs as described above for "<DL" data. Therefore, obtaining as complete a data set as possible is always an important consideration during project design.

2.4 Some "typical" problems encountered when editing a laboratory data report file to a DAS+R file

A data file as received from a laboratory will usually contain a lot of text that is necessary only for general information purposes. Figure 2.3 shows a screen snapshot of one of the Kola Project result files as received from the laboratory. This kind of file cannot be easily imported by most data analysis software. The file will probably contain information on the analytical technique(s) used for the determination of each element, the detection limit, the unit of the measurement for each variable, and more. Some of these text lines contain the information DAS+R can hold in the first 11 comment rows, others may need to be deleted or edited. Files may contain "accessory" information or comments at the beginning as well as at the end of the data array (check the bottom line of any file – see Figure 2.3), or this information may be given in other worksheets (check for multiple work sheets!) within the file. If the laboratory has invested some time in table design and merged cells somewhere, e.g., at the top of the file, this will create severe problems when trying to import such a file into data analysis software.

In contrast an "ideal" file for most data analysis packages will only contain one row identifying the variables (with the first variable being the "Identifier" for the sample (i.e. the sample number), and then row by following row will contain the identifier and the analytical results

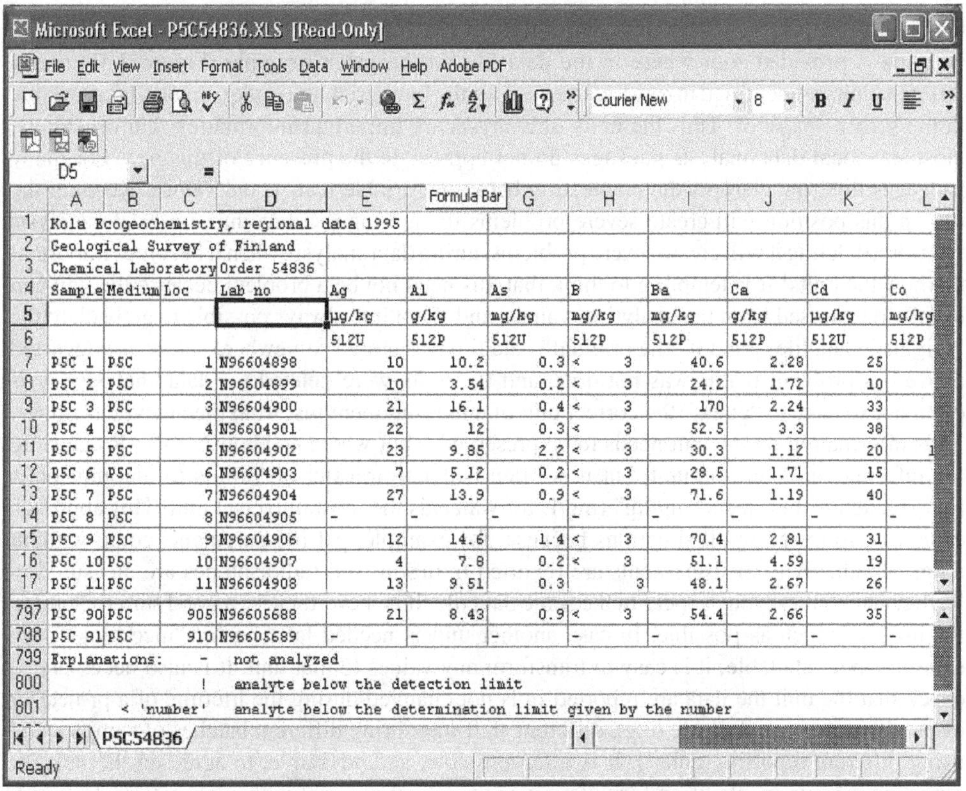

	A	B	C	D	E	F	G	H		I	J	K	L
1	Kola Ecogeochemistry, regional data 1995												
2	Geological Survey of Finland												
3	Chemical Laboratory			Order 54836									
4	Sample	Medium	Loc	Lab no	Ag	Al	As	B		Ba	Ca	Cd	Co
5					µg/kg	g/kg	mg/kg	mg/kg		mg/kg	g/kg	µg/kg	mg/kg
6					512U	512P	512U	512P		512P	512P	512U	512P
7	P5C 1	P5C	1	N96604898	10	10.2	0.3	<	3	40.6	2.28	25	
8	P5C 2	P5C	2	N96604899	10	3.54	0.2	<	3	24.2	1.72	10	
9	P5C 3	P5C	3	N96604900	21	16.1	0.4	<	3	170	2.24	33	
10	P5C 4	P5C	4	N96604901	22	12	0.3	<	3	52.5	3.3	38	
11	P5C 5	P5C	5	N96604902	23	9.85	2.2	<	3	30.3	1.12	20	1
12	P5C 6	P5C	6	N96604903	7	5.12	0.2	<	3	28.5	1.71	15	
13	P5C 7	P5C	7	N96604904	27	13.9	0.9	<	3	71.6	1.19	40	
14	P5C 8	P5C	8	N96604905	-	-	-	-		-	-	-	-
15	P5C 9	P5C	9	N96604906	12	14.6	1.4	<	3	70.4	2.81	31	
16	P5C 10	P5C	10	N96604907	4	7.8	0.2	<	3	51.1	4.59	19	
17	P5C 11	P5C	11	N96604908	13	9.51	0.3		3	48.1	2.67	26	
797	P5C 90	P5C	905	N96605688	21	8.43	0.9	<	3	54.4	2.66	35	
798	P5C 91	P5C	910	N96605689	-	-	-	-		-	-	-	-
799	Explanations:		-	not analyzed									
800			!	analyte below the detection limit									
801			< 'number'	analyte below the determination limit given by the number									

Figure 2.3 One of the "original" laboratory data files as received for the Kola Project C-horizon results

for all variables listed in the first row in the same sequence (see above). If possible such a file format should be agreed upon with the laboratory at an early stage of the project. The "optimal" data report file from a laboratory will look like the files in Figures 2.1 or 2.2 rather than like a "nice table".

2.4.1 Sample identification

Laboratories often add their own "laboratory number" to all samples received for analysis according to the laboratory's standards (see Figure 2.3). This laboratory number is used as a "unique" number for storage in the laboratory's database. Make sure that the laboratory reports the results according to the original sample numbers and not with these "new" laboratory numbers only. In the case where samples were randomised and given new numbers prior to submission (see Chapter 18), it is necessary to ensure that the laboratory analyses the samples in the exact sequence of the new numbers and not in the sequence of the laboratory's own numbers. This requires that the "submitter" must keep and safely back-up the table that relates the original sample numbers to the new randomised sequence numbers that are on the physical samples submitted to the laboratory.

2.4.2 Reporting units

Units used to report the values are often quite different from variable to variable and information on units is provided somewhere in the data file. In the given example (Figure 2.3) several different units as identified in row five are used for the few variables (μg/kg, mg/kg, g/kg) visible in the screen snapshot. Thus the units of analysis are important information. Unfortunately, most statistical data analysis packages do not appreciate the presence of this row. For many packages this row disrupts the connection between variable names and values. Keeping the row in this position will create severe problems in entering the data into most data analysis software, deleting it will create severe problems during data analysis when it becomes important to know the units! It is tempting to think that this need not be a problem because the unit can usually be guessed from the analytical values, and if not it is always possible to go back to the "original" data file (good if "the original" data file was stored somewhere safely as suggested above – a problem if this was not done and the units were deleted for data analysis). It is often not so easy to "guess" the correct unit of a measurement just from the analytical values. Thus information on the unit needs to be preserved – but where and how? DAS+R can store the information about the unit, but it is often better to transfer all results to one unit only. For geochemical projects "mg/kg" (mg/L for water) is the most universal unit. This approach cannot be followed in all situations because, for example, pH measurements come without a unit, conductivity measurements are reported in mS/m, and radioisotopes are measured in Bq/kg, and all these may appear in the same data file. It is, nevertheless, a good start to "unify" the units as much as possible. In case another unit is needed later on for "nice looks", for example in a data table, it is easy to transform any values to that unit. It is also necessary to check that the unit the data are reported in is not changed during the lifetime of a project or even within just one data file (e.g., different staff measuring different batches of samples and using different reporting units!). It is often possible, and advisable, to agree on the unit for reporting measurements when formulating an analytical contract with a laboratory. That way a lot of unnecessary and always dangerous file editing can be avoided.

2.4.3 Variable names

Variable names need to be unique. For the laboratory it may be sufficient to use the same variable name if the method code or extraction is identified as different in another row of the data file – for data analysis this is not sufficient! If one and the same element is analysed by a variety of techniques or in a variety of different extractions, this information needs to be incorporated into the variable name (e.g., Ca_aa, Ca_ar and Ca_tot for Ca determined in an ammonium acetate extraction, Ca determined in an aqua regia extraction and Ca determined by XRF or INAA, giving total concentrations). The same applies when files with different sample materials are linked (e.g., Ca_MOSS, Ca_OHOR, Ca_BHOR, Ca_CHOR). Note that most data analysis programs cannot handle variable names containing a space – it is thus advisable to use another sign as separator and an underscore is a common solution. If material and method need to be included in a variable name (e.g., Ca_CHOR_ar, Ca_CHOR_tot), very long variable names may result. Note that lengthy variable names may not be convenient during data analyses (e.g., when the variable names are displayed in graphics).

2.4.4 Results below the detection limit

The laboratory will have marked results below the detection limit (Section 2.2) somehow – either via a "<" sign followed by the value of the detection limit or by the actual reading of the instrument marked in some specific way as "below detection" (e.g., via an exclamation mark before or after the result or by a "−"). Such markers are acceptable in a printed table, however, a computer or a data analysis package will not know what to do with such special symbols. In the example file (Figure 2.3) the laboratory has managed to use different ways to identify measurements that fall below the lower limit of detection. For some variables a "!" marks such values, for other samples a "<" was used. As most data analysis software cannot handle such special signs they need to be removed. However, these signs cannot just be deleted because the file will most likely contain true measurements with exactly this value, and then these true values could no longer be differentiated from values that are lower. It may also be tempting to just delete the contents of the whole cell marked as "lower than the detection limit". However, an empty cell denotes a "missing value", a value "below detection" denotes a value too small to measure with the chosen analytical method, which is an important difference when making statistical estimations. It is also no solution to replace these values with a "0" – for some variables "0" may be a true data value, importantly "0" will often provide a far too low estimate of the real concentration, and in case there are any "0" in a data file, logarithmic transformations become impossible. One standard solution to the situation (the solution chosen for the Kola data) is to replace all values marked as "lower than detection limit" by a value of one half of the detection limit. Therefore, in the example data file from the Kola Project all values marked with a "!" or a "<" need to be replaced by a value of one half of the detection limit. Before starting to replace the "<" sign, it is necessary to check that the laboratory is operating with one detection limit per variable only. It may happen that the laboratory has operated with different detection limits, depending on the matrix of the samples (a common situation in neutron activation analyses (NAA or INAA ("I" for "instrumental")). When replacing all "<" values with one half of their value in such a situation some of these replaced values may suddenly be in the range of "real" values. Thus this situation requires special care (one possible solution is to replace all the "<" values with one half of the lowest DL value reported – see Section 2.2). Afterwards it is easy to use the "find" and "replace" functions in spreadsheet software to get rid of all the "<" signs (and of the "!").

When actually doing this for the example file, it turned out to be more difficult than antici-pated – the laboratory had used blanks (sometimes none or one, two, three, four, and even five blanks) between the actual number and the sign. Each single one of these possibilities needs thus to be checked until in the end there are no more "<" or "!" in the data file. Thus great care is necessary when replacing all values below detection with a value of one half of the detection limit. To avoid confusions (e.g., when a file contains values <1 and <10) this should be done variable by variable when using "find" and "replace" options of the spreadsheet software rather than for the whole data file at once. It is good practice to agree with the laboratory beforehand exactly how values below (or above) detection limit are to be reported. As can be seen this editing process can be error prone; this just emphasises the importance of making a back-up copy of the original file received from a laboratory, so it can always be referred to in case it is thought that an editing error may have occurred.

2.4.5   Handling of missing values

Empty fields may exist in a data file, indicating "missing values" (Section 2.3), e.g., a sample that got lost during transport or where there was not enough material available for all the different analytical techniques that were ordered. Never replace empty fields by a number (for example it could be tempting to use a "0"). Empty fields indicate "no information", if these fields are replaced with zeros, which may be valid for a measurement, their original meaning will be lost. The "empty" information thus needs to be preserved or replaced by a term that the software recognises as "missing value". In Figure 2.3 in row 14 (and row 798) the special sign "−" was used by the laboratory to identify whole samples or single variables for a sample that were not analysed. Here the "−" thus marks "missing values", a directly dangerous approach. Such a "−" should be immediately replaced by an empty cell. When doing this replacement, it is wise to use the "find entire cells only" option to ensure that no real "−" sign, identifying possibly existing negative values, is replaced. Also, it is not a good idea to just delete a whole row with "missing values". At some point it may either be important to identify the samples with missing values, to know the absolute number of missing values, or another data file will be linked with the first data file, and this new file may contain data for the sample that contains missing values in the first file. The standard in R to identify missing values is to use NA (not available) as marker for a missing value, NA is automatically inserted when a data file containing empty fields is imported to R (or DAS+R). Thus empty fields should be left empty when editing the data file.

2.4.6   File structure

In case the file contains a column identifying the variables and the results follow in columns rather than in rows, it is possible to use the "copy" and "paste special" – "transposed" - features in a spreadsheet to produce a transposed file that follows the above standard. Again it is essential to ensure that the original data file is safely stored somewhere before attempting to transpose the file so that it cannot be destroyed by this procedure.

Data tables as received from different laboratories as spreadsheets may have a very dissimilar structure, and thus there is no simple way around a substantial amount of file editing, no matter whether these files are to be incorporated into a database or directly imported into a data analysis package.

For R (outside DAS+R) there exists the possibility to mark several leading (top) rows with a "#". The software will ignore lines marked this way (just like comment lines in a computer program) when importing data, and the information is at least not deleted. Data files prepared with all auxiliary information for DAS+R can thus be directly read into other R-routines using this facility without having to re-edit the file.

2.4.7 Quality control samples

However, before the file is really ready for import into R or DAS+R, more work may still be required. Hopefully the file contains Quality Control (QC) samples (i.e. duplicates and standards inserted among the "real" samples). These QC-samples must be removed prior to data analysis (and used for quality control (see Chapter 18)). The "key-file" identifying QC-samples needs to be used to identify and remove the QC samples from all other samples.

2.4.8 Geographical coordinates, further editing and some unpleasant limitations of spreadsheet programs

For geochemical mapping and thus identifying the location of unusual results the samples also need coordinates. Thus these coordinates need also to be linked to the data file. Furthermore, analytical results may be received from more than one laboratory, and all these different results will be needed in just one large file for data analysis. A "cut" and "paste" job in a spreadsheet has to be carried out with great care. If very many variables exist, or if results for different materials are to be linked into just one file, it is possible to reach the limits of the most common spreadsheet software packages. These are usually limited to 256 columns. In such a case it is also possible to link the different files in R (or DAS+R). Alternatively, those users familiar with database software, e.g., Microsoft Access™, may choose to import the different field, laboratory, and other tables into the database manager and undertake the editing and final table construction in that environment.

Many packages used for data analysis will not be able to read the special spreadsheet file format (e.g., ".xls" files). Furthermore, software manufacturers sometimes change their formats with each upgrade of their spreadsheet software with the consequence that older versions of the program are no longer able to read files that were saved in a newer version. This can turn out to be a real nuisance when data files are to be exchanged between colleagues. Thus the "final" file should be saved separately in both the spreadsheet format (for easy handling and editing) and in a simple text or ascii format. For this it is best to use the ".csv" format, which is a simple file format where the variables and values are separated by commas. To use the ".csv" format, it is necessary to make sure that the comma was not used as decimal sign in the file. This may actually even depend on the "country" setting of the spreadsheet used for editing. Note that the ".csv" format does not support multiple worksheets; information stored in different worksheet will be lost when storing a former Microsoft Excel™ file in ".csv" format. The major advantage of this format is that almost all "foreign" software packages are able to import the ".csv" format. This is the file format used to import all Kola data into R (and DAS+R).

2.5 Appending and linking data files

The Kola Project was a typical multi-element multi-medium regional geochemical mapping project. In consequence there is not only one regional data set, but there are four such regional data sets, one for each sample material (there are actually even more data sets, one for Topsoil (0–5 cm) and one for lake water (Russia only)). Thus similar files with analytical results were received for the C-, B- and O-horizon of Podzol profiles and for terrestrial moss, all collected at the same sample sites. All these files needed to be prepared according to the above example. The prepared files are provided in ".csv" format on the web page `http://www.statistik.tuwien.ac.at/StatDA/R-scripts/data/`.

However, it may be required to do more with the data than just look at them layer by layer. A direct comparison between the different sample materials will be of interest, it may thus be necessary to be able to produce diagrams and statistics using all four layers at the same time. For that it is necessary to merge the different data sets in some sensible way. This can actually be done in two different ways, and depending on how it is done, different comparisons and diagrams can be produced.

For example, it is possible to just add additional rows to the "first" file (Moss or C-horizon) for each additional material. Hopefully as many variables as possible are identical (i.e. the same analytical program was used for all materials), otherwise this is a substantial data-editing job, which needs a lot of care. Now the sample material needs to be identified in one of the variables, e.g., as "MOSS", "OHOR", "BHOR" and "CHOR". Using this variable it is then possible to prepare "data subsets" according to sample material and compare results for these subsets in tables and different diagrams. It is possible to append data directly in DAS+R. However, it may also be necessary to be able to make a direct comparison sample for sample. This is not easily achieved with such a file structure.

Hopefully this possibility was considered when designing the fieldwork and sample-numbering scheme (data analysis starts with project design). Many organisations may require that a "unique sample number" per sample should be used because organisations such as a Geological Survey often build a project-independent database and want to minimise the chance of mixing up sample numbers. However, working with unique sample numbers can create a nightmare during data analysis because it is necessary to somehow tell the data analysis software which samples are related. In cases where unique sample numbers were used, this will have to be done sample by sample – an unnecessary and time-consuming task that can waste a lot of time for a large data set.

For the Kola Project this issue was considered during project design as it was recognised that at some stage it would be informative to be able to compare results for the different sample materials. The sample number thus included a unique component that identified each catchment where all samples that belong together were taken. An additional unique code identified the sample material. Care was taken that these codes were obviously different from one another in order to minimise the chance of mixing up different materials (e.g., C-, B-, and O-horizon soil samples) in the field or later during sample preparation in the laboratory. For example, parts of the number can disappear from a sample bag during transport and handling of the samples and it should still be possible to identify the material and the identity of the sample beyond doubt. Thus "CHOR", "BHOR" and "OHOR" (as in the data files) was not used in the field but P5C, P4B and "HUM" (for "humus"). In consequence, all samples that were taken at one and the same sample site can then be easily linked via the sample number (ID in the data files). It is only necessary to add on the data from the additional materials as additional columns.

Data analysis software is often not able to handle a file containing several variables with the same name (R will recognise such a situation and "number" such variables, e.g., Ca1, Ca2, Ca3). For the Kola data set, in many instances, four variables exist with the same name, e.g., the variable silver (Ag) in the C-, B- and O-horizons, and moss. Thus the sample material needs to be indicated in the variable name. Again the problem arises that variable names tend to get too long when a lot of information needs to be incorporated (e.g., different extractions and/or analytical techniques used, different grain size fractions analysed, different sample materials analysed). It may thus be beneficial to limit the number of variables included in this file to the "critical variables" where results between the materials are really comparable and most of the samples returned values above the detection limit.

As mentioned above (Section 2.4.8), spreadsheet software has clear limitations as to the number of variables that one file can contain. At present this limit is 256 – with four different sample materials with 60 variables determined, this limit is easily reached when additional field information needs to be incorporated into the file. To append data to a spreadsheet requires that at the very beginning a complete set of "sample numbers" (and coordinates) existed for all materials and that no rows in the data files were deleted just because only "missing values" were present for one of the samples in one of the different layers sampled. It is also necessary to treat the field and analytical duplicates in the different layers the same way (i.e. always removing the second result), otherwise a considerable amount of extra editing work can result. Instead of doing all this in a spreadsheet where this kind of work requires a lot of attention, files can be directly linked in DAS+R using the link command, which will automatically link all samples with the same ID. Or alternatively, this can be done externally in a database package.

In the end there may exist as many as six separate data files: one for each sample material (e.g., CHOR, BHOR, OHOR, and MOSS), one with all results and the sample material encoded in the variable name for the creation of subsets according to sample material, and a last file with four results per analysed element (e.g., Ag_C, Ag_B, Ag_O, and Ag_M), one for each material. Using six different files can create problems if results of one sample are changed or deleted somewhere in one data file during some point of data analysis – it is thus necessary to consider the possibility of keeping all the files up-dated to the newest version at all times, as is done when using a database.

2.6   Requirements for a geochemical database

A geochemical database can be relatively easily built using existing software solutions like Oracle[TM] or Microsoft Access[TM]. A number of questions need to be addressed when building a geochemical database. A database needs an administrator to guarantee its longevity. It is very difficult to build an "ideal" institutional database for an organisation because this requires an almost inhuman amount of foresight in terms of what kind of work will be carried out in future projects. A database constructed without an in-depth system analysis can easily turn into quite a secure data grave, with the guarantee that the data are there and safely stored but never used because they are too difficult to retrieve in the form needed for further data analysis. Even building a good database for just one project requires a thorough system analysis – what are the data going to be used for, and in which form does one need to recover the data for data analysis are some of the very first questions that need to be answered before starting to construct a project database.

In general the data should be entered into the database as soon as possible after collection to minimise the chance that any are lost or important details are forgotten. On first sight, constructing a database may look like an additional bureaucratic effort; in the long run it will pay back the time spent on database design. These days most data are provided by laboratories in the form of spreadsheets, e.g., Microsoft Excel™ or Quattro Pro™ files. Unfortunately, such files are not a secure medium to archive data. However, they do provide a widely available and familiar tool for original data capture and reporting, and some table re-arrangement functions.

2.7 Summary

For entering data into a statistical data analysis package a file structure is needed that deviates from nicely designed tables. Usually substantial data editing is necessary before being able to start any data analysis. Especially with large data tables there is a high risk that errors may be introduced during data editing. Before starting with any data editing, a copy of the original file as received from the laboratory needs to be safely stored to be able to check that any questions that may arise during data analysis are not related to mistakes introduced during data editing. This file should never be modified in any way. A data file usually contains important auxiliary information (e.g., measurement units and techniques) that need to be taken care of. At the same time variable names should be kept as short as possible. Variable names should not contain any blanks or any language specific special characters. When working with chemical analyses, it is important to consider the handling of values below (or above) the detection limit, and missing values. It is also important to consider how to preserve information on sample preparation, analytical technique used, detection limit reached and the data unit. Much work can be spared if a contract is formulated with the laboratory on how the report file for a project should be constructed.

It is an advantage to work with simple standardised file formats, e.g., "csv" (comma separated values) or "txt" because these are independent of the computer's operating system.

3
Graphics to Display the Data Distribution

As the Kola Project progressed, chemical analyses for c. 50 chemical elements in four different sample materials collected at about 600 sample sites were received. These needed to be summarised, compared and mapped; now we enter the realms of statistical data analysis. For data collected in space, two types of distribution need to be considered: the spatial distribution of the data in the survey area and the statistical data distribution of the measured values. At the beginning of data analysis the focus is on the statistical distribution, temporarily setting the spatial component aside (see Chapter 5, where the spatial data distribution is studied).

The statistical data distribution can be explained using some small artificial data sets. The data are simply plotted against a straight line as a scale for the data values. Depending on the data, the distribution of the values can look symmetrical (Figure 3.1, upper left, the data are distributed symmetrically around the value 20). The data could fall into two groups, i.e. be bimodal, and show a gap around the value 20 (Figure 3.1, upper right). The data could show a single extreme outlier (Figure 3.1, lower right). The data could show an asymmetrical distribution with a high density of points at the left side and a decreasing density towards the right side (Figure 3.1, lower right). All these different characteristics, or a mixture thereof, define the statistical data distribution. For further statistical treatment of the data it is essential to get a good idea about the data distribution. Many statistical methods are, for example, based on the assumption of a certain model for the data distribution (see Section 4.1, Figure 4.1 for a number of different model distributions).

With more than 600 measurements it will not be sufficient to simply plot the measured values along an axis as in Figure 3.1, because many data values will plot at the same point. Other graphics are needed to visualise the data distribution. These will be introduced on the following pages.

3.1 The one-dimensional scatterplot

Plotting the data along a straight line works fine as long as the data set is small and the data do not plot too close to, or on top of, one another. Figure 3.2, upper diagram, demonstrates the problem using the analytical results of Sc in the Kola C-horizon. Many more data points may be hidden behind one point plotted along the line. To view samples that have the same

Statistical Data Analysis Explained   Clemens Reimann, Peter Filzmoser, Robert G. Garrett, Rudolf Dutter
© 2008 John Wiley & Sons, Ltd.

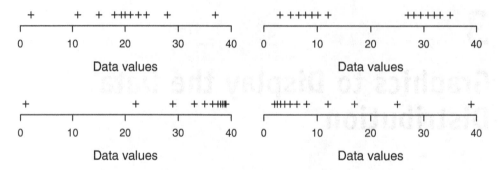

Figure 3.1 Some possible data distributions. Upper left: symmetric; upper right: bimodal with a gap around 20; lower left: left skewed; lower right: right skewed

value it is, however, possible to add a second dimension and add such values as additional symbols against the *y*-axis (Figure 3.2 middle, stacked scatterplot). With many values at the same position the *y*-axis starts to dominate the plot. With some further modifications the one-dimensional scatterplot is obtained (see, e.g., Box *et al.*, 1978); an informative and simple

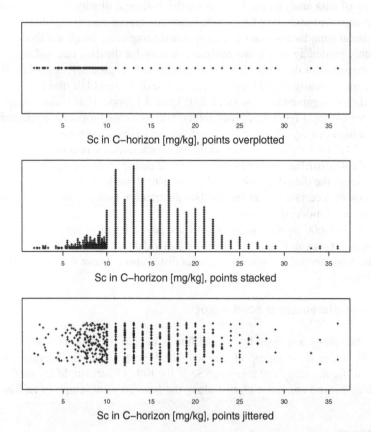

Figure 3.2 Evolution of the one-dimensional scatterplot demonstrated using Sc as measured by instrumental neutron activation analysis (INAA) in the samples of the Kola C-horizon

graphic to study the data distribution. This plot is typically displayed as an elongated rectangle. Each value is plotted at its correct position along the x-axis and at a position selected by chance (according to a random uniform distribution) along the y-axis (Figure 3.2, lower diagram). This simple graphic can provide important insight into structure in the data.

In Figure 3.2 (stacked and one-dimensional scatterplot) a significant feature is apparent that would be important to consider if this variable were to be used in a more formal statistical analysis. The data were reported in 0.1 mg/kg steps up to a value of 10 mg/kg and then rounded to full 1 mg/kg steps – this causes an artifical "discretisation" of all data above 10 mg/kg.

3.2 The histogram

One of the most frequently used diagrams to depict a data distribution is the histogram. It is constructed in the form of side-by-side bars. Within a bar each data value is represented by an equal amount of area. The histogram permits the detection at one glance as to whether a distribution is symmetric (i.e. the same shape on either side of a line drawn through the centre of the histogram) or whether it is skewed (stretched out on one side – right or left skewed). It is also readily apparent whether the data show just one maximum (unimodal) or several humps (multimodal distribution). The parts far away from the main body of data on either side of the histograms are usually called the tails. The length of the tails can be judged. The existence or non-existence of straggling data (points that appear detached from the main body of data) at one or both extremes of the distribution is also visible at one glance. The Kola C-horizon data set provides some good examples.

Figure 3.3 shows four example histograms plotted for the variable Ba (aqua regia extraction) from the Kola C-horizon data set. The x-axis is scaled according to the range of the data, the starting point is usually a "nice looking" value slightly below the minimum value. Intervals along the x-axis are adapted to the number of classes it is required to display, and the number of classes is chosen such that the whole data range of the variable is covered (several rules of thumb exist for the "optimum" interval length or number of classes, one of the easiest is $\sqrt{n}$ for the number of classes, where n is the number of individuals/samples in the data set). The y-axis shows the number of observations in each class or, alternatively, the relative frequency of values in percent.

In theory, when dealing with a very large data set, the length of the intervals could be decreased so much by increasing the number of classes that the typical histogram steps disappear. This results in a plot of a smooth function, the density function. This smooth function is thought to represent the distribution from which the data are sampled by chance. If the data were drawn from a normal distribution the density function would take on the classical bell shape (see Figure 4.1).

One situation that often arises with environmental (geochemical) data is that the distributions are strongly right-skewed. In addition, extreme data outliers occur quite frequently. In such cases a histogram plotted with a "linear" scale may appear as a single bar at the left hand side and a far outlier at the right hand side of the graphic. Such histograms contain practically no information of value about the shape of the distribution (see upper left histogram, Figure 3.3). Statisticians will usually solve such problems via scaling the data differently. In the case at hand it appears advisable to reduce the influence of the high values and to focus on the main body of data that lie in the first bar of the histogram. One solution that meets these requirements is to scale the data logarithmically (lower right histogram, Figure 3.3). This re-scaling is called log-

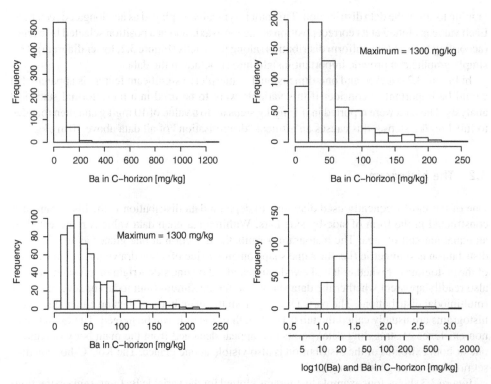

Figure 3.3 Histograms for Ba (aqua-regia extraction) from the C-horizon of the Kola data set. Upper left: original data; upper right: truncated data; lower left: truncated data as upper right but with double the number of classes; and lower right: log-transformed data. The lower right histogram has two scales to demonstrate the logarithmic transformation, the upper scale indicating the logarithms and the lower scale the original untransformed data

transformation of the data; a variety of data transformations are used in statistics to improve the visibility (or more generally, the behaviour) of the data (see Chapter 10). In geochemistry the log-transformation is the transformation that is most frequently used. Note that in the example the log-transformation results in an almost symmetrical distribution of the strongly right-skewed original data and that the range (i.e. maximum value − minimum value) of the data is drastically reduced because the influence of the outliers has been reduced. Plotting a histogram of the log-transformed values has, however, the disadvantage that the direct relation to the original data is missing (upper scale on the x-axis). This can be overcome by using a logarithmic scale for the x-axis (lower scale on the x-axis), this is the procedure adopted by DAS+R.

To permit scaling in the original data units, which are easier to read than a log-scale, it is also possible to plot a histogram for a certain part of the data only, e.g., for a certain data range (upper right histogram, Figure 3.3, range 0 to 250 mg/kg). In that case the outliers are not displayed; and the "real" maximum value should be indicated in the plot so that an unsuspecting reader does not gain an erroneous impression about the complete data distribution from the truncated histogram. Plotting more classes will result in a better resolution of the distribution (lower left) and will often considerably change the histogram's shape. Plotting too many classes will result in ugly gaps in the histogram.

The choice of starting point and class interval will substantially influence the appearance of the resulting histogram. This is shown in the lower left, where the only difference from the histogram in the upper right position is that the number of classes was increased from 13 to 26 (according to the simple $\sqrt{n}$ rule of thumb, 25 would be the "optimum" number of classes for the Kola Project C-horizon data). While the lower right histogram could be taken as an indication that a lognormal distribution is present (i.e. the log-transformed data follow a normal distribution), the more detailed histogram suggests that this may be in fact a multimodal distribution, approaching symmetry if log-transformed. The histogram could clearly be misused to demonstrate a "lognormal" distribution by reducing the number of classes (a statistical test of the distribution – see Chapter 9 – will indicate that even the log-transformed data for Ba do not follow a normal distribution). It may also be possible that the spikes in the distribution only appear because too many classes were chosen for the number of samples – the spikes might then be artefacts of the way the data were reported by the laboratory, e.g., to the nearest 1, 2, 5, or 10 mg/kg. This demonstrates how important the choice of the optimum interval length (number of classes) is when constructing histograms. Modern software packages will automatically use more sophisticated mathematical models than the $\sqrt{n}$ rule of thumb for that purpose (see, e.g., Venables and Ripley, 2002).

It can be concluded that by studying or displaying just one histogram, important aspects of the distribution may be missed. It may thus be necessary to plot a series of histograms for a variable (as above), or the histogram should be augmented with some other graphics displaying the data distribution and showing additional features.

One possibility is to combine the histogram with the one-dimensional scatterplot (Figure 3.4).

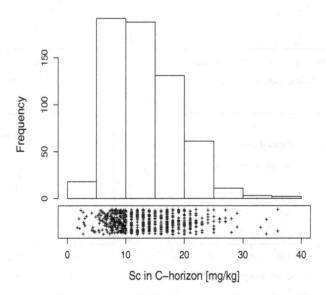

Figure 3.4 One-dimensional scatterplot for Sc (total concentrations as determined by INAA) in the Kola C-horizon data set, combined with the histogram

3.3   The density trace

The density trace (kernel density estimate – Scot, 1992) is on first inspection a "smoothed line, tracing the histogram" (Figure 3.5). It is another approximation of the underlying density function of the data. Each point along the curve is calculated from data within a defined bandwidth using a "weight function" (in most packages; including R, these parameters are chosen by default but can be changed manually). Choice of bandwidth and weight function will crucially determine how the final density trace appears (compare Figure 3.5 density trace upper right with density trace lower left). Although at first glance the density trace is a more "objective" graphic to display the data distribution than the histogram, it can also be manipulated substantially. Density traces are better suited for comparing data distributions than histograms because they can easily be plotted on top of one another (e.g., in different line styles or colours – see Chapter 8). The plots (Figure 3.5) are for the distribution of Ba from the Kola C-horizon data set previously displayed as histograms (Figure 3.3). Compared to the histogram not much information is gained. More or less the same problems are experienced as above when plotting the histograms. Note that for plotting the density trace of the log-transformed data in the lower right display, a log-scale was used to preserve the direct relation to the original data range (Figure 3.5).

A density trace can also be combined with a histogram to give a more realistic impression of the data distribution.

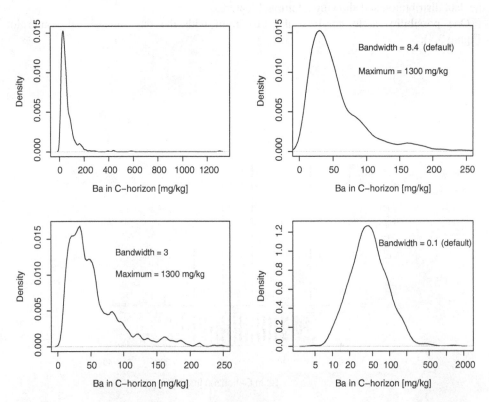

Figure 3.5   Density traces for Ba (aqua regia extraction) from the C-horizon of the Kola data set. Compare to Figure 3.3

3.4 Plots of the distribution function

Plots of the distribution function, e.g., the cumulative probability plot, were originally introduced to geochemists by Tennant and White (1959), Sinclair (1974, 1976) and others. They are one of the most informative graphical displays of geochemical distributions. There exist several different plots of the distribution function that all have their merits.

3.4.1 Plot of the cumulative distribution function (CDF-plot)

Histogram and density trace are based on the density function. However, the percentage of samples plotting above or below a certain data value x could be of interest. The percentage equals an area under the curve of the density function. Percentages can also be expressed as probabilities, i.e. the probability that a value smaller than or equal to the chosen data value x_1 (Figure 3.6) appears is p_1. This probability can be taken as a point in a new plot, where the data values along the x-axis are plotted against probabilities along the y-axis (Figure 3.6). By varying the chosen value x, additional probabilities are obtained that can then be drawn as new

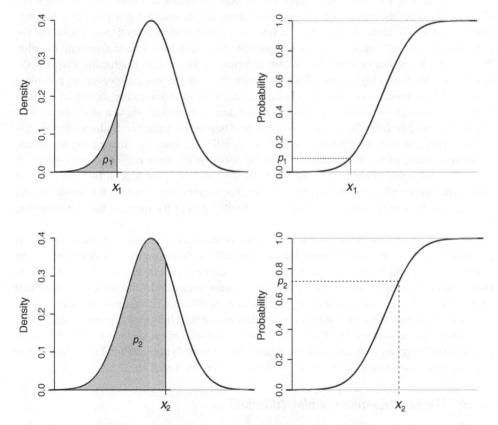

Figure 3.6 Construction of the cumulative distribution function (CDF-plot). The graphic to the left shows the density function and the graphic to the right shows the resulting cumulative distribution function

points in the right hand plot for the corresponding x values (e.g., p_2 for the data value x_2). If this is carried out for the whole x-range, a new plot is generated, the cumulative distribution function or CDF-plot (right hand diagrams).

If the data follow a normal distribution (bell shape in the left hand plot), the distribution function will have a typical symmetrical, sigmoidal S-shape (right hand plot).

3.4.2 Plot of the empirical cumulative distribution function (ECDF-plot)

Just like trying to approach the density function via plotting histograms or density traces, it is possible to approach the distribution function by displaying the empirical cumulative distribution function (the ECDF-plot). The ECDF-plot is a discrete step function that jumps with each data value by $\frac{1}{n}$, where n is the number of data points. As n becomes increasingly large, to infinity, this function will approximate the underlying distribution function.

The ECDF-plot shows the variable values along the x-axis (either scaling according to non-transformed data or following a transformation, e.g., logarithmic). The y-axis shows the probabilities of the empirical cumulative distribution function between 0 and 1 (which could also be expressed as percentages 0–100 percent).

As an example, the gold (Au) results for the Kola Project C-horizon soils are displayed (Figure 3.7). The histogram and density trace demonstrate that the distribution is not normal but extremely right skewed, as a result the typical S-shape mentioned above for the CDF-plot (Figure 3.7, upper right) is not present. Two very high values dominate the plot. The ECDF-plot of the original data is still informative because it graphically displays the distance of these two high values from the main body of the data. However, to provide a more useful visualisation of the main body of data, a log-transformation should be applied. Following log-transformation, the histogram and density trace still show a slight right skew (Figure 3.7, middle left). The resulting ECDF-plot begins to display an S-shape (Figure 3.7, middle right), however, the right skew is still clearly reflected. Instead of using a log-transform, the plotting range of the data could be limited to focus on the main body of data (Figure 3.7, lower left and right). This permits the study of the main body of data in far greater detail than in the upper plot. Displaying the data in this manner requires that the viewer be reminded that the diagrams are displaying only a limited part of the range of the complete data set.

One of the main advantages of the ECDF-plot is that every single data point is visible. A geochemist would now start to search for any unusually high (or low) values and breaks in the distribution. Very high values might, for example, indicate a mineral deposit or anthropogenic contamination source. A break in the distribution could be caused by the presence of different natural factors like geology, weathering and climate, or different contamination sources influencing the data. In a large regional survey, the data may reflect both multiple natural processes and anthropogenic sources. The ECDF-plot can be used with advantage to identify classes for geochemical mapping that have a direct relation to the underlying statistical data distribution via assigning class boundaries to these breaks (see Chapters 5 and 7).

3.4.3 The quantile-quantile plot (QQ-plot)

It is often quite difficult to judge whether the S-shape as displayed in the ECDF-plot indicates a normal or a lognormal (or some other) distribution. When it is necessary to judge the underlying distribution of the empirical data, a different plot is required, and it is advantageous to change

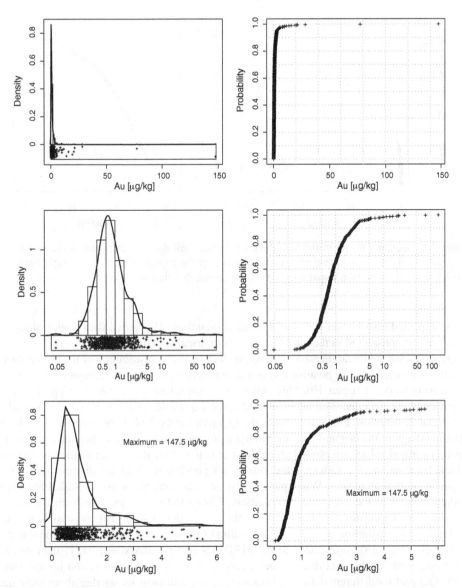

Figure 3.7 Combination of histogram, density trace, and one-dimensional scatterplot (left hand side) and ECDF-plot (right hand side) used to study the distribution of Au in the Kola Project C-horizon soil data set. Upper diagrams: original scale, middle diagrams: log-scale, lower diagrams: truncated plotting range

the y-axis scaling while keeping the x-axis the same. It is easiest to detect changes from an expected distribution if the points fail to follow a straight line. Thus the best approach is to change the y-axis in such a way that the plotted points fall on a straight line if they follow the assumed distribution (normal, lognormal, or any other). To achieve this, the cumulative distribution function must be transformed. A non-linear transformation, the inverse of the expected distribution function of the y-axis, is used for this purpose. The values of the inverse

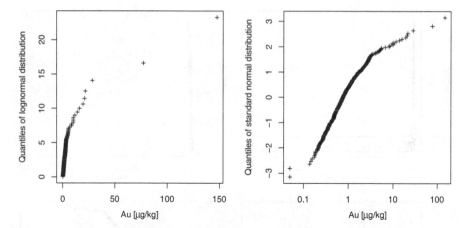

Figure 3.8   QQ-plots for Au in the Kola Project C-horizon soil data set as shown in Figure 3.6. Left hand side: *x*-axis non-transformed data, *y*-axis quantiles of the lognormal distribution; right hand side: *x*-axis log-transformed data, *y*-axis quantiles of the normal distribution

of the expected distribution function are called quantiles. The sorted data values along the *x*-axis can be considered as the quantiles of the empirical data distribution.

Quantiles are the linear measure underlying the non-linear distribution of the probabilities. Quantiles are expressed as positive and negative numbers, they can be compared to standard deviation units (see Chapter 10). This plot is called the quantile-quantile (QQ-) plot because quantiles of the data distribution are plotted against quantiles of the hypothetical normal or lognormal distribution. Figure 3.8 shows the QQ-plot for the Kola C-horizon Au data. In the left hand diagram the original data are plotted along the *x*-axis. In the right hand diagram the log-transformed values are plotted along the *x*-axis to reduce the impact of the two extreme values. By changing the scaling of the *y*-axis, it is possible to check the distribution for log-normality in both diagrams. When plotting the original data, the *y*-axis is scaled according to the quantiles of the lognormal distribution. When plotting the log-transformed data, the *y*-axis is scaled according to the quantiles of the normal distribution. This is possible because the lognormal distribution is simply the logarithm of the normal distribution. For the standard normal distribution 0 corresponds to the MEDIAN and the RANGE $[-1, 1]$ indicates the inner two-thirds of the data distribution. Unfortunately, the construction and interpretation of the QQ-plot for different data distributions requires statistical knowledge about these data distributions and their quantiles that is far beyond the scope of this book. The interested reader can consult Chambers *et al.* (1983).

In general, when checking for other distributions (e.g., Poisson or gamma distributions), the *x*-axis is always scaled according to the original data, and only the *y*-axis is changed according to the quantiles of the expected distribution.

When plotting empirical data distributions in the QQ-plot, it may still be quite difficult to determine whether the data points really follow a straight line. A simple and robust way to plot a straight line into the diagram is to connect first and third quartiles of both axes. In addition to the straight line 95 percent confidence intervals around that line can also be constructed.

The *confidence interval* encloses 95 percent of the data points that could have been drawn from the hypothetical distribution. These limits can be used to support a graphical decision

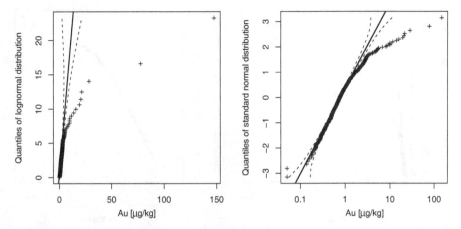

Figure 3.9 QQ-plots as shown in Figure 3.8 with a straight line indicating the hypothetical distribution and the 95 percent confidence intervals as dashed lines

as to whether the empirical data originate from the hypothetical distribution. The width of the confidence interval varies because the "allowed" deviation of data points from the straight line depends on the number of samples. In the example plot (Figure 3.9) the upper tail of the distribution clearly deviates from log-normality.

3.4.4   The cumulative probability plot (CP-plot)

The examples above demonstrate that a scale forcing the points in the plot to follow a straight line is useful. Because in the QQ-plot the scaling of the y-axis is different from distribution to distribution, it would be much easier if the y-axis were expressed in probabilities. In the pre-computer times such a plot was constructed on probability paper that was especially designed for a normal (or lognormal) distribution. This procedure was originally introduced to geochemists by Tennant and White (1959), Sinclair (1974, 1976) and others. The graph in the plot is exactly the same as in the QQ-plot. When using a computer, it is no longer necessary to limit the QQ-plot to a normal (lognormal) distribution. Any other distribution could be introduced for scaling the y-axis. If the scale on the y-axis is expressed in probabilities rather than in quantiles, the plot is generally named the cumulative probability plot (CP-plot) (Figure 3.10). Note that when checking for normality, the probabilities as expressed along the y-axis can never reach zero or one because these values would correspond to quantiles of minus infinity or plus infinity, respectively.

The CP-plot with log-scale for the data (Figure 3.10, right hand side) is especially useful because it allows direct visual estimation of the MEDIAN (50th percentile) or any other value from the x-axis or the percentage of samples falling below or above a certain threshold (e.g., a maximum admissible concentration (MAC)) from the y-axis. It also allows the assigning of a percentage to any break in the curve; in the example several breaks in the Au data are visible, the first one occurs at about 85 per cent (c. 1.5 µg/kg Au), the next at 95 per cent (c. 3 µg/kg Au) of the data. Just as in the QQ-plot, the straight line to judge whether the empirical data follow the hypothetical distribution can be shown.

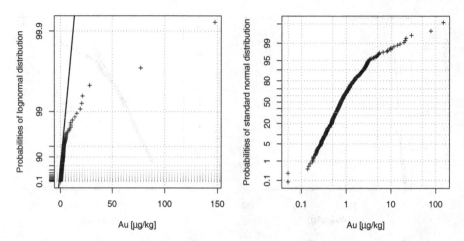

Figure 3.10 CP-plots for Au in the Kola Project C-horizon soil data set as shown in Figures 3.6, 3.7 and 3.8. Left hand side: *x*-axis non-transformed data, *y*-axis probabilities in per cent; right hand side: *x*-axis log-transformed data, *y*-axis probabilities in per cent

3.4.5 The probability-probability plot (PP-plot)

Yet another version of these diagrams is the probability-probability (PP-) plot. Instead of plotting quantiles of the hypothetical distribution against the quantiles of the data distribution at fixed probabilities (QQ-plot), the probability of the hypothetical distribution is plotted against the probability of the empirical data distribution at fixed quantiles (PP-plot). The advantage of the PP-plot is that while the QQ-plot and the CP-plot can be dominated by extreme values, these cannot dominate the PP-plot because of their low probability. The PP-plot will thus focus the attention on the main body of data. In the example plot (Figure 3.11) an additional flexure

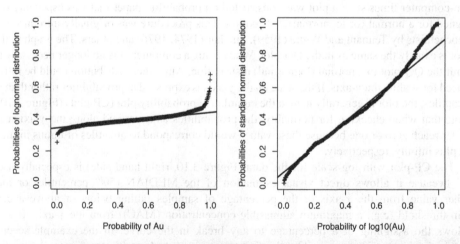

Figure 3.11 PP-plots for Au in the Kola Project C-horizon soil data set as shown in Figures 3.7, 3.8, 3.9, and 3.10. Left hand side *x*-axis: probabilities referring to the non-transformed data, *y*-axis: probabilities for lognormal distribution; right hand side *x*-axis: probabilities referring to the log-transformed data, *y*-axis: probabilities for the normal distribution

at about 20 percent of the Au distribution becomes visible. This is visible as a weak flexure in the ECDF-plot (Figure 3.7) at 0.5 µg/kg Au. The main disadvantage of the PP-plot is that the relation to the original data is completely lost, whereas it is retained in the ECDF- and CP-plot. It is of course possible to use the PP-plot in combination with the CP-plot to identify the data value for 20 percent. A combination of CP-plot and ECDF- or PP-plot may thus be quite powerful in obtaining a more complete picture of the data distribution. Single extreme values are of course hardly visible in the PP-plot though they are in the ECDF-plot. Just as in the QQ- and CP-plot, a straight line can be introduced in the PP-plot to check for agreement with a hypothetical distribution.

3.4.6 Discussion of the distribution function plots

Depending on the empirical data distribution of a given variable, the different versions of these plots all have their merits, especially when the task is to detect fine structures (breaks or flexures) in the data. If only one plot is to be presented, it is advisable to look first at the different possibilities and then select the most informative for the variable under study because this will depend on the actual data distribution of each variable. In applied geochemistry the CP-plot with logarithmic x-axis is probably the most frequently used (Figure 3.10, right). In combination with the less frequently used ECDF-plot and the almost never used PP-plot (Figure 3.11, right), it holds the potential to provide a very realistic picture of the complete data distribution.

As mentioned above, one of the main advantages of these diagrams is that each single data value remains observable. The range covered by the data is clearly visible, and extreme outliers are detectable as single values. It is possible to directly count the number of extreme outliers and observe their distance from the core (main mass) of the data.

When looking at a selection of these plots for As (Figure 3.12), several data quality issues can be directly detected; for example, the presence of discontinuous data values at the lower end of the distribution. Such discontinuous data are often an indication of the method detection limit or too-severe rounding, discretisation, of the measured values reported by the laboratory. The PP-plot corresponding to the log-transformed data shows most clearly how serious an issue this data discretisation due to rounding of the values by the laboratory can become.

Values below the detection limit, set to some fixed value, are visible as a vertical line at the lower end of the plots, and the percentage of values below the detection limit can be visually estimated. The CP-plot with logarithmic scale (middle right figure) displays this best. The detection limit for As was 0.1 mg/kg, about two per cent of all values plot below the detection limit (Figure 3.12). From 0.1 to 1 mg/kg the As values were reported in 0.1 mg/kg steps – obviously a too-harsh discretisation for the data at hand, causing artificial data structures (Figure 3.12). The presence of multiple populations results in slope changes and breaks in the plots (Figure 3.12).

3.5 Boxplots

The boxplot is one of the most informative graphics for displaying a data distribution. It is built around the MEDIAN (see Chapter 4), which divides any data set into two equal halves.

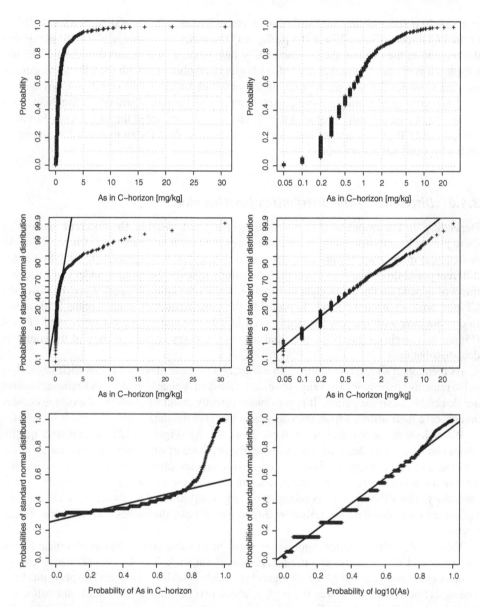

Figure 3.12 Six different ways of plotting distribution functions. Upper row: empirical cumulative distribution function plots (ECDF-plots); middle row: cumulative probability plots (CP-plots); lower row: probability-probability plots (PP-plots). Left half of diagram: data without transformation; right half of diagram: plots for log-transformed data

3.5.1 The Tukey boxplot

Tukey (1977) introduced the boxplot to exploratory data analysis. The construction of the Tukey boxplot is best demonstrated using a simple sample data set, consisting of only nine values:

2.3   2.7   1.7   1.9   2.1   2.8   1.8   2.4   5.9.

The data are sorted to find the MEDIAN:

    1.7   1.8   1.9   2.1 **2.3** 2.4 2.7 2.8 5.9.

After finding the MEDIAN (2.3), the two halves (each of the halves includes the MEDIAN) of the data set are used to find the "hinges", the MEDIAN of each remaining half:

    1.7   1.8 **1.9** 2.1 **2.3** 2.4 **2.7** 2.8 5.9.

These upper and lower hinges define the central box, which thus contains approximately 50 percent of the data. In the example the "lower hinge" (LH) is 1.9, the "upper hinge" (UH) is 2.7. The "inner fence", a boundary beyond which individuals are considered *extreme values* or potential *outliers*, is defined as the box extended by 1.5 times the length of the box towards the maximum and the minimum. This is defined algebraically, using the upper whisker as an example, as

    Upper inner fence (UIF) $= \mathrm{UH}(x) + 1.5 \cdot \mathrm{HW}(x)$.
    Upper whisker $= \max(x[x \le \mathrm{UIF}])$.

where HW (hinge width) is the difference between the hinges (HW = upper hinge–lower hinge), approximately equal to the interquartile range (depending on the sample size), i.e. Q3–Q1 ($75^{th} - 25^{th}$ percentile); and the square brackets, [...] indicate the subset of values that meet the specified criterion.

    The calculation is simple for the example data:

    Hinge width, HW $=$ UH $-$ LH $= 2.7 - 1.9 = 0.8$.
    Lower inner fence, LIF $=$ LH $- (1.5 \cdot \mathrm{HW}) = 1.9 - (1.5 \cdot 0.8) = 0.7$.
    Upper inner fence, UIF $=$ UH $+ (1.5 \cdot \mathrm{HW}) = 2.7 + (1.5 \cdot 0.8) = 3.9$.

By convention, the upper and lower "whiskers" are then drawn from each end of the box to the furthest observation inside the inner fence. Thus the lower whisker is drawn from the box to a value of 1.7, the lower whisker and minimum value are identical, and the upper whisker is drawn to the value of 2.8. Values beyond the whiskers are marked by a symbol, in the example the upper extreme value of (5.9) is clearly identified as an extreme value or data outlier (see Chapter 7) and is at the same time the maximum value.

    Figure 3.13 shows a "classical" Tukey boxplot for Ba in the Kola C-horizon samples. No lower extreme values or outliers are identified. The lower inner fence is lower than the minimum value, and thus the lower whisker terminates at the location of the minimum value. The upper inner fence and termination of the upper whisker fall together in this example, and all values to the right of the upper whisker are identified as upper extreme values or outliers.

    In summary, the Tukey boxplot – one of the most powerful EDA graphics – shows in graphical form:

- the "middle" (MEDIAN) of a given data set, identified via the line in the box;
- spread (see Chapter 4) by the length of the box (the hinge width);
- skewness (see Chapter 4) by the symmetry of the box and whisker extents about the median line in the box;
- kurtosis (see Chapter 4) by the length of the whiskers in relation to the width of the box;
- the existence of extreme values (or outliers – see Chapter 7), identified by their own symbol.

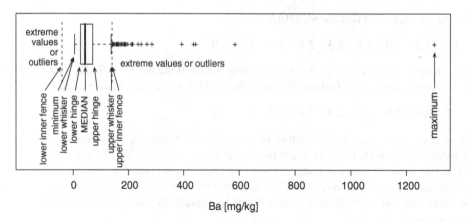

Figure 3.13 Tukey boxplot for Ba in the Kola C-horizon soil samples

Furthermore, because the construction of the boxplot is based on quartiles, it will not be seriously disturbed by up to 25 percent of "wild" data at either end of the distribution. It is not even seriously influenced by widely different data distributions (Hoaglin *et al.*, 2000).

3.5.2 The log-boxplot

It is important to recognise that the calculation of the whiskers in the above formula assumes data symmetry, lack of which is easily recognised by the median line not being close to the middle of the box. The recognition of extreme values and outliers is based on normal theory, and the standard deviation (SD) of the distribution is estimated via the hinge width (HW) or interquartile range (IQR). So for the estimate of SD to be appropriate, there has to be symmetry in the middle 50 percent of the data (see Chapter 4). Thus for strongly right-skewed data distributions, as frequently occur in applied geochemistry, the Tukey boxplot based on untransformed data will tend to seriously underestimate the number of lower extreme values and overestimate the number of upper extreme values.

Figure 3.13 demonstrates that the boxplot detects a high number of upper extreme values for the Ba data and no lower extreme values. The reason for this is due to the right-skewed data distribution, indicated by the MEDIAN falling only one-third of the hinge width (HW) above the lower hinge (LH). This feature was also apparent in the histogram and density trace (Figures 3.3 and 3.5). These figures demonstrate that the Ba distribution approaches symmetry when the data are log-transformed. The Tukey boxplot of the log-transformed data will thus be suitable for providing a realistic estimate of the extreme values at both ends of the data distribution. Figure 3.14 shows the Tukey boxplot for the log-transformed Ba data. As expected, the number of upper extreme values is drastically reduced, and one lower extreme value is now identified (Figure 3.14, upper diagram). It is of course possible to plot a log-scale for the original data to regain the desirable direct relationship (Figure 3.14, middle). Because the boxplot is reliable for symmetric distributions, it is appropriate to calculate the values for the whiskers for the log-transformed distribution and then back-transform the values for the fences to the original data scale (Figure 3.14 lower diagram). Note that the MEDIAN and hinges will not be changed by log- and back-transformation because they are based on order statistics. This version of the boxplot is called the log-boxplot and should be used when the

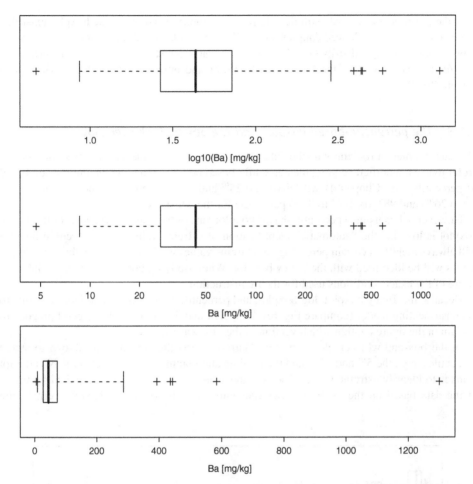

Figure 3.14 Boxplots for Ba. Upper diagram: the boxplot of the log-transformed data. Middle diagram: the same but scaled according to the original data. Lower diagram: log-boxplot of the original data with whiskers calculated according to the symmetrical log-transformed data and back-transformed to the original data scale – compare with the Tukey boxplot in Figure 3.13

original data are strongly skewed and it is still desirable to preserve the data scale. Comparison of Figure 3.13 with Figure 3.14 (lower diagram) demonstrates that the limits of the box are not changed; however, the extent and position of the whiskers and the number of extreme values has changed dramatically.

In conclusion, the Tukey boxplot should not be applied to strongly skewed data distributions without an appropriate data transformation because it will result in a wrong impression about the upper and lower extreme values. Thus for applied geochemical data, the data distribution should always be checked for symmetry before drawing either a Tukey boxplot or the log-boxplot. In the majority of cases the log-boxplot (Figure 3.14, lower plot) will be better suited to identify the number and position of extreme values.

If the log-transformed data still deviate from symmetry, versions of the boxplot exist that can deal with strongly skewed data sets and will still provide useful fences for extreme values at both ends of the data distribution. They are based on a robust measure of skewness (Section 4.4) for the calculation of the fences for the lower and upper whiskers (Vandervieren and Hubert, 2004).

3.5.3 The percentile-based boxplot and the box-and-whisker plot

The data symmetry problems with the Tukey boxplot are probably the reason why some workers prefer to use a modified version, the percentile-based boxplot, where all definitions are based on percentiles (see Chapter 4): MEDIAN and 25th and 75th percentile for the box and 2nd (or 5th to 20th) and 98th (or 80th to 95th) percentile for the whiskers.

However, when using a percentile-based boxplot, one of the major advantages of the Tukey boxplot is lost, i.e. the "automatic" identification of extreme values. The percentile boxplot will always identify a certain percentage of extreme values while it is possible that no extreme values will be identified with the Tukey boxplot. When studying boxplots, it is essential to be aware of the exact conditions used for their construction.

Because the Tukey boxplot, log-boxplot, and percentile-based boxplot all look the same to the unsuspecting reader (compare Figures 3.13, 3.14 and 3.15), it should be good practice to explain in the figure caption which version of the plot was used.

In the box-and-whisker plot (see, e.g., Garrett, 1988) the whiskers are drawn to stated percentiles, e.g., the 5th and 95th, and the minima and maxima plotted as crosses. No attempt is made to identify extreme values. The box-and-whisker plot is simply a graphical summary of the data based on the order statistics (percentiles) with no assumptions concerning the

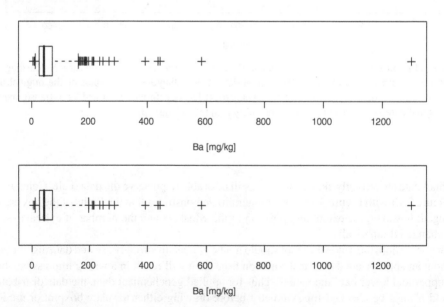

Figure 3.15 Percentile-based boxplot using the 5th and 95th (upper boxplot) and 2nd and 98th percentile (lower boxplot) for drawing the whiskers. Variable Ba, Kola Project C-horizon; compare with Figures 3.13 and 3.14

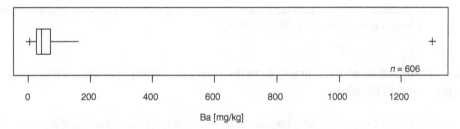

Figure 3.16 Box-and-whisker plot of the variable Ba, Kola Project C-horizon

underlying statistical model. Figure 3.16 is the box-and-whisker plot for Ba in Kola Project C-horizon soils. The resulting plot is the same as the boxplot shown in Figure 3.15, upper boxplot, without identifying all outliers or extreme values.

3.5.4 The notched boxplot

Often the information included in the Tukey boxplot is extended by adding an estimate of the 95 percent confidence bounds on the MEDIAN. This leads to a graphical test of comparability

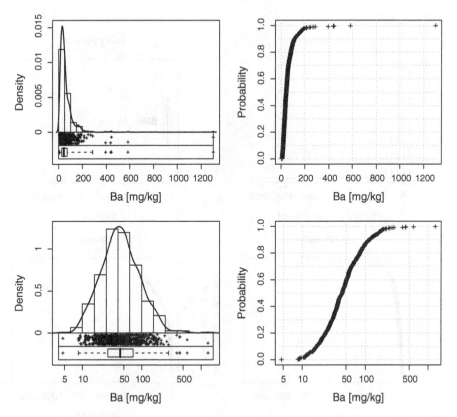

Figure 3.17 Histogram, density trace, one-dimensional scatterplot, and boxplot in just one display, combined with the ECDF-plot. Variable Ba, Kola Project C-horizon. Upper diagrams: original data (with log-boxplot); lower diagrams: log-transformed data

– much like the more formal t-test (see Chapter 9) – of MEDIANS via notches in the boxplot. This use of boxplots is discussed in Section 9.4.1.

3.6   Combination of histogram, density trace, one-dimensional scatterplot, boxplot, and ECDF-plot

Several of the plots named so far – one-dimensional scatterplot, histogram, density trace and boxplot – can advantageously be combined into just one display (Reimann, 1989). Figure 3.17 (left) shows this combined graphic for Ba. The ECDF-plot is another graphic that will reveal interesting properties of the data distribution (Figure 3.17, right). In contrast to the QQ-, CP-, or PP-plot, the ECDF-plot is not based on the assumption of any underlying data distribution. It is thus ideally suited as an exploratory data analysis tool in the first stages of data analysis.

In combination these graphics provide an excellent first impression of the data, as each of them highlights a different aspect of the data distribution (Figure 3.17). Figure 3.17 shows that Ba is strongly right skewed for the original data. It displays an almost symmetrical distribution for the log-transformed data. For the log-transformed data both the one-dimensional scatterplot and ECDF-plot show some minor disturbances at the lower end of the Ba distribution. For further work with most statistical methods requiring symmetrical data, the log-transformed

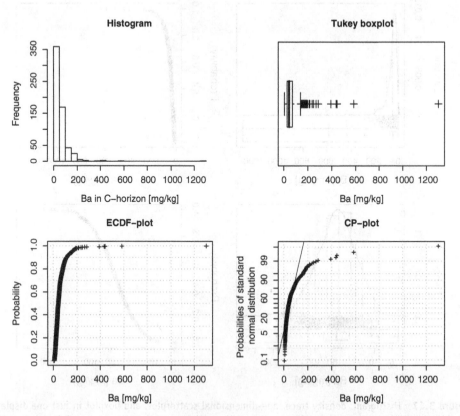

Figure 3.18   Combination of histogram, Tukey boxplot, ECDF-, and CP-plot for the variable Ba, Kola Project C-horizon

values of this variable should obviously be used. One could also conclude that the log-boxplot should be used for further work with this variable for a realistic identification of extreme values whenever the data are kept in the original scale.

3.7 Combination of histogram, boxplot or box-and-whisker plot, ECDF-plot, and CP-plot

A different combination of the plots in this chapter has also proven informative in a single display (Figure 3.18). Using the data for Ba in Kola C-horizon soils, the histogram in the upper left is the same as in Figure 3.17. The upper right display is either a Tukey boxplot or a box-and-whisker plot, the user's choice. The lower left display is an ECDF-plot and the lower right a CP-plot. The ECDF-plot and CP-plot permit easy inspection of the middle and extreme parts of the data distribution, respectively. The choice of Tukey boxplot or box-and-whisker plot depends on whether the user wishes to identify potential outliers or simply have an order statistics based replacement for the histogram. The plotting of the Tukey boxplot above the CP-plot permits easy comparison of the two plots. Figure 3.18 indicates that all plots are dominated by some few extreme values. Thus the whole display should be plotted using logarithmic scaling (Figure 3.19).

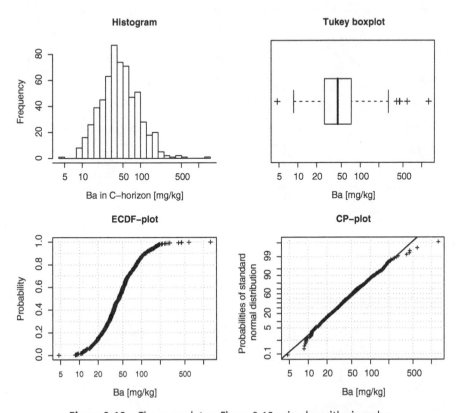

Figure 3.19   The same plot as Figure 3.18 using logarithmic scales

3.8 Summary

When working with a new data file it is highly advisable to study the data distribution in detail before proceeding to all subsequent statistical analyses. Many statistical methods are based on assumptions about the data distribution. A documentation of the data distribution will help to decide which statistical methods are appropriate for the data at hand. A histogram alone is not sufficient to get a good impression of the data distribution. The impression gained from histogram and density trace alone depends strongly on the choice of parameters. A combination of different graphics will often provide greater insight into the data distribution. All variables should thus be documented in a combination of summary plots, e.g., histogram combined with density trace, boxplot and one-dimensional scatterplot and ECDF- or CP-plots.

It is advisable to have copies of a number of distribution graphics for all variables under study at hand for easy reference when more advanced data analysis methods are applied.

4
Statistical Distribution Measures

When studying the data distribution graphically, it is apparent that the distributions of the variables can look quite different. Instead of looking at countless data distributions, it may be desired to characterise the data distribution by a number of parameters in a table. What is required?

First the central value of the distribution needs to be identified, together with a measure of the spread (variation) of the data. Furthermore, the quartiles and different percentiles of the distribution may be of interest (i.e. above what value fall the uppermost two, five, or ten per cent of the data).

When working with "ideal" data, two further measures are often provided in statistical tabulations: skewness and kurtosis. Skewness is a measure of the symmetry of the data distribution, kurtosis provides an expression of the curvature – the appearance of the density trace can be flat or steep. Such "summary values" are frequently used to compare data from different investigations. For real data there is often the problem that the presence of outliers and/or multimodal distributions will bias these measures. Both can be easily recognised from the graphics described in the previous chapter.

4.1 Central value

What is the most appropriate estimator for the central value of a data distribution? What is actually the central value of a distribution? It could be the "centre of gravity", it could be the most likely value, it could be the most frequent value, it could be the value that divides the samples into two equal halves. Accordingly there exist several different statistical measures of the central value (location) of a data distribution.

4.1.1 The arithmetic mean

The most frequently used measure is probably the arithmetic mean. Throughout this book we will use the term MEAN for the arithmetic mean. All values are summed up and divided by the number of values. If $x_1, x_2, \ldots, x_n$ denotes the values of the data with n indicating the

Statistical Data Analysis Explained   Clemens Reimann, Peter Filzmoser, Robert G. Garrett, Rudolf Dutter
© 2008 John Wiley & Sons, Ltd.

number of samples, the simple formula for calculating the arithmetic mean is:

$$\text{MEAN} = \frac{1}{n} \sum_{i=1}^{n} x_i.$$

4.1.2 The geometric mean

The geometric mean G is often used with advantage for right-skewed (e.g., lognormal) distributions. G is calculated by taking the n-th (n = number of samples) root of the product of all the data values. It requires that all values are positive, negative values or zeros in the data set are not permitted. When dealing with applied geochemical data, this is the usual case, negative concentrations of a chemical element cannot exist, and the concentration is always conceptually greater than zero, even if it is so low that it cannot be measured. An exception may be measurements of organic pollutants that do not occur in nature, where the concentration could actually be zero. The formula is:

$$G = \sqrt[n]{x_1 \cdot x_2 \cdots x_n} \quad \text{or} \quad \sqrt[n]{\prod_{i=1}^{n} x_i}.$$

Using this form of calculation should be avoided as rounding errors may occur due to the arithmetic precision limitations of computers. A preferable method is to use logarithms, then:

$$\text{mean} = \frac{1}{n} \sum_{i=1}^{n} \log(x_i) \quad \text{and} \quad G = 10^{\text{mean}} \quad \text{or} \quad G = e^{\text{mean}}$$

depending on whether logarithms to the base 10 or natural logarithms were used.

4.1.3 The mode

The MODE is the value with the highest probability of occurrence. There is no simple formula to estimate the mode; however, the mode is often estimated from a histogram or density trace – the MODE being the value where the histogram or density function shows a maximum.

4.1.4 The median

The MEDIAN divides the data distribution into two equal halves. The data are sorted from the lowest to the highest value, and the central value of the ordered data is the MEDIAN. In the case that n is an even number, there exist two central values and the average of these two values is taken as the MEDIAN. This may best be demonstrated by a simple example:

   2.3   2.7   1.9   2.1   1.8   2.4   2.0   5.9.

The data are then sorted:

   1.8   1.9   2.0   2.1   2.3   2.4   2.7   5.9.

The two central values are:

   1.8   1.9   2.0   **2.1**   **2.3**   2.4   2.7   5.9,

and the MEDIAN is (2.1 + 2.3)/2 = **2.2**.

For comparison, the MEAN of these eight values is 2.64, the geometric mean, G, is 2.44, and to estimate a MODE does not make much sense when dealing with so few data. Looking at the differences between the MEAN, G, and MEDIAN demonstrates that it may be difficult to define a meaningful central value. These large differences between the estimates of the central value are caused by the one high value (5.9) that clearly deviates from the majority of data. When computing the MEAN, each value enters the calculation with the same weight. The high value of 5.9 thus has a strong influence on the MEAN. The logarithm of the geometric mean G is the arithmetic mean of the log-transformed data. Log-transforming (base 10) the above data:

0.26   0.28   0.30   0.32   0.36   0.38   0.43   0.77;

the MEAN of these log-transformed values is 0.39. When this value is transformed back ($10^{0.39}$), the value of the geometric mean G, 2.44, is obtained. Thus the same problem observed for the MEAN remains, G is attracted by the (still) extreme value of 0.77, though to a lesser extent due the nature of the logarithmic transformation. Using the above example, the difference between the MEDIAN and maximum is 3.7 units, but in logarithmic units the difference is only 0.43. Clearly, when calculating the arithmetic mean, the maximum value will have less leverage when using logarithmically transformed data.

Because the MEDIAN is solely based on the sequence of the data values, it is not affected by extreme values – the extreme value could even be much higher without any influence on the MEDIAN. The MEDIAN would not even change when not only the largest value but also the next two lower values were much higher (or the lowest values much lower). For large n the MEDIAN is resistant to up to 50 per cent of the data values being extreme. Thus the MEDIAN may be the best measure of the central value if dealing with data containing extreme values.

4.1.5   Trimmed mean and other robust measures of the central value

To this point four methods of obtaining a central value have been discussed. However, others exist, such as calculating a trimmed mean. Here a selected proportion of extreme data values are excluded before calculating the mean. Thus the five per cent trimmed mean would be the mean of the data between the 5th and 95th percentiles, i.e., the top and bottom five per cent of the data have been discarded from the calculation. Because the exact proportion that should be trimmed is unknown, the procedure introduces an amount of subjectivity into computing the mean. Graphical inspection of the data distribution, e.g., in the CP-plot, can be very helpful in deciding on the trimming percentage. There are other robust estimators (i.e. estimators that are less influenced by data outliers and deviations from the model of a normal data distribution) of the central value, based on the M-estimator (Huber, 1964). Data points that are far away from the centre of the distribution are down-weighted by these robust estimators. The M-estimator is less robust than the MEDIAN. However, because the MEDIAN uses less information it is less precise in estimating the central value of the underlying distribution.

4.1.6   Influence of the shape of the data distribution

It has been demonstrated how the different measures of the central value behave when the data contain one (or several) extreme value(s). How does the shape of the distribution influence the central value? Figure 4.1 shows, starting with a normal distribution, how the different measures of the central value depend on the shape of the data distribution.

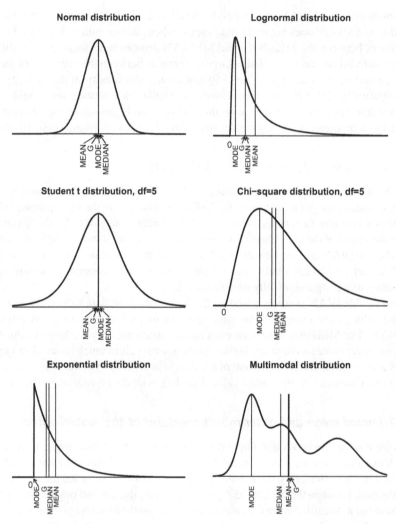

Figure 4.1 Graphical presentation of a selection of different statistical data distributions (normal, lognormal, chi-square, Student t, exponential and multimodal) and the location of the four different measures of the central value discussed in the text

In the case of a *normal distribution*, all four measures will, theoretically, take the same value (Figure 4.1), though there may be slight differences depending on the actual data. For a normal distribution the MEAN is the best (most precise) central value.

For a *lognormal distribution* important differences occur – the arithmetic mean is strongly influenced by high values (Figure 4.1). The MEDIAN and G are (theoretically) equal. The reason is that the logarithm of the lognormal distribution is a normal distribution. Thus the geometric mean G of the lognormal distribution corresponds to the MEAN of the normal distribution. For a normal distribution the MEAN is the best measure of the central value. Thus G can be considered as the best measure of the central value of a lognormal distribution. The position of the MEDIAN will not be influenced by a logarithmic transformation. Thus we get the same value for G and MEDIAN. The MODE is lower than all other measures and identifies the peak of the lognormal distribution. Still, the MODE may not be the best measure of the central value because far fewer data occur below the mode than above. The MEDIAN still separates the data into two equal halves and is, together with the geometric mean G, a good measure of the central value for a lognormal distribution. The most precise measure is again the MEAN, calculated for the log-transformed data and then back-transformed to the original data scale.

The *Student t distribution* is a symmetrical distribution with a different shape than the normal distribution (Student (W.S. Gosset), 1908). The likelihood that it contains values that are far away from the centre of the distribution is much higher; it has "heavy tails". The "heaviness" of the tails depends on a property named the degrees of freedom (df) (see, e.g., Abramowitz and Stegun, 1965). A small value of df results in very heavy tails. With ever increasing df the shape of the normal distribution is approached. Because the distribution is symmetrical, in theory all four measures of the central value are again equal (Figure 4.1). However, in practice the Student t distribution describes data with many extreme values, and thus the measures can strongly deviate from each other. Of the measures discussed here, the MEDIAN is the only reliable measure of the central value because it is resistant to a high proportion of extreme values.

The *chi-square distribution* is a right-skewed distribution (see, e.g., Abramowitz and Stegun, 1965). As for the t distribution, the parameter df (degrees of freedom) determines the shape. With increasing df the distribution becomes more and more symmetric and will in the end approximate the normal distribution. All four measures of the central value will usually differ (MODE < G < MEDIAN < MEAN – Figure 4.1). The MEAN is influenced by the high values (right-skewed distribution). The geometric mean G will be influenced by the low values because the log-transformed values of a chi-square distribution are left skewed. The MODE identifies the peak, but in a right-skewed distribution fewer values will occur below the MODE than above. The MEDIAN is thus the best measure of the central value.

The *exponential distribution* is again right skewed, and the same sequence of the measures of the central value as for the chi-square distribution is observed. The reasons are the same. Here the MODE is zero and uninformative. Again the MEDIAN is the best measure of the central value.

A *multimodal distribution* has more than one peak. All four measures of the central value will generally provide different locations. In the case shown in the example plot (Figure 4.1), it is very difficult to decide which measure is the best indication of the central value because three data populations are clearly present where the distribution functions are superimposed on one-another. It would be best to separate the three distributions and estimate the central value of each distribution individually. This is impractical because of the extent of the area of overlap and one central value for the whole distribution has to be estimated, even knowing that the central value will not be meaningful for any of the three underlying distributions. The MODE will be meaningful as a central value for one population. MEAN and G can both be strongly influenced by outliers or skewness. The MEDIAN still divides the resulting total distribution into two equal halves and may thus again be the best measure of the central value. Note that it is impossible to transform this distribution to approach symmetry.

In conclusion, it can be stated that the MEDIAN is the most suitable measure of the central value when dealing with distributions with different shapes (i.e. when working with real data). It is also preferable because it is robust against a sizeable proportion of extreme values.

4.2   Measures of spread

The central value can be used to compare data. However, even with the same central value, the data can exhibit completely different distributions. An additional measure of variation is required for a better description of the data.

How can the variation of data be characterised? Statisticians define the variance as the average squared deviation from the central value. For practical purposes variance may not be the most usable measure because it is expressed in squared units. A measure expressed in the same unit as the data may thus be desired. Such a measure is called a "measure of spread" rather than the measure of variance.

One possibility to estimate a measure of spread is to look at the interval into which the data fall, i.e. maximum value minus minimum value. To make this measure more robust against (i.e. less influenced by) extreme values one could choose to look at the interval defined by the "inner" 50 per cent of the data. Another possibility is to consider the variation of the data points around the central value.

4.2.1   The range

The RANGE is defined as maximum (MAX) minus the minimum (MIN) and thus gives directly the interval length from the smallest to the largest values:

$$RANGE = MAX - MIN.$$

However, real data often contain extreme values (data outliers). The RANGE is extremely sensitive to outlying data and thus not usually a good descriptor of the data variation when working with applied geochemical or environmental data.

4.2.2   The interquartile range (IQR)

A more robust equivalent of the RANGE is the interquartile range, IQR. Instead of calculating MAX − MIN, the difference between the 3rd and the 1st quartile of the data is calculated. Thus the upper and lower 25 per cent of the data are not used in the calculation; the IQR measures the range of the inner 50 per cent of the data. To reach conformity with the spread measure of the underlying data distribution, the IQR needs to be multiplied by a constant. For an underlying normal distribution the value is 0.7413. The resulting formula for the spread-measure IQR of a normal distribution is:

$$IQR = 0.7413 \cdot (Q3 - Q1),$$

where Q1 and Q3 are 1st and 3rd quartiles of the data.

The IQR is robust against 25 per cent of data outliers or extreme values, and thus considerable deviations in the tails of the normal distribution can be tolerated. Because the IQR is robust, it can be used as a measure of spread even for real data that are skewed and have a high proportion of outliers.

4.2.3 The standard deviation

The standard deviation, SD, considers (just like the MEAN) each single data point. The average of the squared deviations of the individual values from the arithmetic mean is computed. This describes the average spread (n is replaced by $n - 1$ to correct for the estimated mean) of the data around the central value. To report the measure in the original data units, the square root of the results is defined as the SD:

$$\text{SD} = \sqrt{\frac{1}{n-1} \sum_{i=1}^{n} (x_i - \text{MEAN})^2}.$$

When calculating the standard deviation for real data, the problem again is that each data value has the same weight. If the data are skewed and/or extreme values are present, not only will the MEAN be biased (see above), but the SD will be even more biased (because squared deviations are used to estimate it). Because of this fact, the SD should never be calculated for data without a prior check of the distribution. If the data do not show a symmetrical distribution, it is most prudent to not estimate and provide a SD.

To get a meaningful estimate of the standard deviation, the distribution must approach normality before any calculations (see Chapter 10 on transformations). In the case of data with a normal distribution, the SD is the best measure of spread. In applied geochemistry and the environmental sciences, the data are sometimes strongly right skewed, and then the standard deviation is not a useful measure of spread. The MEAN can be calculated for the log-transformed data and then re-transformed to the original data scale. The spread (expressed as SD), however, cannot be calculated for the log-transformed data and then back-transformed to the original data scale, because the scale is changing under transformation (see Figure 3.3, histogram, and compare range for the original data (RANGE = MAX(Ba) − MIN(Ba) = 1300 − 4.7 = 1295.3) and for the log-transformed data (RANGE = MAX(log10(Ba)) − MIN(log10(Ba)) = 3.11 − 0.67 = 2.44)).

4.2.4 The median absolute deviation (MAD)

The robust equivalent of the SD is the median absolute deviation, MAD. It is also a measure of average deviation from the central value. In contrast to the SD, for calculating the MAD the MEDIAN is taken as the central value. The absolute deviations are computed and their average estimated with *their* MEDIAN. Unlike the standard deviation, and like the IQR-based estimate of spread, it is not necessary to take the square root because the MAD is already in the same unit as the data. The median of the absolute deviations provides a value that again has to be multiplied by a constant so that it conforms to the underlying distribution, to provide the MAD. For the normal distribution the constant is 1.4826.

$$\text{MAD} = 1.4826 \cdot \text{median}_i(|\text{MEDIAN} - x_i|).$$

The MAD is robust against up to 50 per cent of extreme values in the data set because it is based solely on MEDIANS. Due to the robustness of the MAD considerable deviations from the normal distribution are tolerable.

When looking at the above example data set (Section 4.1.4):

1.8   1.9   2.0   **2.1**   **2.3**   2.4   2.7   5.9,
MEDIAN = 2.2;

to estimate the MAD, the difference between the values and the MEDIAN is calculated, namely:

−0.4   −0.3   −0.2   −0.1   +0.1   +0.2   +0.5   +3.7.

Then the differences are arranged in order of magnitude (neglecting the sign) and the MEDIAN of these ordered values is determined:

0.1   0.1   0.2   **0.2**   **0.3**   0.4   0.5   3.7,
MEDIAN = 0.25.

The MEDIAN of the deviations is 0.25. Note that this value is again not influenced by the absolute value of the outlier. The MAD is calculated by multiplying the MEDIAN of the deviations with the "normal distribution" factor (1.4826, see above).

$$MAD = 1.4826 \cdot 0.25 = 0.37.$$

Just for comparison the RANGE of our simple data set is 4.1, the SD is 1.35, and the IQR is:

$$IQR = 0.7413 \cdot (2.55 - 1.95) = 0.44.$$

Note that if we use R to calculate the IQR, we will get a slightly different result (IQR = 0.37) because R uses a better approximation of the quartiles. As mentioned above, just like IQR, the MAD can be applied as a useful measure of spread for real data even when working with the untransformed (original) data.

4.2.5   Variance

Range, IQR, SD, and MAD all have a direct relation to the unit of the data. All these measures of spread can be used to estimate the variance by simply taking the square of the obtained values. The "sample variance" is defined as

$$VAR = \frac{1}{n-1} \sum_{i=1}^{n} (x_i - MEAN)^2,$$

and the SD is calculated as the square root of VAR.

4.2.6   The coefficient of variation (CV)

The coefficient of variation (relative standard deviation) is independent of the magnitude of the data and thus of the data measurement units. It is usually expressed in per cent (CV values times 100 in percent). It is well suited to compare the variation of data measured in different units, or the variation in data sets where the MEDIANS are quite different. The coefficient of variation (CV), also known as the relative standard deviation (RSD), is defined as:

$$CV\ \% = 100 \cdot \frac{SD}{MEAN}.$$

4.2.7 The robust coefficient of variation (CVR)

The robust coefficient of variation, CVR, is defined as:

$$\text{CVR \%} = 100 \cdot \frac{\text{MAD}}{\text{MEDIAN}}.$$

Just as the CV, it is usually expressed in per cent.

4.3 Quartiles, quantiles and percentiles

Quartiles, quantiles and percentiles have already been used in the previous chapter when discussing QQ-, CP- and PP-plots and constructing a boxplot. Quartiles divide the data distribution into four equal parts. One-quarter of the data occur below the first quartile, three-quarters of the data above the first quartile. The MEDIAN has been defined in Section 4.1.4 – half of the data occur above the MEDIAN, accordingly the other half of the data occur below the MEDIAN. Three-quarters of the data occur below the third quartile and one-quarter above.

For studying the tails of a given data distribution, one could be interested in other fractions than quartiles. Generally, an α-quantile is defined as the value where a fraction α of the data is below and the fraction $1 - \alpha$ is above this value. Values of $\alpha = 0.02, 0.05, 0.1, 0.9, 0.95$, and 0.98 are often of special interest to identify extreme values of a given data distribution.

Percentiles are defined analogously to quantiles. For percentiles the sought fractions of α are expressed in percent, i.e. 2, 5, 10, 90, 95, and 98 per cent of the data distribution.

The position of quartiles, quantiles, and percentiles in a given data set is not changed by strictly monotone transformations like the log-transformation. This fact was used when constructing the log-boxplot (Section 3.5).

4.4 Skewness

Skewness is a measure of the symmetry of the data distribution, it is computed as the third moment about the mean (the variance is the second moment about the mean) (see, e.g., Abrramowitz and Stegun, 1965). Skewness is zero if a distribution is symmetric, it takes on negative values for left-skewed data and positive values for right-skewed data. The value in itself conveys little. Skewness should be standardised with a constant, depending on the sample size. The standardised skewness should be in the range ± 2, otherwise the skewness must be considered to be extreme.

4.5 Kurtosis

Kurtosis is a measure that will indicate whether the data distribution is flat or peaked, it is computed as the fourth moment about the mean (see, e.g., Abrramowitz and Stegun, 1965). For a normal distribution the normalised kurtosis is zero, the "raw" kurtosis is three. It takes negative values for peaked (most values in the centre of the distribution) and positive values for flat (many values towards the tails of the distribution) distributions. Just as skewness, it should be standardised with a constant depending on the number of samples. If the standardised kurtosis is outside the limit ± 2, it is considered to be extreme.

4.6   Summary table of statistical distribution measures

As a first general description of a data set, a summary table of some (or all) of the above distribution measures is often needed. Table 4.1 shows such a table for selected variables of the Kola Project Moss data set. It is up to the individual scientist to decide which parameters, and especially how many percentiles, are included in such a table. According to the above discussions, the variable name, minimum and maximum values, a central value (the MEDIAN will usually be the most suitable choice when working with "real" data), and a measure of spread (if the MEDIAN is provided as the central values MAD or CVR will be the best measures of spread) should be provided as a minimum. Number of samples, 2nd and 98th percentile, and the IQR are also of considerable interest. For environmental or geochemical data, the unit of the variable and the lower detection limit should also be included.

Table 4.1 shows that for the Moss data set a detection limit problem exists for quite a number of variables (minimum value < value of the detection limit, e.g., Ag, B, Bi, Cr, Na, Tl, U). For these variables the number (or percentage) of samples below the detection limit is important to know – if it is not provided in the summary table, it is important to consult the CP-plot of these variables to determine the actual percentage. MEDIAN and G (also called "MEAN-log", as it is equal to the back-transformed arithmetic mean of the log-transformed data) are often quite comparable and in the majority of cases considerably lower than the MEAN. This is a clear indication of right-skewed distributions. For the same reason the values for SD are usually much higher than those for MAD and IQR. The table clearly indicates that MEAN and SD do not provide realistic measures of location and spread for most of the data. MEAN and SD are used to calculate CV. It is thus no surprise to note the major difference between CV and its robust counterpart CVR. For the data at hand, the CVR is the most reliable estimate of the coefficient of variation. The elements Ni, Na, and Co show the highest CVR. The distributions of Ni and Co are strongly influenced by the emissions from the Russian nickel industry, the distribution of Na is strongly influenced by the steady input of marine aerosols along the coast of the Barents Sea. Variables with a high CVR may thus suggest the existence of an unusual process in a survey area.

In Table 4.1 all variables are reported in the same unit (mg/kg). Many scientists use different units to provide a more comparable data range for the variables. This is quite convenient for tabulating data because all numbers can be shown in the same format and kept short. However, for comparisons between variables, it is always necessary to consider two different pieces of information, the data value and the unit, at the same time. The direct comparability of variables is thus limited. For computing and graphical data analysis, the "beauty of the number" is of no importance. Many data analysis systems do not permit storing and using both the unit and the value of the variables. Unit and values can thus become separated, and it can be quite difficult to "guess" a unit for a variable. For practical purposes and direct comparability of the values it is thus a good idea to reduce all values to just one unit wherever possible. In applied geochemistry the most convenient data unit is mg/kg, and to avoid confusion in case the unit is lost at some stage of data processing, it may be a good idea to keep all values in all data files in just this one unit (see discussion in Chapter 2).

4.7   Summary

A number of statistical parameters are available to characterise a distribution. For the "classical" normal distribution arithmetic mean and standard deviation are best suited. Environmental

Table 4.1 Summary table providing a selection of important statistical parameters to describe selected variables of the Kola Project Moss data set.

	DL	MIN	$Q_{0.05}$	Q1	MEDIAN	MEAN-log	MEAN	Q3	$Q_{0.95}$	MAX	SD	MAD	IQR	CV %	CVR %
Ag	0.01	< 0.01	0.014	0.023	0.033	0.036	0.050	0.052	0.144	0.824	0.061	0.019	0.022	121	58
Al	0.2	34	92	141	193	215	300	285	768	4850	458	90	106	152	46
As	0.02	0.037	0.08	0.126	0.173	0.195	0.260	0.268	0.71	3.42	0.301	0.085	0.105	116	49
Bi	0.004	< 0.004	0.005	0.012	0.018	0.018	0.027	0.029	0.076	0.544	0.041	0.012	0.013	151	66
Ca	20	1680	2040	2360	2620	2677	2740	2940	3764	9320	681	415	430	25	16
Cd	0.01	0.023	0.042	0.069	0.089	0.095	0.115	0.121	0.259	1.23	0.111	0.036	0.039	96	40
Co	0.03	0.11	0.17	0.24	0.40	0.51	0.92	0.89	3.4	13.2	1.478	0.304	0.48	161	77
Cr	0.2	< 0.2	0.28	0.43	0.60	0.68	0.93	0.99	2.7	14.4	1.132	0.334	0.415	121	56
Cu	0.01	2.6	3.6	4.7	7.2	9.70	17	16	60	355	28	4.65	8.35	167	65
Fe	10	47	86	145	212	252	386	384	1396	5140	545	128	177	141	60
Mg	10	518	773	918	1090	1100	1132	1270	1691	2380	282	260	261	25	24
Mn	1	29	126	290	433	388	444	577	797	1170	204	214	212	46	49
Mo	0.01	0.016	0.039	0.058	0.080	0.087	0.107	0.120	0.27	1.08	0.096	0.037	0.046	90	46
Na	10	<10	24	42	71	77	107	136	305	918	99	53	69	92	75
Ni	0.3	0.96	1.46	2.27	5.39	7.12	20	18	81	396	41	5.43	11.84	209	101
Pb	0.04	0.84	1.59	2.29	2.98	3.02	3.34	3.84	5.8	29	2.06	1.13	1.15	62	38
S	15	543	708	792	863	877	888	952	1140	2090	154	119	119	17	14
Sb	0.01	0.011	0.021	0.030	0.041	0.043	0.052	0.056	0.125	0.623	0.045	0.018	0.019	87	43
V	0.02	0.28	0.67	1.08	1.60	1.75	2.58	2.54	7.5	84	4.58	0.91	1.08	178	57
Zn	1	11.7	19	26	32	32	34	39	54	82	11	9.4	9.5	32	29

DL: value of the lower detection limit; MIN = minimum value; Q = quantile; Q1= 1st quartile, Q3 = 3rd quartile; MEAN-log = mean of the log-transformed data back transformed (equals G, the geometric mean); MAX = maximum value reported; SD = standard deviation; MAD = median absolute deviation standardised to conform with the normal distribution; IQR: interquartile range standardised to conform with the normal distribution; CV = coefficient of variation (per cent); CVR: robust coefficient of variation (per cent). $n = 594$, all values in mg/kg, no missing values.

data will usually deviate from the normal distribution and it is better to routinely use robust counterparts, e.g., MEDIAN and median absolute deviation (MAD).

A good summary table of the variables contained in a data set will provide auxiliary information in addition to the "usual" statistical parameters, e.g., number of values, number of missing samples (if any), the measurement unit(s), the detection limits. A comparison of "classical" and "robust" parameters will prove informative. If these clearly deviate there is good reason to take great care during any subsequent statistical analysis of the data. Such deviations indicate the presence of data outliers and/or skewed data distributions.

5

Mapping Spatial Data

As mentioned in the introduction, most data sets in applied earth sciences differ from data collected by other scientists (e.g., physicists) because they have a spatial component. They present data for individuals/samples that were taken somewhere on Earth. In addition to analytical results, the samples are characterised by coordinates. During data analysis of environmental and earth science research results, this spatial component must be included. To effectively study the spatial component, the data's statistical structure must be shown in appropriate maps (Reimann, 2005).

Many studies investigating the regional distribution of chemical elements in different sample materials have been prepared during the last 40 years. Geochemical atlases or collections of regional distribution maps of selected chemical elements are available for many countries (see, e.g., Webb *et al.*, 1964, 1978, Webb and Howarth, 1979, Weaver *et al.*, 1983, Shacklette and Boerngen, 1984; Fauth *et al.*, 1985; Bølviken *et al.*, 1986; Jianan, 1989; Thalmann *et al.*, 1989; Lahermo *et al.*, 1990; Kolijonen, 1992; McGrath and Loveland, 1992; British Geological Survey, 1993; National Environment Protection Agency of the People's Republic of China, 1994; Rühling, 1994, Lis and Pasieczna, 1995; Lahermo *et al.*, 1996; Mankovská, 1996; Rapant *et al.*, 1996; Rühling *et al.*, 1996; Reimann *et al.*, 1998a; Rühling and Steinnes, 1998; Siewers and Herpin, 1998, Čurlík and Šefčík, 1999; Kadunas *et al.*, 1999; Li and Wu, 1999; Rank *et al.*, 1999; Ottesen *et al.*, 2000; Siewers *et al.*, 2000; Gustavsson *et al.*, 2001, Skjelkvåle *et al.*, 2001a,b; Reimann *et al.*, 2003; Salminen *et al.*, 2004, 2005).

The resulting maps from some of these investigations may look informative whereas others are featureless. Is this a question of the area or element mapped, or the area's size? Is it just a question of whether there are anomalies, outliers, present in the mapped area? Has it something to do with the way the map was designed, and is it thus a question of art? Or is it possible to make technical mistakes when preparing a geochemical map that will result in a severe loss of information? There is of course, also, the underlying additional question of whether this might be a consequence of poor survey design and data quality.

Considering the costs invested in sample collection and chemical analyses to make a geochemical map, and particularly for producing a multi-element geochemical atlas, it is surprising that so little attention appears to have been paid to geochemical map preparation. Reliable and informative geochemical maps are needed in mineral exploration, in environmental studies, and for communicating relevant geochemical findings to environmental regulators and policy makers. Studying many existing examples, it appears that practitioners in applied geochemistry

Statistical Data Analysis Explained   Clemens Reimann, Peter Filzmoser, Robert G. Garrett, Rudolf Dutter
© 2008 John Wiley & Sons, Ltd.

and environmental sciences have not evolved a common effective methodology. Sometimes one has the impression that the geochemical data just have been dropped onto an existing base map. Many different mapping techniques have been suggested and tried over the past 40 years, but no standard has emerged. In contrast, geographers pay great attention to map design and production (see, e.g., Monmonier, 1996).

In this chapter different mapping techniques suggested over the years will be reviewed and the resulting maps illustrated. Advantages and disadvantages of these maps for understanding the geochemical process(es) creating the distribution patterns and for highlighting geochemical anomalies will be discussed. In the following it is demonstrated that several techniques are well suited to present geochemical data, while others fail.

5.1 Map coordinate systems (map projection)

Map projections are needed to transform the three-dimensional surface of the planet Earth onto a flat, two-dimensional plane, the map. Many different map projections, and thus many different coordinate systems, are in use in geographical mapping. Due to the fact that a three-dimensional globe is projected onto a two-dimensional plane, all resulting maps have some kinds of scale distortion. It is important to know in what coordinate system the locations for the samples were recorded – and preferably that system should match the coordinates of any background maps to be employed. Most Geographical Information System (GIS) programs are able to transform coordinates between the different systems, but it is essential that those systems are known so that the correct transformations may be applied. In many parts of the world the UTM-system (Universal Transverse Mercator projection) is widely used, and co-ordinates are reported in metres easting (X-coordinates) and metres northing (Y-coordinates) as in the example data sets in this book. The world is divided into 60 UTM-zones, each 6° of longitude wide. Thus it is possible for survey areas to span two or more zones, and it is essential that prior to map preparation, the location coordinates are all transformed to a single zone. When a survey area spans several UTM-zones it may be preferable to use a different projection to minimise the distortions that occur in displaying a three-dimensional surface on a plane, in such instances Lambert Conic Projections and Albers Equal Area may be used. When using projections for small scale maps, the projection property of equal area is important so that biased impressions of spatial extent of patterns are not introduced. In the field the GPS-systems now widely used for determining location will automatically switch from one zone to another when crossing the boundary between two zones. Thus it is necessary to record the UTM-zone of the location in addition to the coordinates themselves, this leads to a 15 digit coordinate that with a leading minus sign for locations in the southern hemisphere uniquely locates any point on the Earth's surface to one metre. Thus a geochemical database often requires three fields to record the location of a sample site, UTM-zone, easting, and northing. Only in surveys within a single UTM-zone could the zone information be omitted, but the zone must be recorded in the metadata for the project, and if possible, be recorded in the variable definitions table of the database, if one is being built. It must be remembered that it is important to record the spheroid and datum used for field coordinate acquisition and the display map projection. For example, in North America a change has been made from North American Datum 1927 (NAD27) to North American Datum 1983 (NAD83). As a result a GPS receiver may be set up to generate NAD83 coordinates using the WGS84 spheroid, but old base maps may be for NAD27 on the Clarke 1866 spheroid; this can lead to errors of up to 100 metres. Some organisations working internationally and in countries with a wide east–west span may choose to record locations directly in geographic coordinates, i.e. latitude and longitude. With the advent of GIS, the conversion to another coordinate system and the use of an appropriate projection for map presentation has become a simple task.

5.2 Map scale

One of the first questions arising when producing a map or an atlas is related to choice of scale. In geography a long tradition of using well-defined mapping scales exists (e.g., ratio scales, called "Representative Fractions", like 1:10 000, 1:50 000, 1:100 000, 1:250 000, 1:500 000, 1:1 000 000, 1:10 000 000 where one unit of distance on the map relates to the stated unit of distance on the ground), depending on the size of the mapped area and the purpose of the map. A scale of 1:10 000 is larger than a scale of 1:250 000 because $\frac{1}{10\,000}$ is a larger fraction than $\frac{1}{250\,000}$. Public authorities commonly use large scales, in urban areas 1:5000 or larger, because they need to show much detail in the maps (e.g., the exact location of power and water lines). The correct scale of a map depends on the amount of detail that needs to be shown and the amount of generalisation, either smoothing out or omissions, that is acceptable. Note that often several scales may be needed to provide a realistic impression of both local detail and the broad overall picture. Scale should be optimised for purpose.

Geographical atlases are often large-format books because whole countries or even continents are fitted on a single page, yet the maps still contain a wealth of information. Geochemists appear to have somewhat misunderstood that concept and concluded that "an atlas has to have a large format". As a consequence many geochemical atlases are large-format books or map collections that are awkward to use. The key factor in designing geographical maps is scale; the map should contain just enough information so that the user can easily grasp the inherent facts at the selected scale. Too much information and the map is hard to read and assimilate, too little information, and it is uninteresting and not fit for purpose. Sample density is an important factor that should determine the scale of a geochemical map. Is it, for example, important to see the exact location of every single sample point, or rather should the representation reflect the processes influencing the regional distribution of the mapped variable?

Geochemical surveys usually result in many maps (at least as many maps as analysed elements and additional parameters). Although handling many maps has become much easier with the advent of GIS, many maps still have to be compared with one another and with maps containing additional information that may explain some of the observed patterns. Experience has shown that A4 (210 × 297 mm) or letter (8.5 × 11 inches) size is a good practical format for a geochemical map. This size also allows the easy use of the maps for other purposes such as readable slide presentations. The A4 (or letter) format may not be possible for high-density surveys where it is important to show the results for each single sample location, such as in some detailed mineral exploration or contamination studies, or in case of unfortunate survey design (e.g., clusters of locations arising from a high spatial sampling density versus large tracts of land with no samples collected). Geochemists need to be aware that usually the information value of a geochemical map increases when the scale is decreased to give the smallest map possible that still shows all the required information.

5.3   Choice of the base map for geochemical mapping

A geochemical map could be plotted without any geographical (or other) background information. Sample sites with coordinates and the analytical results are sufficient to start mapping and identifying spatial structures in the data. However, the samples were collected somewhere, and users more often than not will want to see the geochemical structure in an informative geographical context. The identification of the geographical location is often essential to aid interpretation of the identified spatial data structures. Thus a suitable base map for plotting the geochemical results is required.

A geochemical map will thus usually be superimposed on some kind of topographical background. The choice of base map can be the reason why geochemical atlases are often so large. However, if the base map is primarily a geographical (or geological) map, well designed for that purpose, it contains already the maximum possible information density for that scale. The addition of geochemical information on such a map will frequently result in a cluttered appearance, and such maps are hard to read. As one consequence, it may be necessary to map at a considerably larger scale than the presentation of the geochemical data alone would require.

Figure 5.1 shows two examples of likely base maps for mapping the Kola Project data. The geological map from the Kola Project area (from Figure 1.2) is one possible choice (Figure 5.1, upper left). The other map (Figure 5.1, upper right) offers minimal topographical information to provide orientation about the location of the samples. In terms of interpreting the geochemical results in a geological context, it is tempting to opt for the geological map as base map.

However, there is already so much information in the geology map that adding geochemical data on top results in a thoroughly cluttered map where neither geology nor geochemistry remain clearly visible (Figure 5.1, lower left). Furthermore, when choosing geology as the background, the underlying assumption is that geology is the most important geochemical process determining the element distribution in the area. This may not be the case, as several other important geochemical processes in the area exist that can influence the regional distribution of the mapped element. In the example map it is not the Caledonian sediments along the coast of the Barents Sea that determine the Mg concentrations in the O-horizon, but rather the steady input of marine aerosols along the coast of the Barents Sea. When selecting an unsuitable base map the chance of recognising such processes unrelated to the base map will be severely limited. The right-hand maps in Figure 5.1 (upper and lower right) are thus the much better and more objective choice for geochemical mapping. The geological background can, of course, still be a very informative addition in an advanced stage of data analysis and mapping.

When producing geochemical maps or atlases, the most important information is the spatial distribution of the chemical elements studied. This statement implies that it is the geochemistry that should govern the mapping and not geography, geology, or other ancillary information. Geochemical maps are primarily produced to understand the processes causing the spatial distribution patterns of chemical elements and to determine the sources that account for those patterns, such as an unusual rock type, a mineral deposit, or contamination from human activities. When producing a set of multi-purpose geochemical maps, information not related to geochemistry should be avoided because it distracts attention from the primary focus of the maps. However, a minimum of geographical information is necessary to allow for easy orientation. More background information, such as the geology, may be added later if required for interpretation. Geology underlaid on a geochemical map right at the beginning may distract the eye from other important patterns in the data. It may also result in a biased map

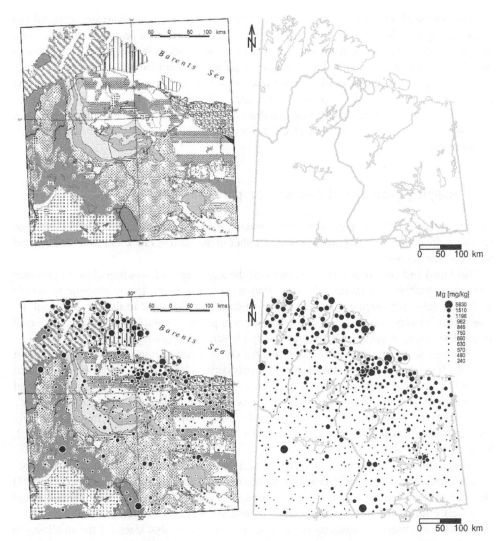

Figure 5.1 Geological map (upper left, see Figure 1.2 for a legend) and simple topographical map (upper right) of the Kola Project area used as base maps for geochemical mapping (lower maps). Lower maps: distribution of Mg in O-horizon soils added to base maps. A colour reproduction of this figure can be seen in the colour section, positioned towards the centre of the book

interpretation because it suggests that the features on the geological map should determine the distribution of the elements. For instance, if the map shows only the geological units and not the fault structures, geochemical patterns related to faults may not be discerned. If the main process determining the distribution of an element is non-geological, it may not even be noticed. Such maps should not be the first product of data analysis and mapping, but are more appropriate for an advanced stage of mapping, making full use of the power of modern GIS. A good first geochemical map should contain enough geographical information for basic orientation.

A map must also have a way to determine scale and orientation (north arrow), a legend and a text stating the element mapped, and probably some information on the sample medium, sample preparation, and method of analysis. In some applications it may also be important to add the date of map generation. Geochemical maps are often copied and used out of context; thus some basic geochemical information like element name, unit, method of analysis, and sample material may be quite useful on the same sheet so that it will not be lost during copying. Again, the amount of information that can be presented for assimilation depends on the map's purpose. Note that a map appropriate for a printed atlas is often too cluttered for a slide presentation.

5.4 Mapping geochemical data with proportional dots

For presenting geochemical data in a map, usually the element values are grouped, and the resulting classes are then displayed on a map, either in the form of black and white or coloured symbols (see Section 5.5).

Björklund and Gustavsson (1987) discussed the advantages of avoiding classes altogether and using symbols that increase in size continuously in relation to the absolute measured analytical value for placing geochemical data into a map. This led to the "proportional dot map", that is today one of the most popular ways of preparing black and white geochemical maps. The resulting maps are often graphically pleasing and are easy to read.

However, an interesting-looking proportional dot map depends crucially on the scaling of the dots. If the dots just grow linearly with the analytical values, it soon becomes obvious that depending on the data distribution, some maps will result in clear differences in the spatial data distribution while on others most of the dots will have almost the same size. Depending on the distance of the maximum value from the main body of data, there may appear only one really large dot on the map – or a whole lot of dots of almost the same size (Figure 5.2, left).

To overcome the problems with different data distributions, an exponential dot size function determining the relative proportions of the dots was introduced. Gustavsson *et al.* (1997) describe in detail an exponential dot size function, which is monotonically increasing and empirically fitted to two percentiles (e.g., 10 and 99 per cent) of the empirical cumulative distribution of observed values on the map area. Thus the symbol sizes for the smallest (10 per cent) and largest (1 per cent) data points are constant. In between the two selected percentiles the dots grow exponentially. Thus the exponential growth of the dot sizes is slow near the detection limit (where there is often large variation due to poor analytical quality, see Chapter 18) and becomes steeper towards larger values. The size function is purposely designed for geochemical variables following a lognormal or positive skew distribution. It does, however, have a disadvantage: it over-emphasises the highest and the lowest values, indirectly indicating that in-between there is nothing of much importance. However, about 90 per cent of all data fits into this "in-between" category. In general, the size function will down-weight the low values and put more weight on the high values.

Using this technique, proportional dot maps showing clear regional patterns can be prepared for almost any data set (Figure 5.2, right). However, because some users appear unaware of the potential of these maps, and others may not have the software for constructing the dot size function, proportional dot maps can be uninformative. A general problem with proportional dot maps is that dots sized on a continuous scale are often difficult to distinguish from

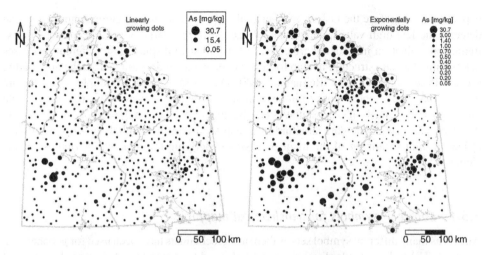

Figure 5.2 Two maps showing the distribution of As in C-horizon soils of the Kola Project area. A map based on a linear relationship between analytical value and proportional dot size (left) and a map of the same data using an exponential weighting function (right)

one another in a map unless they occur in close proximity. A further disadvantage may be that due to their optical weight, the attention of the reader is completely drawn to the high values.

5.5 Mapping geochemical data using classes

It is often assumed that geochemical data follow a single (log)normal distribution, which is "disturbed" by a limited number of extreme values, the outliers. Traditionally the high extremes have been of greatest interest. The problems with this concept are discussed in a number of recent papers (see, e.g., Reimann and Filzmoser, 2000a; Reimann and Garrett, 2005b, Reimann *et al.*, 2005c) and in several chapters of this book. In general, geochemical data do not follow a normal or lognormal data distribution but are poly-populational. The resulting distribution may mimic a lognormal distribution due to the superimposition of a number of separate distributions related to different geochemical processes. These processes are reflected in separate data populations, i.e. sets of data with unique statistical parameters, such as means and standard deviations.

Usually, the most dominant distributions in regional geochemical surveys are data related to the geochemically distinct bedrock lithologies present in the survey area. Superimposed on these rock type-related distributions are effects of secondary processes, such as anthropogenic contamination, sea spray, or enrichment or depletion of elements due to a wide variety of causes (pH, grain size, Fe- and Mn-oxyhydroxides, presence and amount of organic material, to name just a few). All these processes are location dependent. Sometimes non-bedrock factors can become so dominant that the "original" geochemical signature of different bedrock lithologies is lost.

Thus, *geochemical data do not consist of independent samples* as assumed in classical statistics, but the samples related to certain processes are linked additionally by a spatial dependence. Therefore the task in geochemical mapping and interpretation cannot be to just detect some few high values, which may be indicative of a large mineral occurrence or extreme contamination in a survey area. The task is rather to display these different processes determining the data structure in map form and to detect local deviations from the dominating process in any one sub-area (Reimann, 2005). Due to the fact that multiple processes are involved, this may appear close to impossible at first glance. However, by splitting the data into groups on the basis of order statistics (see Sections 5.5.2 and 5.5.3) it is possible to display the spatial aspects of the data structure in a map such that the symbols or isopleths reflect at least a limited number of the main processes underlying the regional distribution of the elements.

5.5.1 Choice of symbols for geochemical mapping

Almost as many different symbol sets as there are geochemists have been used for geochemical mapping. The lack of standardisation for symbols used for geochemical mapping has resulted in maps that are not directly comparable, and the preferred use of proportional dot maps. These do, however, focus on very high values and do not facilitate detecting the data structure in a map.

It is EDA approaches that have permitted the acceptance of a standard set of symbol class boundaries. However, within this context different practitioners have used different symbol sets for mapping (Figure 5.3). These aim to provide an even optical weight for each symbol in a map in order to be able to focus on data structure and not on "high" or "low" values. The original EDA symbol set is based on the boxplot, reflecting the data structure, and permits the mapping of five classes. Limiting the number of classes results in a rather "quiet" and relatively easy to grasp distribution on a map. Experience teaches that seven classes is the maximum number that should be shown in a black and white map.

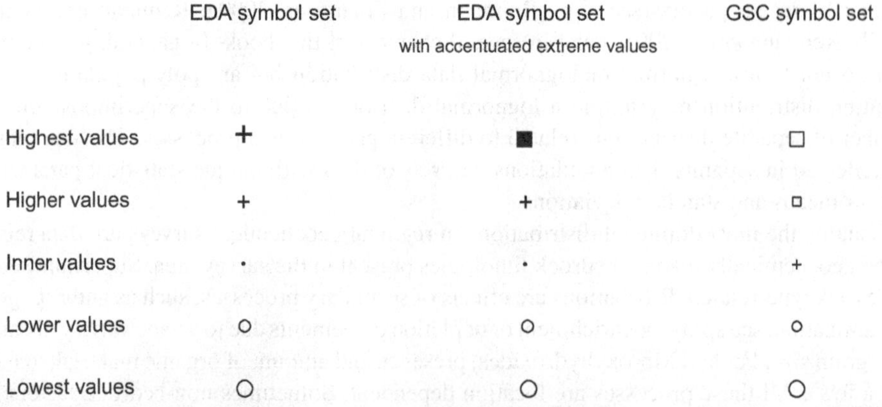

	EDA symbol set	EDA symbol set with accentuated extreme values	GSC symbol set
Highest values	+	■	☐
Higher values	+	+	▫
Inner values	·	·	+
Lower values	o	o	o
Lowest values	O	O	O

Figure 5.3 Three possible symbol sets of use for geochemical mapping: original EDA symbol set; EDA symbol set with accentuated upper extreme values; an alternative symbol set as used by the Geological Survey of Canada (GSC). If desired these symbol sets can be easily extended from five to seven classes by using an additional size-class for the outer symbols

Although today colour is most widely used for mapping, black and white geochemical maps still have the advantage of being inexpensively copied and printed. In addition, black and white maps introduce less perceptual bias than colour maps. This may have substantial advantages when trying to detect geochemical processes with such a map. As a further advantage, colour-blind individuals are able to work with black and white maps. However, for a black and white map the choice of symbols is crucial. A black and white map can also be drawn as a contour, isopleth map, avoiding the use of discrete individual symbols altogether or as a smoothed surface map (see Section 5.6).

Note that the choice of symbols and their size in relation to the size of the map are critical to map appearance. It is not a trivial task to find the optimal symbol type and size for a map. Too many different symbols will usually result in a cluttered and thus hard to read map. Because the symbol size must have a relation to the scale of the map, it is a clear advantage if plotting software permits scaling a whole group of symbols at once so that the relative size proportions between the symbols stay constant once a good set of symbols has been determined.

5.5.2   Percentile classes

Percentiles are based on order statistics and their use does not assume any underlying data distribution. This is a major advantage when dealing with geochemical data. However, the geochemist is left with the decision as to how to distribute the classes over the range of percentiles from 0 to 100. If the task is to identify geochemical processes from regional patterns in a map, this data structure must be revealed in the map. One possibility is to spread the symbols or colours almost evenly across the range of values, e.g., via using the 20th, 40th, 60th, 80th percentiles (Figure 5.4, lower left). However, geochemists are often more interested in the tails of the data distribution. To highlight the tails the 5th, 25th, 75th, and 95th percentiles can be used (Figure 5.4, upper left and upper right). Additional classes can easily be accommodated. A logical choice may be to include further class boundaries at the 50th and 98th (and possibly 2nd) percentiles (Figure 5.4, lower right). An extended version of the EDA symbol set as well as the Geological Survey of Canada symbol set as introduced above (see Figure 5.3) can easily handle up to seven classes for mapping (Figure 5.4, lower right). Maps constructed with percentile classes and one of these symbol sets will usually facilitate the direct recognition of a number of major processes determining the distribution of the mapped variable in space. A remaining problem is that when using percentiles, there is no satisfactory way to identify extreme values or true outliers. The percentile-based map thus includes the assumption that the uppermost, or lowermost, two, five, or ten percent of the data are outliers. This can be justified as identifying an appropriate number of samples for further inspection (Reimann et al., 2005c).

5.5.3   Boxplot classes

The boxplot is based on order statistics and is almost free of any assumption about data distribution (see Section 3.5). More than 15 years ago Kürzl (1988) realised that it is also well suited to define classes for geochemical mapping. O'Connor and Reimann (1993), O'Connor et al. (1988) and Reimann et al. (1998a) subsequently used this technique with great success. It should be recognised, however, that the normal distribution sneaks into the definition of the

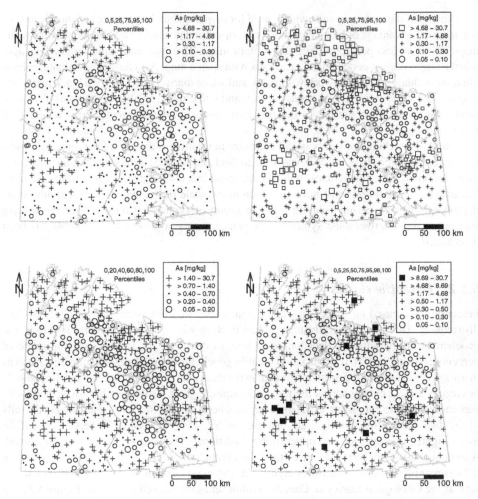

Figure 5.4 Geochemical maps of the variable As in the Kola C-horizon based on different percentile classes and using EDA (upper left), GSC (upper right), extended EDA (lower left), and accentuated EDA (lower right) symbol sets

boxplot when the whiskers, the borders for extreme values, are computed. It was demonstrated above (see Section 3.5) that the boxplot will not recognise lower extreme values and will seriously overestimate the number of upper extreme values when the data distribution is right skewed (the most usual case in applied geochemistry). If identifying "too many" upper extreme values is desirable, one can continue using the original Tukey boxplot with the "raw" data. The boxplot then provides a simple and fast method to define class boundaries for mapping. If the task is to really study the data structure in a map, it may be preferable to use percentiles, a version of the boxplot based on percentiles, or a version of the boxplot that takes care of the vulnerability of the original boxplot to skewed distributions (e.g., the log-boxplot).

Five main classes result from using the boxplot for class selection for mapping: lower extreme values to lower whisker, lower whisker to lower hinge, the box, containing the inner

50 per cent of data (the box can be divided into two classes if needed), upper hinge to upper whisker, and upper extreme values. A set of black and white symbols (see Figure 5.3) based on this EDA approach can be used to display these five classes on a map. These symbols are based on the concept that to show the data structure on a map objectively, each class should have an even optical (graphical) weight. This is achieved by using large symbols for the lower and upper extreme values. These symbols will not dominate the map, because there are usually not a large number of extreme values. At the same time, the symbols of circle ("○") and cross ("+") or square ("□") are almost intuitively interpreted as low and high. The next classes at both ends of the distribution, consisting of ≤ 25 per cent of the data each, employ smaller versions of the extreme symbols. The inner 50 per cent, and thus the majority, of the data have the smallest symbol, a dot. However, in some instances the dots are hard to see, and in such instances a small cross can be used to advantage (GSC symbol set, Figure 5.3).

Maps constructed using these symbol sets look unusually "calm" at first glance, maybe even lacking in information content, to someone used to reading maps where the eye is automatically drawn to the high values. Their real power lies in the fact that the spatial data structure becomes visible (Reimann, 2005). The underlying geochemical processes usually determine the spatial data structure. Such maps can thus be used to understand and interpret the main processes governing the geochemical distribution in space. In addition, even locally unusual data behaviour which is not marked as "extreme" can be easily recognised in such a map in the form of the occurrence of different symbols in an otherwise uniform area. Furthermore, via the boxplot an automated check for extreme values is performed (based on an assumption of log-normality), which is the main advantage of the boxplot over percentiles. Note that for this to be effective, a symmetrical boxplot that has been adjusted to handle right-skewed data distributions should be used, for example by employing a log-transform, or a large number of upper extreme values will be displayed. Instead of the EDA symbols, colour classes can also be used. With the exception of the rare situations where no outliers exist, colour maps constructed using boxplot classes will in general look similar to maps constructed with percentile classes with a symmetrical distribution of symbols about the MEDIAN. Figure 5.5 shows the As distribution in the C-horizon of the Kola Project area based on boxplot classes. Figure 5.5 (right) uses the original EDA symbols with an accentuated outlier symbol. To satisfy the geochemists quest for the highest value, this symbol grows continuously in direct relation to the analytical result.

Considering the major advantages of boxplot classes in combination with EDA symbols for black and white mapping, it is surprising that the technique has found so little application in applied geochemistry. One reason is probably that other maps, such as proportional dots, look much simpler at first glance. Another may be the lack of available software to prepare such maps. Many geochemists still think in terms of "high" (= interesting, may indicate a mineral occurrence, or contamination in environmental sciences) and "low" (= useless, no mineralisation in these areas, or background in contamination studies), and they seek a map that simply displays the range of the data in a relative manner. However, EDA maps achieve the task of spatially locating high and low extreme values even better, by displaying symbols indicative of the extent of "extremeness" in terms of the data distribution itself. Other reasons may be that EDA symbols need appropriate scaling in relation to the other features or symbols displayed on the map (Kürzl, 1988) and that hardly any software is available that permits this approach. An additional reason may be the overestimation of the number of upper extreme values by the original Tukey boxplot when care is not taken to use a transform that brings the data into symmetry.

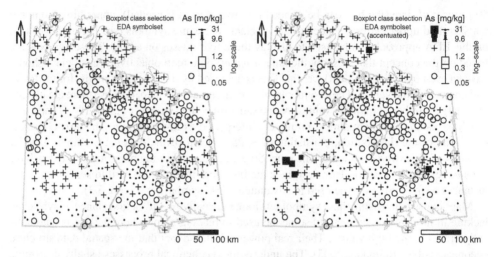

Figure 5.5 Boxplot class-based maps showing the distribution of As in C-horizon soils of the Kola Project area. Tukey boxplot (with log-scale) classes and original EDA symbol set (left), Tukey boxplot classes (log-scale), and accentuated EDA symbol set (right)

5.5.4 Use of ECDF- and CP-plot to select classes for mapping

One of the best procedures to study geochemical distributions graphically is to plot the empirical cumulative distribution function (ECDF-plot) and cumulative probability plot (CP-plot) (Sections 3.4.2 and 3.4.4) and seek breaks or changes of slope in the plots that can be used to identify useful classes for geochemical mapping. The procedure is discussed in detail in Reimann *et al.* (2005c). Working with ECDF- and CP-plots requires experience and detailed study of data distributions prior to mapping. It is thus a technique that an experienced applied geochemist will use in a second step of data analysis, once the first set of maps using a standard default technique (preferably percentile or boxplot classes with EDA symbols) as described above has been prepared.

5.6 Surface maps constructed with smoothing techniques

To construct surface maps, different techniques can be used. Some form of moving average (or median) and more general interpolation techniques or kriging (Section 5.7) are most often applied. Surface maps are needed because most users (and regulators and policy makers) want maps that look "pleasing". For a trained geochemist, hoping to extract knowledge about geochemical processes governing the distribution of any one element in space, surface maps have, however, several disadvantages. Firstly, information on local variability is lost. Secondly, depending on the algorithm (e.g., search radius, distance weighting), maps having different appearances will result. Thirdly, the untrained user may gain from surface maps the impression of a much greater "precision" than the few samples it is based on will support.

Most applied geochemical data are spatially discontinuous, exceptions being anthropogenic contaminants that do not occur naturally and are dispersed from point sources, with levels controlled by geological lithologies and other environmental factors. Very detailed mapping shows step discontinuities associated with these features, not smooth gradients as are expected from geophysical maps where the properties of a "unit" are exerted beyond its physical extent. Therefore it is usually best to prepare surface maps at scales four to five times smaller than a survey's working scale so that the generalisation is acceptable in the eyes of users knowledgeable in the underlying discontinuities. This is particularly important in surveys employing drainage basin sediments, where the recorded sample site location is not the location that data conceptually represent. It must always be remembered that surface maps based on sparse data can be misleading as their preparation requires extrapolation based on some mathematical model that may not reflect the geochemical reality between the widely spaced data points.

The simplest method to construct a surface map is based on moving averages (for geochemical data the moving median is the preferable choice). For constructing a moving median map a window of a fixed size is defined. This is quite similar to smoothing lines (Section 6.4), but the window is now defined by the x- and y-coordinates and is thus two-dimensional. The size of the chosen window will have a major influence on the degree of smoothing. The window is moved across the map, and all data points falling into the window are used to estimate a value at the centre of the window. This value could simply be the MEDIAN of the points within the window or some kind of weighted median, taking the distance of each point from the centre of the window into consideration. The data points on the irregular sampling grid are thus used to estimate new values on a regular grid. The resolution of the resulting map will depend on the chosen raster size. The choice of the raster size, i.e. the intervals in the x and y directions at which the estimator is computed, may be limited by the complexity of the algorithm, i.e. computing time.

More complex smoothing methods use polynomial functions to interpolate irregularly spaced data onto a regular grid (see, e.g., Akima, 1996). These methods will as a rule result in smoother surface maps than moving median techniques. The selection of classes for smoothed surface maps is a crucial step in providing meaningful geochemical maps. Theoretically, an endless grey (or colour) scale could be used for a surface map, with its own grey (or colour) value for each data value, avoiding classes altogether and thus adopting the philosophy of the "proportional dot maps" (Section 5.4). In practice, the same problems as discussed for the proportional dots (Section 5.4) occur, and the appearance of such a map is highly dependent on the data distribution. Thus some kind of scaling to avoid a too-strong influence of outliers and extreme values is needed (just like the "dot size function" – Section 5.4). This can be achieved by using, e.g., 100 percentile classes, each with its own grey value (Figure 5.6, left). If percentiles are used for defining a preset number of classes, the chosen classes must have a relationship to the data structure (Reimann, 2005).

This can be achieved in the majority of cases when the boxplot is used to define the class boundaries (Kürzl, 1988), resulting in five (or seven) classes for the map (see above, Section 5.5.3). Of course, instead of using the boxplot for class selection, it is also possible to directly use percentiles, e.g., the 5th, 25th, 50th, 75th, 90th and 95th, as class boundaries (Figure 5.6, right). Surface maps will usually look better the more classes are applied (Figure 5.6, left); using percentiles to define a more limited number of classes may sometimes facilitate the detection of data structure and thus processes in the maps. It is of course also possible to carefully select class boundaries on the basis of the inspection of CP- and ECDF-plots as done for symbol maps. When a limited number of fixed percentiles are used, care must be taken not to use only one class for the lowermost 50 per cent, or even 75 per cent, of the data, and then to use all the remaining classes for the high end of the distribution. Such an approach will make it impossible to discern the data structure, though such a map might be attractive to some

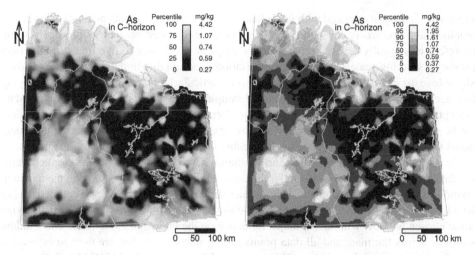

Figure 5.6 Smoothed surface maps of the variable As in the Kola C-horizon; left: constructed using a "continuous grey scale" (100 percentiles), right: the As element distribution map constructed using seven selected percentile-related classes

mineral prospectors, or with data that have a very large proportion of "less than detection limit" values.

5.7 Surface maps constructed with kriging

In general, kriging is a more informative approach for generating surface maps from point source data than smoothing (Section 5.6) because it will deliver a statistically optimised estimator for each point on the grid selected for the interpolation (see, e.g., Cressie, 1993). As an additional advantage, kriging will provide an approximation of the prediction quality at each point within the grid. It is possible to either estimate the values for the blocks defined by the intersections of the grid lines (block kriging), or for any location in space (point kriging). The basic idea behind spatial analysis and kriging is that because samples taken at neighbouring sites have a closer relationship to each other than to samples collected at more distant sites the closer samples should have greater influence on the interpolated estimate. The question then becomes, how to set the weights proportional to distance.

5.7.1 Construction of the (semi)variogram

The basis of kriging is the (semi)variogram (see, e.g., Cressie, 1993), which visualises the similarity (variance) between data points (y-axis) at defined distances (x-axis). The distance plotted along the x-axis in the (semi)variogram is the Euclidean distance (distance of the direct connection) between two points in the survey area. Usually about 30 to 50 per cent of the maximum distance in a survey area is used as a maximum for the x-axis of the (semi)variogram. One reason for this choice is that the variability of the difference in the element concentrations of very distant samples can be expected to be high (approaching the total variability of the

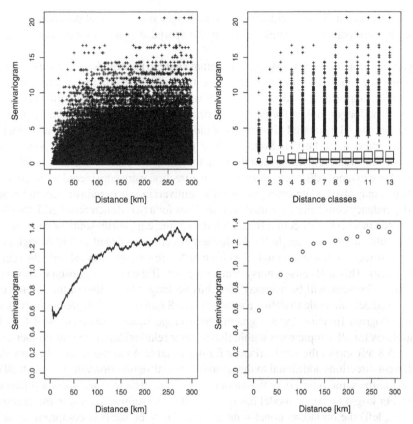

Figure 5.7 Visualisation of the basic information for the computation of the semivariogram for the variable As in the Kola Project C-horizon

element in the survey area). For the Kola Project area the maximum distance is about 700 km. For plotting the (semi)variograms, a maximum distance of 300 km was chosen (43 per cent) when preparing the Kola Atlas (Reimann *et al.*, 1998a) and is thus used here, although a shorter distance, e.g., 200 km, would in many cases be more appropriate.

The basis for the calculation of the (semi)variogram is the *semivariogram cloud* shown in Figure 5.7 (upper left). The semivariogram cloud displays the squared difference of the element concentrations between all possible pairs of data points. The pairs are plotted according to their separation distance along the *x*-axis. The *y*-axis records the semivariance, which is half of the value of the squared differences between the values of the pairs. The semivariance is used because it allows an easier visualisation of the total variance in the semivariogram. The values along the *x*-axis can then be summarised in distance classes. Such a summary is shown using boxplots in Figure 5.7 upper right. The MEDIANS displayed in the boxes increase with distance. The plot shows many outliers, i.e. pairs of data points that have an unusual large variability. Instead of summarising in boxplots, it is also possible to smooth the data along the *x*-axis. Figure 5.7 lower left shows the smoothed line for the semivariogram cloud in Figure 5.7 upper left. Again the increase of the semivariance with increasing distance is visible. At a certain distance the smoothed line flattens out, i.e. the variance becomes constant.

From this distance on, the concentrations will no longer show a spatial dependency. For small distances the semivariance is small and thus the spatial dependency between the points is high.

Progressing from the very noisy semivariogram cloud to the boxplot presentations, the trend is becoming smoother and smoother. However, for finally constructing a semivariogram, an even smoother appearance is desirable. For this it is necessary to leave the semivariogram cloud and to no longer look at the pairs of all data points but to summarise more points using distance intervals. The squared differences of the element concentrations for all points within the distance interval (20 km) are averaged. This procedure results in the few points shown in Figure 5.7 lower right, which are then the base for fitting the semivariogram model.

In the above example distance intervals in any direction were considered (*omnidirectional*). In practice, it may also be interesting to study the semivariance in certain directions because the spatial dependency could change with direction. Thus for a two-dimensional grid, the distances are calculated in a certain direction from each data point, e.g., north–south or east–west. If the grid is irregular, a tolerance angle for the defined direction is needed to find enough points in the selected direction. If a small angle is chosen only a few points may fall into the segment at small distances. This will cause a noisy semivariogram. If the angle is chosen to be very large, directional differences will be averaged and thus no longer be visible in the semivariogram. An often-used default angle to define a segment is $\pi/8$ radians (22.5 degrees). For calculating the semivariogram function for a segment, the average squared differences of the element concentrations for all sample pairs within that angular relationship from one another are used.

Figure 5.8 left shows the semivariances for the variable As in the Kola C-horizon data for four different directions additional to the omnidirectional semivariogram. Although all curves start at about the same point, they flatten out at different distances and show a different total variance. For kriging a single model has to be fitted to the semivariances. In the example plot (Figure 5.8, left) the omnidirectional semivariogram may be the best compromise for fitting the data (Figure 5.8, right).

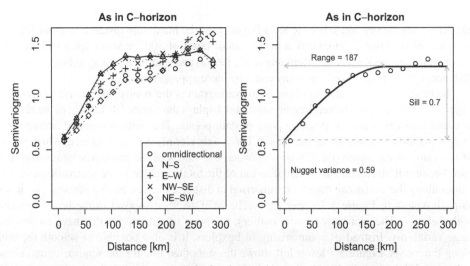

Figure 5.8 Directional semivariograms (left) for the variable As in the Kola Project C-horizon samples and spherical semivariogram model for the omnidirectional semivariogram (right)

For *fitting a model* to the semivariogram there are different options. The curve can be approximated by a spherical model (most common case) as shown in Figure 5.8 (right). Linear, exponential, Gaussian, and several other models can also be used to fit the semivariogram function. The final fitting should be based on a visual inspection of the semiovariogram.

5.7.2 Quality criteria for semivariograms

There are a number of central terms used in describing a semivariogram. The *range* (187 km) defines the distance where the samples can be considered as independent from each other. This is the distance to the point where the semivariogram model flattens out in *x*-direction (Figure 5.8 right). This is also the point where the model reaches the *total variance* of 1.2 (in *y*-direction). The range relative to the maximum distance is an indication of the data quality. A very short range is an indication that the area is undersampled or that the data quality is very poor. The prediction quality improves with an increasing range because the dependency between the samples increases, and thus a larger number of samples are used in the interpolation.

The *nugget variance* (often named the nugget effect) is the variance at the distance zero. In theory it should be zero. However, even for two samples taken at very close distance there will always be several causes for a certain variation, e.g., the sampling error and the analytical error, but also that no two absolutely identical samples can in practice be taken. The name "nugget effect" originates from the assays in gold deposits, where a very high variation between neighbouring samples can be caused by the existence of single gold nuggets. If the range is zero, the semivariogram consists only of the nugget effect, indicating completely independent samples, and no sensible spatial prediction for generating a surface map is possible. In applied geochemistry it will be an indication that the sampling density was too low (sparse) or of very poor analytical quality.

The *sill* (Figure 5.8 right) is the difference between total variance and nugget variance. The proportion of the sill to the total variance should be large because this will indicate a well-defined spatial dependency, leading to a high prediction quality.

5.7.3 Mapping based on the semivariogram (kriging)

The fitted semivariogram model can now be used for kriging, i.e. to either estimate the values at the intersections of the grid lines, or any other point (point kriging) or for each block defined by the grid (block kriging). Based on these estimated values, a smoothed surface map is finally constructed. The variogram determines the weight of neighbouring samples used for the prediction. More distant observations will in general obtain less and less weight out to the range. For each element an individual function, the semivariogram, is estimated. All observed values, weighted according to the semivariogram, are then used for the prediction. This is a major advantage over other methods (e.g., moving medians), which consider only samples in a defined search radius.

For constructing isoline maps point kriging is used, and the gridded point estimates are passed to a contouring package. For the Kola Project data block kriging was used, permitting the direct construction of a surface map. Figure 5.9 shows two kriged maps for As in the Kola Project C-horizon samples. Depending on the block size chosen for kriging, the blocks will be visible (20 × 20 km blocks, Figure 5.9, left), or the map display becomes smoother with decreasing block size (e.g., 5 × 5 km blocks, Figure 5.9, right).

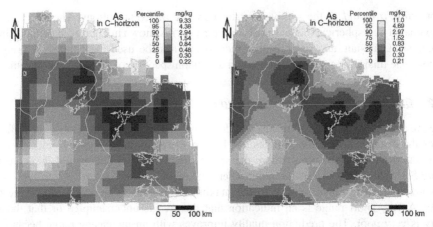

Figure 5.9 Block kriging used to construct surface maps of the distribution of As in the Kola Project C-horizon. Left map, 20 × 20 km blocks; right map, 5 × 5 km blocks. The semivariogram in Figure 5.8, right, is used for kriging

The major advantage of this approach over all other available interpolation techniques is that the semivariogram permits judgement of the data quality, or better, their suitability for constructing a colour surface map. Via the estimated *kriging variance*, the reliability of the obtained data estimate for each cell can be assessed. In terms of temporal monitoring, it thus becomes possible to give a statistical estimate of how likely it is that observed changes in element concentrations over time are significant. The kriging variance can also be mapped. If regions within the survey area show an unusually high variance, they would need to be sampled at a higher density if an improved map was required. However, highly localised geochemical processes in an area could also be marked by an unusual local variance. Although a combination of the geochemical map with a map of the kriging variance can clearly guide interpretation, many projects will not support such an intensive activity.

Class selection and scaling are again important considerations when constructing kriged maps. The data structure should become visible and thus the classes should have a direct relation to the data structure. As discussed above (Sections 5.5.2 and 5.5.3), this is best achieved via boxplot or percentile classes. The approach of "avoiding" classes via using a more or less continuous grey scale (see Section 5.6 and Figure 5.6, left) is also a possibility.

5.7.4 Possible problems with semivariogram estimation and kriging

As mentioned above (Section 5.7.2), it is possible that the nugget effect is almost as large as the total variance. Figure 5.10 shows this effect for Hg measured in the Kola O-horizon samples. Only completely independent samples would allow for a pure (linear) nugget effect model. In regional geochemistry it is very unlikely that the samples are completely independent. The variogram thus suggests that the sample density was not high enough to detect the regional data structure of Hg. It is of course also possible that such a result is caused by an insufficient analytical quality (see Chapter 18) or that the sample material O-horizon is not suited to map

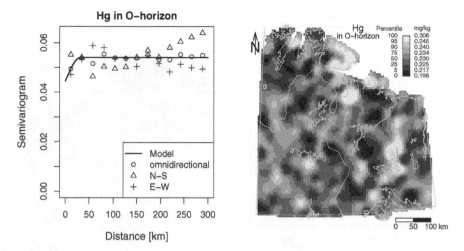

Figure 5.10   Semivariogram (left) and resulting kriged map (right) for Hg in Kola Project O-horizon soils

the regional distribution of Hg. To be able to map, a spherical model with a small range was fitted to the semivariogram. The resulting map (Figure 5.10, right) is very noisy. A map for a variable with a very high nugget effect may even become less informative than a properly prepared point source map.

A common assumption for constructing a semivariogram is that the variance in the selected direction depends only on the distance between two data points but is independent of the location of the data points. The variance at a fixed distance is thus assumed to be the same throughout the survey area. This is not the case if there is a systematic directional trend of increasing or decreasing values in the data (see Figure 6.7 for some examples). The Kola data show such trends for some selected elements in some materials (moss and O-horizon) with distance from coast. This would theoretically call for the use of more advanced kriging methods (e.g., universal kriging, see Cressie, 1993). However, the standard methods can often still be successfully applied to the data. To visualise the existence of such trends, it can be advantageous to show the directional and not only the omnidirectional semivariograms.

Figure 5.11 demonstrates this using the variable Pb measured in Kola O-horizon soils. The Pb concentrations show a strong north–south trend, as demonstrated in the directional semivariogram. In north–south direction independence of the samples is not reached within the survey area (Figure 5.11, upper left). In contrast, in the east–west direction the spatial dependency is rather low and the nugget effect dominates (Figure 5.11, upper left). However, it is possible to construct a kriged map based on the omnidirectional "compromise" (Figure 5.11, upper right) or on the east–west-model alone (Figure 5.11, lower left). The common choice would probably be the omnidirectional compromise. For comparison the smoothed surface map (Section 5.6) is shown in Figure 5.11 lower right. The maps suggest that the omnidirectional compromise results in excessive smoothing, hiding important details of the spatial variation of Pb.

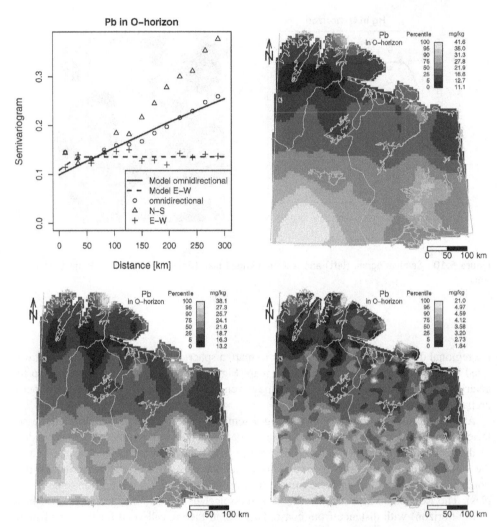

Figure 5.11 Directional (east–west and north–south) and omnidirectional semiovariograms for the variable Pb measured in Kola Project O-horizon soils (upper left). A model was fitted to the east–west and to the omnidirectional semivariogram (upper left). Kriged maps based on the omnidirectional model (upper right), based on the east–west model (lower left) and a smoothed surface map (lower right) are shown

5.8 Colour maps

What is the difference between colour and black and white maps? At first thought, there is no real difference other than that most people will be attracted by colour and thus feel that the map is "easier" to read. As a first approach, different black and white symbols may be replaced by just one symbol that is displayed in different colours (e.g., a filled circle or a filled square). Point source colour maps will result. However, surface maps are the true realm of colour mapping.

Colour is usually employed to make a map visually attractive. Because there are many different colours that can be easily distinguished, it is much less demanding to produce maps with a lot of information in colour than in black and white. However, colour can both help and hurt a map, and the most effective use of colour is not easy. Many colour maps look merely pretty but are not particularly informative. In geochemistry black and white maps may be more readily and reliably decoded than colour maps. It is often beneficial to investigate simple black and white mapping as the first step of spatial investigation.

One of the first problems with using colour in geochemical maps is that the colours have to be sorted somehow from low to high. There exists no logical, widely accepted colour scale. If ten different people are given a stack of colour cards and asked to sort them from "low" to "high" the chance is good that ten different scales will emerge. Some may order from green to red, some from blue to red, again others may attempt to order according to the rainbow sequence of colours, and some may use similar scales but in different directions.

In geochemistry a scale from deep blue for low to red and purple for high values is frequently used (the "rainbow" scale), similar to a global topographic map spanning from ocean depths to mountain peaks. Often variations arise in the "middle" with various choices for the use of yellows and greens. When looking through a selection of geochemical atlases, it is obvious that different colour scales have been employed, this is an unfortunate situation for the untrained reader. Preparing maps for the same data using the full rainbow scale, firstly ordered from blue to red, and then secondly, an inverse map, where the values are ordered from red (low) to blue (high), a scale that is also sometimes used for pH measurements, demonstrates some of the dangers that lie in colour mapping.

A simple, consistent grey scale, from light grey for low to black for high values may actually be far better suited to displaying concentration differences in a contoured surface map than a more pleasant colour scale (see Figure 5.6). Sometimes simple colour scales are used, e.g., from light yellow over light and dark brown to black which provides a consistent, logical and readily comprehended ordering from light to dark or low to high. Such a map may be a compromise between a "dull" black and white or grey-scale map and a "pretty" full colour map. A further advantage of a grey-scale map is that it will have the same appearance to those with colour blindness as to the rest of the population.

Dark red and dark blue on a colour map, a scheme often used in global topographic maps, will look very similar when copied in black and white. The information in a colour map is thus easily lost when copying or faxing a map. Colours are also bound to emotions, and colour significance may vary culturally. Colours can thus be used to manipulate the reader and may not be the optimal choice in an international world.

The colour red is a good case in point, in the western cultures it automatically gets more attention than all other colours, being a signal for attention or danger. However, in eastern cultures red signals happiness. Thus colour maps are not universally objective and attempts to draw most attention to the high values (which are usually plotted in red) may not always succeed with an international audience.

If a rainbow scale is used for mapping, it is also very important where, in relation to the data set, the main colour changes occur, and class selection becomes even more important than in black and white mapping. Colours may increase the visibility of data structure in a map – or destroy it completely. Usually a colour map based on the rainbow scale and with the major colour changes carefully spread over the full range of percentile classes will provide the most informative impression. When studying such maps the reader should at least always check where in relation to the data the break from yellow to red occurs.

The "no classes" approach can also be transferred to colour mapping because so many colours exist that virtually any single analytical value can have its own value in a map (see, e.g., Lahermo *et al.*, 1996). This has been demonstrated above using a continuous grey scale (Figure 5.6, left) and is shown for the rainbow scale in Figure 5.12 (lower left).

In a GIS environment the use of colour on maps is especially tempting. In that case the alert reader has to watch for inadvertent camouflage. In some cases poor contrast between the background and geochemical colours can result in severe difficulties in extracting useful information from a pretty coloured map.

5.9 Some common mistakes in geochemical mapping

5.9.1 Map scale

As discussed in Section 5.2, one of the most widespread mistakes is to scale a geochemical map according to some existing base map or to some nice looking scale (e.g., 1:1 000 000). A geochemical map needs to be scaled such that the inherent geochemical information is optimally displayed. The geochemistry and not the base map information should govern the map scale. Usually, the smallest possible map will be best suited for geochemical interpretation.

5.9.2 Base map

A common mistake discussed in Section 5.3 is to use an existing topographical or geological map that was already designed (and scaled) for purpose. This will lead to "cluttered" and hard to read geochemical maps (see Figure 5.1). It may also lead to misinterpretations in instances where the chosen background information is not the main factor determining the geochemical distribution.

5.9.3 Symbol set

Two (three with proportional dots) "standard symbol sets" (EDA and GSC) have been introduced above (Section 5.5.1, Figures 5.3 and 5.4). What can be read from a black and white map depends crucially on the symbol set and unfortunately there are about as many different symbols sets as there are geochemists. Figure 5.13 demonstrates how different maps can look when the symbol set is changed and how information can actually be lost (or be manipulated) by the choice of the symbol set (see also Figure 5.4 for different impressions gained from the EDA and GSC symbol sets). A symbol set needs to include a certain easy to grasp inner order as provided by both GSC and EDA symbol sets. Here readability of the map for different users will in the end just be a question of personal taste and training. If there is no "inner order" in the symbol set, it will be close to impossible to gain an impression of the regional data structure in a map.

5.9.4 Scaling of symbol size

The size of the symbols in relation to the size of the map will have an important influence on the appearance of the map (Figure 5.14). Even the relative proportions of the size of the

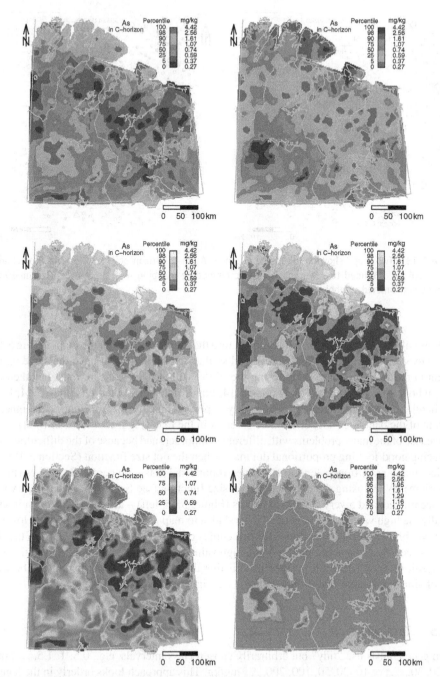

Figure 5.12 Colour smoothed surface maps for As in Kola C-horizon soils. Different colour scales are used. Upper left, percentile scale using rainbow colours; upper right, inverse rainbow colours; middle left, terrain colours; middle right, topographic colours; lower left, continuous percentile scale with rainbow colours; and lower right, truncated percentile scale. A colour reproduction of this figure can be seen in the colour section, positioned towards the centre of the book

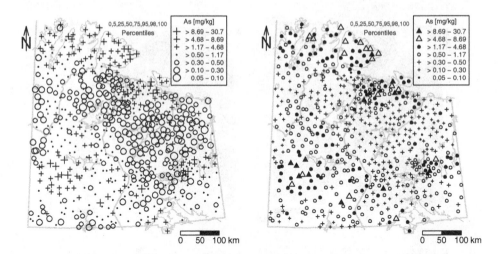

Figure 5.13 Two maps showing the distribution of As in C-horizon soils of the Kola Project area. EDA symbol set extended to seven classes to accommodate mapping with percentiles (left map) and arbitrary, non-ordered symbol set (right map)

symbols will influence the appearance, and in turn the interpretability, of the map (Figure 5.14). The EDA symbol set was carefully designed so that each symbol will have the same optical weight in the resulting map. When using the EDA symbol set, great care must be taken that the symbols do not get too small (Figure 5.14, upper right) or too large (Figure 5.14, lower left) in relation to the map scale. It is also very important that the dot representing the inner 50 per cent of the data does not get too much weight in the map (Figure 5.14, lower right).

Because of the many problems with different symbol sets and because of the difficulties with producing good-looking proportional dot maps when the dot size function (Section 5.4) is not applied, many practitioners tend to combine proportional dots with percentile classes (Figure 5.15, right). When using the exponential dot size function (see Section 5.4), the high values are accentuated, and a very clear map results (Figure 5.15, left) – as long as the main interest is really the high values. It is in general a good idea to map using percentile classes. However, when combining proportional dots with percentiles classes, the visual impression of the data structure as well as the accentuation of the high values is lost, and a rather uninformative map may result (Figure 5.15, right). The reason is that the eye is not able to distinguish between dots of almost the same size if they do not occur right beside one another.

5.9.5 Class selection

Often geochemists use "tidy" but arbitrarily chosen class intervals, e.g., 0.5, 1, 1.5, . . . ; or 5, 10, 20, 30, . . . ; or 10, 20, 50, 100, 200, . . . mg/kg. This approach looks orderly in the legend, but the legend itself is not the most important part of a geochemical map. Such arbitrarily picked classes bear no relation to the spatial or statistical structure of the mapped data. Such classes often result in uninformative, "dull" geochemical maps that fail to reveal the underlying geochemical processes causing regional geochemical differences. Arbitrary classes should therefore be avoided. As a historical note, arbitrary classes were extensively used when optical

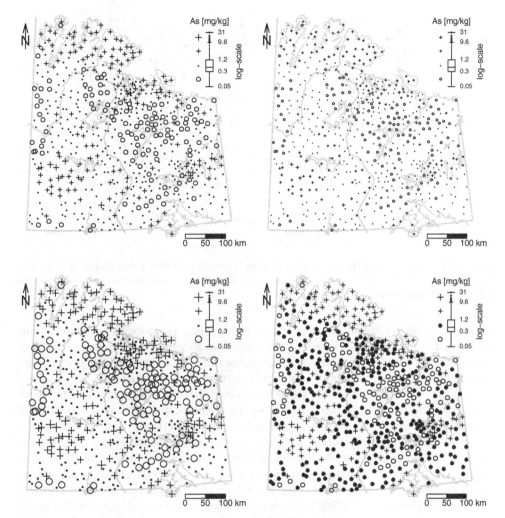

Figure 5.14 Four maps showing the distribution of As in C-horizon soils of the Kola Project area. Percentile classes and the EDA symbol set are used for mapping. To get a clear visual impression of the data distribution, the size of the symbols needs to be adjusted to the scale of the map

spectroscopy was employed as a geochemical analysis tool and the plates or filmstrips were read by eye with a comparator. This tended to lead to clustering of the reported values around the standards used for comparison. Thus the class boundaries were set at the mid-points between the standards, often on a logarithmic scale, e.g., with standards at 1, 5 and 10 units, class boundaries were set at 1.5, 3 and 7.

Another approach that often fails to reveal meaningful patterns is to base the classes in the mean plus/minus multiples of the standard deviation. This approach is based on the assumption that the data follow a (log)normal distribution. In this case – and only in this case – the approach provides an easy method of identifying the uppermost 2.5 per cent of the data as "extreme values". However, as has long been recognised, most applied geochemical survey

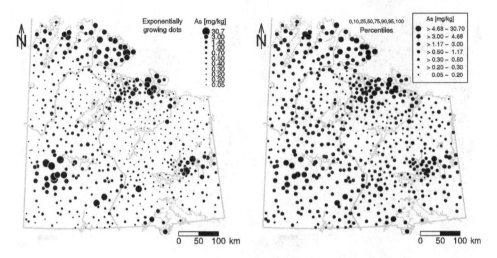

Figure 5.15 Maps showing the distribution of As in C-horizon soils of the Kola Project area. Proportional dots according to the exponential dot size function (left) in direct comparison to growing dots scaled according to percentile classes (right)

measurements do not follow a lognormal distribution. Some practitioners choose to assume that geochemical data follow a lognormal distribution because this facilitates the use of classical statistics by taking the logarithms of the univariate values (Reimann and Filzmoser, 2000d). Regrettably, modern (often "robust") statistical procedures, that are better suited for geochemical data, are often omitted from basic statistics courses, and this may explain why the "mean plus standard deviation" approach is still so popular in geochemistry Reimann *et al.*, 2005c). The problem is that for data (or their logarithms) that are not normally distributed, but reflect a whole range of different processes and thus are multimodal, the mean, and especially the standard deviation, are not good measures of central tendency and spread (variability). For example, all class boundaries defined by this approach are strongly influenced by the outliers. Yet, to identify statistical outliers is often one of the reasons why the samples were collected at the outset Reimann *et al.*, 2005c). Geochemists should recognise the important distinction between the extreme values of a normal (or lognormal) distribution (detected satisfactorily by the mean ± standard deviation approach) and outliers due to multimodal distributions (e.g., several "overlapping" geochemical processes), where this approach is unfounded.

Even when using percentile-based classes, many geochemists accentuate differences among the high numbers at the expense of variations at lower concentrations (Figure 5.12, lower right). This results in a map focussing completely on the high values and neglecting the lower end of the distribution. In some studies, e.g., for trace element deficiencies, this is completely inappropriate. Focussing on the uppermost end of the distribution again neglects (as with arbitrarily chosen classes) the fact that there exists something called the "data structure".

5.10 Summary

Environmental data are characterised by their spatial nature. Mapping should thus be an integral and early part of any data analysis. Once the statistical data distribution is documented (Chapters 3 and 4) the task is to provide a graphical impression of the "spatial data structure". There are many different procedures for the production of maps, only a few provide a graphical impression of the spatial data structure. As a first step the focus should not be on the high values alone but rather on an objective map of the complete data distribution of all studied variables. Classes need to be carefully spread over the whole data range, the log-boxplot or percentiles are well suited for class selection. They need to be combined with a suitable symbol set (e.g., EDA symbols).

If only "highs" and "lows" are of interest proportional dot maps provide nice, and apparently "easy", maps. Just as the statistical data distribution should be documented in a series of summary plots the spatial data distribution needs to be documented in a series of maps of all variables before more advanced data analysis techniques are applied. Black and white maps are best suited for this purpose. Map scale and a north arrow should always accompany a map. An easily legible legend is a further requirement.

More advanced mapping techniques will be used in a later stage, when surface, contour or isopleth maps are designed for "presentation". For this purpose point data need to be inter-polated to surface data. Surface maps can be highly manipulative, the choice of parameters and the choice of colour are important considerations. When kriging is used for constructing surface maps a solid understanding of the method is required. Maps prepared by smoothing are not based on statistical assumptions in the same way as kriged maps. If the additional information from kriging (semivariogram, kriging variance) is not required, smoothing maps may be the better choice.

6

Further Graphics for Exploratory Data Analysis

6.1 Scatterplots (xy-plots)

Once there is more than one variable, it is important to study the relationships between the variables. The most frequently used diagram for this purpose is a two-dimensional scatterplot, the xy-plot, where two variables are selected and plotted against one another, one along the *x*-axis, the other one along the *y*-axis. Scatterplots can be used in two different ways. Firstly, in a truly "exploratory" data analysis approach, variables can be plotted against one another to identify any unusual structures in the data and to then try to determine the process(es) that cause(s) these features. Secondly, the more "scientific" (and probably more frequently used) way of using scatterplots is to have a hypothesis that a certain variable will influence the behaviour of another variable and to demonstrate this in a plot of variable *x* versus variable *y*. When working with compositional data the problem of data closure needs to be considered. It will often be a serious issue whether simple scatterplots show the true relationships between the variables (consult Section 10.5).

Both methods have their merits, the exploratory approach facilitates the identification of unexpected data behaviour. In pre-computer times this approach was not widely used as it was time-consuming with the risk of time and resources expended with no reward. Even with computers many data analysis procedures require considerable effort to draw such displays. The process must be simple and fast so that it becomes possible "to play with the data" and to follow up ideas and hypotheses immediately without any tedious editing of data files before the next plots are displayed.

Figure 6.1 shows a scatterplot for Cu versus Ni in the moss samples from the Kola Project. These two elements are of special interest because approximately 2000 tons of Ni and 1000 tons of Cu are emitted to the atmosphere annually by the Russian nickel industry, situated in Nikel, Zapoljarnij, and Monchegorsk (see Figure 1.1 for locations).

Figure 6.1 reveals that a positive, sympathetic relationship exists between Cu and Ni. However, there are a substantial number of samples with high analytical results for both Cu and Ni that mask the behaviour of the main body of data. Plotting the same diagram with logarithmic axes (Figure 6.1, right), decreases the influence of the high values and increases the visibility of the main body of data. It can now also be seen that the spread (variance) of

Statistical Data Analysis Explained   Clemens Reimann, Peter Filzmoser, Robert G. Garrett, Rudolf Dutter
© 2008 John Wiley & Sons, Ltd.

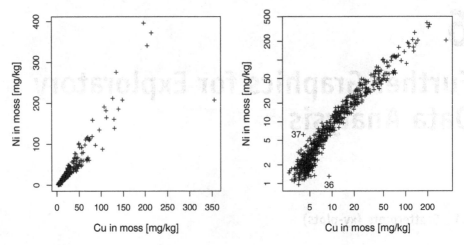

Figure 6.1 Scatterplot of copper (Cu) versus nickel (Ni) in moss samples from the Kola Project. Left plot: original data scale; right plot: log-scale. In the right plot two unusual samples were identified

the data is relatively homogeneous over the whole range of the data. It is also apparent that Cu and Ni are strongly related over the whole range of data. In addition, attention is drawn to a number of unusual samples in both diagrams that deviate from the general trend displayed by the majority of samples. It should be possible to identify these individuals by just "clicking" on them on the screen. Samples like numbers 36 and 37 (Figure 6.1, right) should undergo an additional "quality check" to ensure that they are truly different and that the discordant data are not due to a sample mix-up or poor analytical quality.

6.1.1 Scatterplots with user-defined lines or fields

It is often desirable to include a line or one or several fields in a scatterplot. The simplest case is probably a line indicating a certain proportion between the data plotted along the two axes. One example would be two variables where the same element was analysed with two different analytical techniques. Here the results might be expected to follow a strict one to one relationship. Plotting a 1:1 line onto the diagram allows for the easy detection of any deviations from the expected relationship, and to see at a glance whether the results obtained with one of the methods are higher than expected.

Figure 6.2 shows two such diagrams. In the left plot La was determined with two different analytical techniques. It takes but a glance to notice that although the analyses are similar, by and large, the analytical results obtained with instrumental neutron activation analysis (INAA) are clearly higher than those from an acid extraction. Only INAA provides truly total La-concentrations in the samples and thus this result is no surprise. In the right hand plot, results for Ag in two different soil horizons are depicted (Figure 6.2). The analytical techniques are comparable (both strong acid extractions), but the plot indicates that Ag is highly enriched in the O-horizon and does not show any relation to the C-horizon results. This is surprising; many applied geochemists or environmental scientists would argue that the C-horizon, the mineral or parent soil material, will provide the "geochemical background values" (see Chapter 7) for

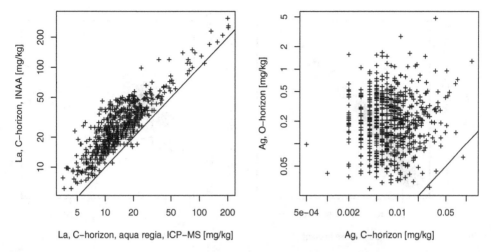

Figure 6.2 Scatterplots with 1:1 line added. Left plot: La determined with two different analytical techniques; right plot: Ag determined with comparable techniques in two different sample materials. Log-scales used for both graphics

the upper soil horizon. The diagram indicates that the relationship is not that simple. One could be tempted to argue that then all the Ag in the O-horizon must be of anthropogenic origin. There is, however, no likely major Ag-emission source in either the survey area or anywhere near. Thus the conclusion must be that there exist one or more processes that can lead to a high enrichment of Ag in the organic layer – independent of the observed concentrations in the soil parent material.

Petrologists frequently use xy-plots where they mark predefined fields to discriminate between different rock types. To include such "auxiliary" background information in a scatterplot is desirable and feasible, but the actual definition of the field limits in the diagrams is a highly specialised task. One software package, based on R, that can plot these lines and fields for many pre-defined plots used by petrologists is freely available at http://www.gla.ac.uk/gcdkit/ (Janousek *et al.*, 2006).

6.2 Linear regression lines

The discussion of Figure 6.2 demonstrated that it is informative to study dependencies between two or more variables. In Figure 6.2 a 1:1 relation was a reasonable starting hypothesis and thus the line could be drawn without any further considerations. However, many variables may be closely related but do not follow a 1:1 relation (e.g., Cu and Ni as displayed in Figure 6.1). In such cases it might aid interpretation if it were possible to visualise this relationship. In most cases the first step will be to look for linear rather than non-linear relationships. To study linear relationships between two or more variables, linear regression is a widely used statistical method (see also Chapter 16). The best-known approach for fitting a line to two variables is the least squares procedure. To fit a line, it is important to take a conscious decision which variable is the *y*-variable, i.e. the variable to be predicted. The least squares procedure determines the line where the sum of squared vertical distances from each *y*-value to this line is minimised.

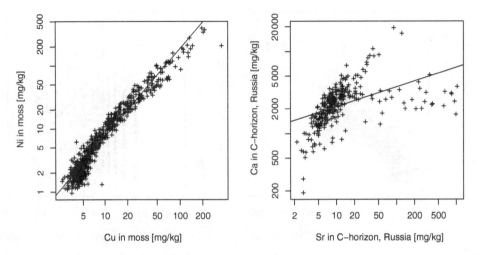

Figure 6.3 Two scatterplots with least squares linear regression lines. Left plot: Cu versus Ni, Kola Project Moss data set with log-scale as used in Figures 6.1 and 6.2; right plot: Sr versus Ca (log-scale), Kola Project C-horizon, Russian samples

This situation can be avoided by using less common procedures for estimating the linear trend, e.g., total least squares or orthogonal least squares regression (Van Huffel and Vandewalle, 1991) which minimise a function of the orthogonal distances of each data point to the line.

Figure 6.3 (left) shows the relationship, regression line, between Cu and Ni in moss (Figure 6.1). When plotting a regression line onto the diagram using the log-scale it is visible that the relation is not really linear but slightly curvilinear. However, a straight line is quite a good approximation of the relationship between the two variables. The other diagram in Figure 6.3 shows the relation between Ca and Sr for the C-horizon samples collected in Russia (log-scale). A number of samples with unusual high Sr-concentrations disturb the linear trend visible for the main body of samples. These samples exert a strong influence (leverage) on the regression line, such that it is drawn towards the relatively few high Sr values. This example demonstrates that the least squares regression is sensitive to unusual data points – in the example the line follows neither the main body of data nor the trend indicated for the outliers (Figure 6.3, right).

It is thus necessary to be able to fit a regression line that is robust against a certain number of outliers to the data. This can be achieved by down-weighting data points that are far away from the linear trend as indicated by the main body of data. The resulting line is called a robust regression line.

Figure 6.4 shows the two plots from Figure 6.3 with both the least squares and a robust regression line. For the left hand plot, Cu versus Ni in moss, both methods deliver almost the same regression line, the difference in the right hand plot (Ca versus Sr in the C-horizon of the Russian samples) is striking. The robust regression line follows the main body of data and is practically undisturbed by the outlying values. For environmental (geochemical) data it is advisable to always use a robust regression line, just like MEDIAN and MAD are better measures of central value and spread than MEAN and SD.

Regression analysis can be used not only to visualise relationships between two variables but also to predict one variable based on the values of one or more other variables (see Chapter 16).

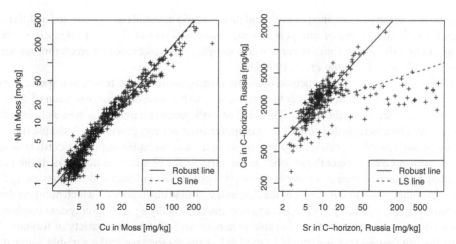

Figure 6.4 Two scatterplots with least squares (LS) and robust regression lines. Left plot: Cu versus Ni, Kola Project Moss data set with log-scale as used in Figure 6.3 (left); right plot: Sr versus Ca (log-scale), Kola Project C-horizon, Russian samples

Relationships between two variables do not need to follow a linear trend. In fact, any non-linear function could be fitted to the data, or the data points could be simply connected by a line. However, to be able to discern the trend, the function will often need to be smoothed. Many different data-smoothing algorithms do exist. Usually a window of specified width is moved along the x-axis. In the simplest case one central value for the points within the window is calculated (e.g., the MEAN or MEDIAN). When connected, these averages provide a smoothed line revealing the overall data behaviour. Depending on the window width, the resulting line will be more or less smoothed. More refined methods use locally weighted regression within the window.

In environmental sciences the most usual application of smoothed lines will be in studying time or spatial trends as described below (Sections 6.3, 6.4).

6.3 Time trends

In environmental sciences "monitoring" is often an important task; it is necessary to determine whether analytical results change with time. The large test data set used here comes from the regional part of the Kola Project. It permits the study of the distribution of the measured elements in space via mapping (Chapter 5) but not to study changes in element concentration over time. It would become possible if the regional mapping exercise was repeated at certain time intervals. This would be a very expensive undertaking, and it would be more effective to collect the required data in a monitoring exercise at selected locations more frequently than to repeatedly re-map the whole area.

For interpreting the regional Kola data set, more detailed information on local variability versus regional variability and on changes of element concentrations with the season of the year was needed. These data were obtained in a catchment study, which took place one year before the regional mapping program (see, e.g., de Caritat *et al.*, 1996a,b; Boyd *et al.*, 1997; Räisänen *et al.*, 1997; Reimann *et al.*, 1997c). For the catchment study eight small catchments were sampled at a density of one site per km^2, and most sample materials were collected at different times through all seasons of one year. Stream water was one of the most intensely studied sample materials, and at some streams water samples were taken once a week throughout a hydrological year (de Caritat *et al.*, 1996a,b).

These data are used here to demonstrate the investigation of time trends in a special form of the scatterplot where the *x*-axis is the time scale and the measured concentration of diverse variables is plotted along the *y*-axis. This kind of work requires careful planning as to how the time of sampling is recorded so that the computer software can process it. A simple trick to overcome such problems could be to give all samples a consecutive number according to the date of collection and select the number as the variable for plotting the time axis. If the field sampling has not occurred at regular intervals, this procedure will lead to distortions along the *x*-axis. Another procedure is to record the date as the "Julian day number", which numbers days consecutively from January 1, 4713 BC and is an internationally agreed-upon system employed by astronomers. Fortunately, R is also able to handle "usual" dates in a variety of formats. To easily discern the trends, a line should be used to link results for a particular variable. Often it is advantageous to be able to estimate a smoothed line running through the results by a variety of different algorithms (see Section 6.4). In the simple case of about 50 water measurements per catchment, spread almost evenly over the year, a simple line directly connecting the results will suffice. However, where multi-year data are acquired, some systematic periodicity, seasonal effect, may be expected and smoothing may help identify such effects.

In Figure 6.5 it is demonstrated how pH and EC change in the course of one year in catchment two, which is in close proximity to the Monchegorsk smelter. A drop in pH and conductivity during May is obvious (Figure 6.5). This indicates the influence of the snow melt period on the stream water samples.

If an abundance of data exists that exhibits high variability, it can also be informative to summarise days, weeks, or months (maybe even years) in the form of boxplot comparisons (see Section 8.4). It is then possible to study differences between the MEDIANS and differences in

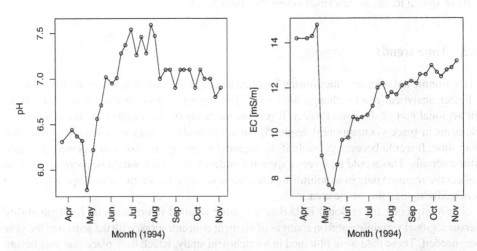

Figure 6.5 Time trends of acidity (pH) and electrical conductivity (EC) in stream water collected on a weekly base in catchment two, Monchegorsk

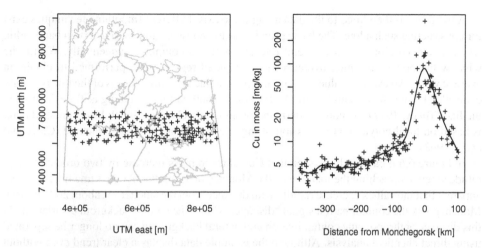

Figure 6.6 Right plot: east–west transect through Monchegorsk, showing the decrease of the concentration of Cu in moss (y-axis, log-scale) with distance from Monchegorsk (x-axis). The map (left plot) shows the location of the samples used to plot the transect

variation (and skewness) at the same time, and trends may be easier to discern when looking at such a graphical summary instead of looking at all the data points and a trend line.

6.4 Spatial trends

Instead of looking at a geochemical map (see Chapter 5), it may also be informative to study geochemical transects, another special case of the xy-plot, where the x-variable is linked to the coordinates of the sample site locations. With the Kola Project data it is, for example, informative to look at an east–west-transect running through Monchegorsk from the eastern project border to the western project border. This facilitates a study of the impact of contamination and the decrease of element concentrations with distance from Monchegorsk. Another interesting study would be a north–south-transect along the western project boundary, in the "background" area, to investigate the impact of the input of marine aerosols at the coast and of the influence of the change of vegetation zones from north to south on element concentrations in moss or soils.

For constructing such a transect, it is necessary to have a tool to define the samples belonging to the transect subset. It is of great advantage if this can be done interactively directly in the map. Such samples could then, for example, be identified as belonging to the subset "EW_Monchegorsk" and "NS_Background". Once the subset is selected, it is easy to study the distribution of all the measured variables along such a transect via selecting the x- (for the east–west-transect) or y-coordinate (for the north–south-transect) for the x-axis of the plot and any other variable for the y-axis and plotting these as xy-plots. In instances where the coordinates of the coastline or of Monchegorsk are known, new variables for the x-axis can be defined "distance from coast" and "distance from Monchegorsk" via a simple subtraction. If the transects are not parallel to the coordinate system, distances can be estimated by plane geometry.

A trend line is then fitted to the data along the x-axis. Different smoothing techniques exist for constructing such a line. The basic principle is to choose a window of a certain bandwidth, which is moved along the x-axis and relevant statistics computed sequentially. Within the window the points are either averaged or a weighted regression is performed to obtain an average value. All average values are connected by a line. The smoothness of the line is mainly controlled by the chosen bandwidth. A very simple method for smoothing is Tukey's moving median (Tukey, 1977). A more modern and prominent technique, which has been used here, is based on local polynomial regression fitting and known as the "loess" method (Cleveland *et al.*, 1992).

In Figure 6.6 it can easily be seen that Cu values in moss increase by two orders of magnitude when approaching the location of the Monchegorsk smelter. At the same time it is also apparent that the values decrease rapidly with distance from the smelter. At about 200 km from Monchegorsk the contamination signal "disappears" into the natural background variation. At this distance anthropogenic contamination and natural background can no longer be separated using direct chemical analysis. Although the example data display a clear trend even without a smoothed line, the line provides a more informative picture of the overall trend without distortion by local variation.

A north–south transect at the western project border in a pristine area without influence from industry shows other processes (Figure 6.7). The steady input of marine aerosols from the coast of the Barents Sea has an important influence on moss and O-horizon chemistry. Elements like Na and Mg, which are enriched in sea water, show a steady decrease from the coast to a distance of about 200 to 300 km inland (Figure 6.7). This indicates the important influence of a natural process, the steady input of marine aerosols along the coast, on the element concentrations observed in moss and O-horizon samples on a regional scale. An "exploratory" surprise was the opposite trend as detected for Bi (and several other elements – see Reimann *et al.*, 2000b) (Figure 6.7). This trend follows the distribution of the major vegetation zones in the survey area (tundra–subarctic birch forest–boreal forest) and is interpreted as a "bio-productivity" signal: the higher the bio-productivity, the higher the enrichment of certain elements in the O-horizon.

In the examples shown so far, the trends were visible when studying the distribution of the sample points in the figures (Figure 6.6 and 6.7) by eye. Trend lines will be even more interesting when the data are far noisier and trends are not visible at first glance. The distribution of the pH values in the O-horizon samples along the transects provides such an example (Figure 6.8). Along the north–south transect, pH is clearly influenced by distance from the coast. The steady input of marine aerosols along the coast (see Figure 6.8) leads to a replacement of protons in the O-horizon and results in a steadily increasing pH towards the coast. The replaced protons end up in the lakes, resulting in lake water acidification along the coast (Reimann *et al.*, 1999a, 2000a). Along the east–west transect one would expect clear signs of acidification (lower pH) when approaching Monchegorsk with its enormous SO_2 emissions. Figure 6.8 (right) suggests the opposite: pH increases slightly when approaching Monchegorsk, the anthropogenic effect of one of the largest SO_2-emission sources on Earth on the pH in the O-horizon is clearly lower than the natural effect of input of marine aerosols along the coastline. The fact that pH actually increases towards Monchegorsk is explained by co-emissions of alkaline elements like Ca and Mg which more than counteract the acidic effects of the SO_2 emissions. The diagram suggests that it would be a severe mistake to reduce the dust emissions of the smelter before the SO_2 emissions are reduced because then the positive environmental effects of these co-emissions would be lost.

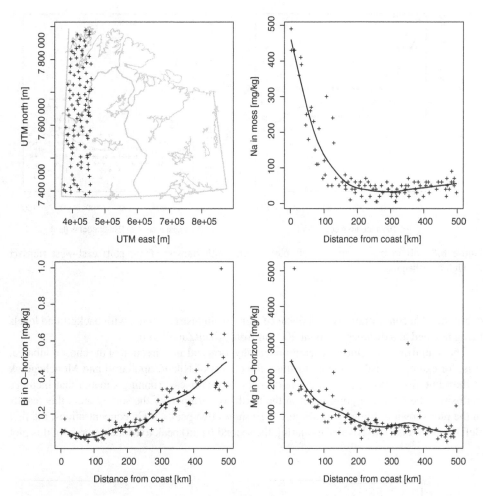

Figure 6.7 North-south transects along the western project border, the most pristine part of the Kola Project survey area. The map (upper left) indicates the samples selected for plotting as transects. Upper right: Na in moss; lower left: Bi in O-horizon soil; and lower right: Mg in O-horizon soil

6.5 Spatial distance plot

There may be cases where it is difficult to define a sensible transect. In such instances it could still be interesting to study systematic changes of a variable with distance from a defined point. This can be done in the spatial distance plot where a subarea (of course the whole survey area can also be used) and a reference point are defined.

The calculated distance of each sample site from the reference point is plotted along the *x*-axis. Along the *y*-axis the corresponding concentrations of the variable at the sample sites are plotted. To better recognise a possible trend a smoothed line is fitted to the points.

Figure 6.9 shows an example of such a spatial distance plot using the variable Cu in moss and the location of Nikel/Zapoljarnij as the reference point. The plot clearly demonstrates the

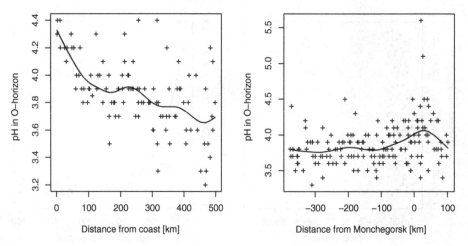

Figure 6.8 pH in the O-horizon. Left plot: north–south transect. Right plot: east–west transect through Monchegorsk

decreasing Cu concentrations with distance from the industrial centre, with background levels being reached at a distance of about 200 km from Nikel/Zapoljarnij.

The spatial distance plot can accommodate any form and any direction of the chosen subarea. It is, for example, easily possible to study the effects of Nikel/Zapoljarnij and Monchegorsk at the same time. When choosing a north–west to south–east running subarea that includes both sites, the reference point is set in the north-western part of the survey area; this results in the plot shown in Figure 6.10. Figure 6.10 shows two peaks of Cu concentrations, the first (left) peak is related to Nikel/Zapoljarnij, the second (right) peak to Monchegorsk. In this plot

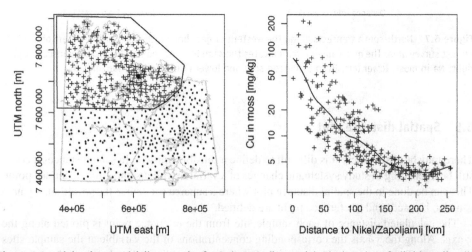

Figure 6.9 Right plot: spatial distance plot for Cu in moss (log-scale). The location of Nikel/Zapoljarnij (dot on left map) is chosen as reference point, the outline indicates the sample sites included in the spatial distance plot (right)

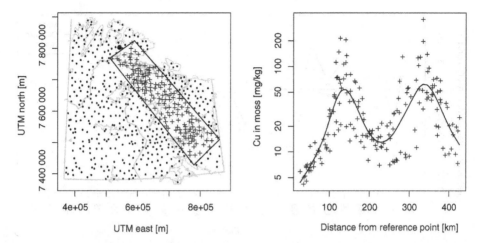

Figure 6.10   Right plot: spatial distance plot for Cu in moss (log-scale). The subarea (left map) includes Nikel/Zapoljarnij and Monchegorsk, and the reference point (dot) is placed in the north-western part of the survey area; the outline indicates the sample sites included in the spatial distance plot (right)

it is clearly visible that the Monchegorsk Cu emissions travel further than the emissions from Nikel/Zapoljarnij, as indicated by the width of the peaks. One likely explanation could be that the emissions at Nikel/Zapoljarnij have a higher proportion of particulates.

6.6 Spiderplots (normalised multi-element diagrams)

Spiderplots, where a selection of elements are plotted along the x-axis against the ratio of the element to the concentration of the same element in some reference value (e.g., "chondrite", "mantle", "shale", "upper continental crust") on the y-axis, are popular in petrology (see, e.g., Rollinson, 1993). The technique was originally developed to compare the distribution of the Rare Earth Elements (REEs) between different rock types. REEs with an even atomic number show in general higher concentrations in rocks than REEs with uneven atomic numbers (Taylor and McClennan, 1985). Through normalising REEs to a reference value, the resulting saw-blade-pattern can be brought to a more or less straight line by plotting the ratio against the elements. Deviations from the line are then immediately visible. Usually the elements are sorted according to atomic number, but other sorting criteria (e.g., ionic potential or bulk partition coefficient) are also used.

Only a few samples can be plotted before the graphic becomes increasingly unreadable; it is thus not really suited for large data sets. The appearance of the plot will strongly depend on the sequence of elements along the x-axis and, of course, on the selection of the set of reference values used for normalisation. Except for the REEs, there are no standardised display orders, a fact that limits the usefulness of these diagrams.

To display the Kola Project results in such a spider plot, the MEDIAN concentration of selected variables per layer (moss, O-, B-, or C-horizon) could be shown. It could be argued at length as to what reference value the results should be compared against. "Upper crust" or "world soil" might be the most appropriate candidates at first glance. However, the average

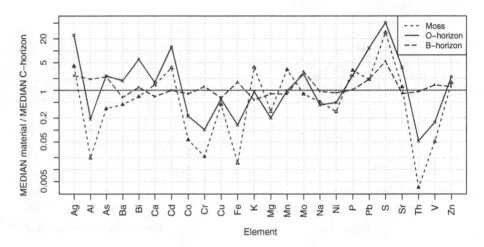

Figure 6.11 Spiderplot showing element concentrations in the B-horizon, O-horizon and moss sample materials relative to the element concentration in the C-horizon from the Kola Project plotted in chemical abbreviation order (from Ag to Zn)

values quoted for "upper crust" or "world soil" are true total concentrations, which are not really comparable to the aqua regia and concentrated HNO_3-extraction data of the Kola Project. Thus there are problems in selecting the most appropriate reference values for the display. The upper soil horizons could be referenced to the C-horizon results from the Kola Project as the best estimate of parent material geochemistry for the survey area (bearing the "cost" of being "only" able to directly compare B-horizon with O-horizon and moss). It would also be possible to use another large-scale data set where comparable analytical techniques were used (e.g., the Baltic Soil Survey data – Reimann *et al.*, 2003). For the example plot the MEDIAN of each variable in moss, O- and B-horizon was divided by the corresponding MEDIAN in the C-horizon. The line at "1" in the plot is the reference line of the C-horizon against which the other sample materials are compared. In the first example (Figure 6.11) the elements have been ordered alphabetically (according to chemical abbreviation from Ag to Zn).

The second example (Figure 6.12) shows the same plot where the elements are ordered according to a steadily decreasing ratio in moss. Although the same data are plotted, the resulting diagram looks very different. Here it is easier to compare the differences of the other two sample media (B- and O-horizon) from the two extremes, C-horizon and moss.

It is immediately apparent that the average composition of the B-horizon – with the exception of very few elements – closely follows that of the C-horizon. Furthermore, it is clear that a number of elements show unusually high concentrations in the moss, while others have much lower values in the moss than in the C-horizon. It is also quite easy to identify those elements that show a different behaviour in the O-horizon than they do for the moss or B-horizon.

6.7 Scatterplot matrix

The scatterplot matrix is a collection of scatterplots of all selected variables against each other (Figures 6.13 and 6.14). The scatterplot matrix is one of the most powerful tools in graphical

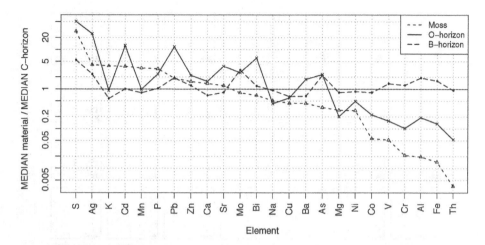

Figure 6.12   The same spiderplot as Figure 6.11, but the elements are sorted according to the continuously decreasing ratio for the moss samples relative to the underlying C-horizon

data analysis, and is an entrance point to multivariate data analysis. It can, for example, be used to identify those pairs of elements where a more detailed study in xy-plots appears to be necessary.

When using the scatterplot matrix with a multi-element data set, one of the major problems is screen and paper size. The individual plots become very small on the screen if more than 10 to 12 variables are displayed. If an A0 (or 36 inch) printer or plotter is available, up to a 60 × 60 matrix may be plotted. An important consideration before plotting a scatterplot matrix is the handling of outliers. Extreme outliers in several variables will greatly disturb the plot (Figure 6.13). Thus transformations need to be considered prior to plotting in order to down-weight the influence of outliers (Figure 6.14). Alternatively samples that exhibit extreme outliers can be removed (trimmed) prior to plotting a scatterplot matrix. Although the scatterplot matrix is a graphic and not a "formal" correlation analysis, the special problems of the closure of geochemical data still remain (see Section 10.5), especially for major and minor elements. The scatterplot matrix may help, however, to visualise some of these problems right away. Furthermore, by additive or centred logratio transformation of the major and minor element data (see Sections 10.5.1 and 10.5.2), the "opened" data may be displayed.

6.8   Ternary plots

Ternary plots show the relative proportions of three variables in one diagram (see, e.g., Cannings and Edwards, 1968). They have been used in petrology since the early 1900s to study the evolution of magmas or to differentiate between certain rock types. When three variables for a single observation sum to 100 per cent, they can be plotted as a point in a ternary plot. To construct an informative ternary plot, the three variables should have analytical results within the same order of magnitude, or all points will plot into the corner of the variable with the highest values. If there are large differences between the variables, it is necessary to multiply one or two of the variables by a factor to bring the values into the same data range as that of the other variables. The values for all three variables are then re-calculated to 100 per cent and

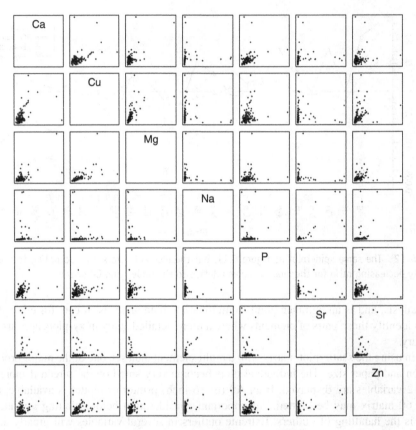

Figure 6.13 Scatterplot matrix of some selected elements, Kola Project C-horizon data. Axes could be labelled, but that will result in even smaller scatterplots, which contain the important information of this graphic. Some extreme values dominate the plot

plotted into the ternary space. In a ternary plot only two variables are independent, the value of the third variable is automatically known; the data are closed (Aitchison, 1986), and thus these diagrams should not be used to infer "correlations".

However, they can be used in environmental and other applied geochemical studies to aid "pattern recognition" and comparison with other studies.

Figure 6.15 shows two ternary plots constructed with the Kola Moss data set. The scale of the three axes varies between 0 and 100 per cent for each element (100 per cent in the corner of the element). The grey dashed lines in Figure 6.15 show the percentages. Just as in the scatterplot, it should be possible to identify interesting samples (usually outliers) via a simple click of the mouse. The Cu-Ni-Pb plot demonstrates that Pb is not an important component of the smelter emissions, which are dominated by Cu and Ni. Sample 907 was taken close to the Monchegorsk smelter. The Mn-Al-Fe ternary plot was constructed because both Al and Fe concentrations will be strongly influenced by the input of dust to the moss, while Mn is a plant nutrient. It is apparent that there are samples where Al dominates the dust component and others where Fe is more important. In addition, it appears that the majority of samples tend to the Mn corner of the diagram.

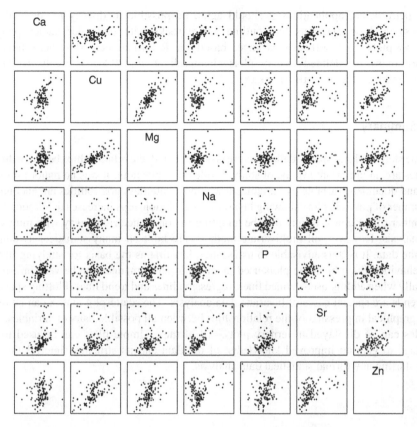

Figure 6.14 The same scatterplot matrix as above (Figure 6.13) but with log-transformed data to decrease the influence of extreme values

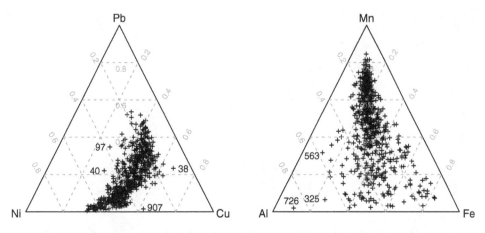

Figure 6.15 Two ternary plots for selected elements of the Kola Project Moss data set

For petrological applications it should again be possible to display pre-defined fields (or lines) in these diagrams that outline where rocks with particular names plot. Just as for xy-plots, one software package, based on R, which can plot these fields and lines for many pre-defined ternary plots (Janousek *et al.*, 2006), is freely available at http://www.gla.ac.uk/gcdkit/.

6.9 Summary

Once the statistical and spatial data structure is documented, the relationships between different variables are an important avenue of study. While up to this point it was no major problem to study and document variable by variable, now almost countless possibilities are encountered. The number of possible plots showing bivariate data relationships increases quadratically with the number of variables. It may thus be tempting to commence right away with multivariate data analysis. However, all multivariate methods are built on many statistical assumptions about the data. It is thus advisable to use scatterplots first as the basis for studying bivariate data behaviour. Such bivariate plots used in an exploratory manner can be very informative, especially when adding user-defined lines, regression lines and trend lines. Plotting spatial and time trends are special cases of bivariate data analysis. The scatterplot matrix can be used to gain a graphical impression of the relationships between all possible pairs of variables. Three variables can be displayed in ternary plots. All of these simple plots can be used to great advantage to gain an improved knowledge of the data and its structure, interrelationships, before starting any formal statistical data analysis.

7

Defining Background and Threshold, Identification of Data Outliers and Element Sources

The reliable detection of data outliers and unusual data behaviour is one of the key tasks in the statistical analysis of applied geochemical data, and has remained a core problem since the early days of regional geochemistry (see, e.g., Hawkes and Webb, 1962; Matschullat *et al.*, 2000; Reimann and Garrett, 2005; Reimann *et al.*, 2005).

In Chapters 3 and 4 several methods to visualise the data distribution and to calculate distribution measures have been introduced. Statistically, it should now be no problem to find the univariate extreme values (for multivariate outlier detection see Chapter 13) for any determined variable. To identify the extreme values in a normal distribution, statisticians usually calculate boundaries via the formula MEAN $\pm 2 \cdot$ SD. The calculation will result in about 4.6 percent of the data belonging to a normal distribution being identified beyond the boundaries as extreme values, 2.3 percent at the lower end and 2.3 percent at the upper end of the distribution. The inner 95.4 percent of the data will represent the "background" variation in the data set. In the early days of regional geochemistry, this formula was frequently used to identify a certain proportion of "unusually high" values in a data set for further inspection (Hawkes and Webb, 1962). The upper border calculated by the "MEAN $\pm 2 \cdot$ SD" rule was defined as the "threshold", the boundary between background variation and extreme values. The terms "background" and "threshold" have often been intermingled as identifying the boundary between "normal" and "extreme" values, but should be kept strictly separate because "background" identifies the main range of variation in the data set while "threshold" is a single value.

However, are these extreme values of a normal distribution, as defined by a statistical rule, really the "outliers" sought by applied geochemists and environmental scientists? Will this rule provide reliable-enough limits for the background population to be of use in environmental regulations? Can it be used to reliably separate geogenic, natural background variation from anthropogenic element inputs?

Extreme values are of interest in investigations where data are gathered under controlled conditions. A typical question would be "What is the 'normal' range of weight and length among newborns?" In this case extreme values are defined as: *values in the tails of a statistical*

Statistical Data Analysis Explained   Clemens Reimann, Peter Filzmoser, Robert G. Garrett, Rudolf Dutter
© 2008 John Wiley & Sons, Ltd.

distribution, and can be calculated using the "MEAN $\pm$ 2 $\cdot$ SD" rule once enough data (usually $n > 30$ is recommended for a first estimate) have been gathered. The estimates for 'normal' range and extreme become increasingly reliable as more data become available; the boundaries will no longer change dramatically when the size of the statistical sample is increased (see Chapter 9).

In contrast, geochemists are typically interested in outliers as indicators of rare geochemical processes. In such cases, these outliers are not part of one and the same distribution. For example, in exploration geochemistry samples indicating mineralisation are the outliers sought. In environmental geochemistry the recognition of contamination is of interest. *Outliers* are statistically defined (Hampel *et al.*, 1986; Barnett and Lewis, 1994; Maronna *et al.*, 2006) as: *values belonging to a different population because they originate from another process or source, i.e. they are derived from (a) contaminating distribution(s)*. In such a case the "MEAN $\pm$ 2 $\cdot$ SD" rule cannot provide a relevant threshold estimate. Outliers could occur anywhere within a given data distribution, not only at the extremes, and may be quite difficult to identify – at least in the univariate case (see Chapter 13). The reason that the "MEAN $\pm$ 2 $\cdot$ SD" rule appears to function in practice is that fortunately the data representing the sought-after distributions are often very high (or low) when compared to the majority of data. In practice, it is often quite difficult to differentiate between true data outliers and extreme values and the term "outlier" is often used for any kind of value deviating from the majority of the data.

Furthermore, because of the inherent spatial component of applied geochemical and environmental data and the difference between "extreme values" and "outliers", the estimated threshold values separating background from outliers will usually change when size and/or location of the survey area are changed. This is an unacceptable situation for regulators needing to define "action levels", "maximum admissible concentrations (MAC-values)", or simply to decide whether a measured value is "natural" or due to "contamination".

From the above discussion it is obvious that procedures that do not consider the spatial component of applied geochemical data will not be able to provide a reliable estimate of background and threshold. A combination of statistics and geochemical mapping is needed. Defining action levels or MAC-values should probably be based more on ecotoxicology than on statistics drawn from regional geochemical data. For toxicologically defined "action levels", the discussion of anthropogenic versus geogenic origin of the measured concentration becomes irrelevant. The question is rather "too high", "OK", or even "too low" for a certain use. On a continental scale natural element variations will play the dominating role, while on a local scale contamination from anthropogenic sources may become increasingly important. It may still be necessary to reliably identify the source of high values in cases where somebody is required to carry the cost of a remediation effort. The best chance for identifying a source lies in appropriate geochemical mapping at a suitable scale (see Chapter 5).

7.1 Statistical methods to identify extreme values and data outliers

7.1.1 Classical statistics

Statisticians use mean and standard deviation to identify extreme values of the normal distribution (see Chapter 7). This is a well-established and very reliable method if independent data are drawn from a normal distribution.

Table 7.1   Lower and upper limits for extreme values estimated using the "MEAN $\pm$ 2 $\cdot$ SD" rule for Cu in the Kola Project O-horizon data. Results are shown when using the non-transformed and the log-transformed values for the calculation. The log-transformed values have been back-transformed to the original data scale

	[MEAN $-$ 2 $\cdot$ SD]		[MEAN $+$ 2 $\cdot$ SD]	
	original	log-based	original	log-based
Cu	-451	1.8	540	98
Cu-Finland	2.3	3.9	12	13
Cu-Norway	-31	2.6	56	32
Cu-Russia	-626	2.5	790	213

The distribution of geochemical data has been repeatedly discussed in the literature (for a recent paper see, for example, Reimann and Filzmoser, 2000). Many of the classical statistical methods (not only "MEAN $\pm$ 2 $\cdot$ SD") require that the data follow a normal distribution or, as a minimum, approach symmetry. In the previous chapters it was demonstrated that geochemical data are usually strongly right skewed and are characterised by the existence of outliers. Skew and outliers will have a strong influence on location (central value) and spread of a data set. Table 7.1 shows the estimated threshold values calculated for the variable Cu as measured in the Kola Project O-horizon data set based on the classical statistical parameters. Because the data distribution is strongly skewed, the resulting values are clearly biased. To overcome the right skew, the data could be log-transformed. Values for MEAN and SD can then be calculated for the log-transformed data; the boundaries can be calculated using these estimates, and back-transformed to the original data scale. All the statistical parameters have changed dramatically, and the estimate of threshold is now much lower. But is it correct? A statistical test for data normality (see Chapter 9) will show that the variable Cu in the Kola Project O-horizon data set follows neither a normal nor a lognormal distribution (see Table 9.1). This provides a strong reason to be suspicious of calculated threshold values. Due to the strong right skew of the data, the calculation of the limits for the lower outlier boundaries, "thresholds", without the use of a log-transformation results in negative estimates. But negative concentrations do not exist in geochemistry, which further highlights the problem with this approach.

The problem of using the above statistical formula to identify extreme values and thus the upper and lower boundaries of the background variation for spatial (applied geochemical and environmental) data is obvious when treating the three countries as data subsets. Calculating MEAN, SD, and outlier boundaries for each of the countries using the above approach (non-transformed versus log-transformed values) results in totally different estimates for each country. What is the most reliable estimate of "background" for the survey area? Is it justified to base regulatory action or guidance values on any of these estimates?

7.1.2   The boxplot

As demonstrated in Chapter 3 the boxplot as defined by Tukey (1977) will automatically identify extreme values. However, it has also been discussed that the calculation of the Tukey outlier boundaries requires data symmetry and is based on assumption of underlying normality (Section 3.5.2). Thus, before using the boxplot to identify data outliers, the data

Table 7.2 Lower and upper boundaries for extreme values for Cu in the Kola Project O-horizon data as obtained with the Tukey boxplot (non-transformed data) and the log-boxplot (data log-transformed to approach symmetry and resulting boundaries for extreme values back-transformed into the original data scale)

| | Lower whisker | | Upper whisker | |
	original	log-based	original	log-based
Cu	2.7	2.7	35	76
Cu-Finland	2.7	3.9	11.9	13
Cu-Norway	4.5	4.5	17	24
Cu-Russia	3.1	3.1	72	185

distribution needs to be studied in some detail and the required data transformation to approach distributional symmetry carried out. When this is done, the boxplot should be better suited to identify data outliers in geochemical data. Table 7.2 compares outlier boundaries as obtained when using the boxplot for the original and for log-transformed Cu data. Using the log-transformed data and back-transforming the resulting boundaries to the original data scale is equivalent to using the log-boxplot (Section 3.5.2) on the original data. Due to the construction rules for the boxplot, negative lower boundaries for extreme values are now impossible (the lower whisker is taken to the lowest original data value). The values for the upper whisker are often considerably higher when using the log-boxplot (see Section 3.5.2). Compared to the boundaries calculated above using the "MEAN ± 2 · SD" rule, only the upper threshold for the country subset "Finland" remains the same when using log-transformed data in both cases. Again the three country subsets all provide different results.

7.1.3 Robust statistics

To better accommodate the special properties of applied geochemical data it is possible to replace MEAN and SD in the "MEAN ± 2 · SD" formula by robust estimators for the location (central value) and spread (see Sections 4.1 and 4.2). When, for example, using MEDIAN and MAD, the resulting formula for identifying extreme values is "MEDIAN ± 2 · MAD". However, the definition of the constant used in estimating the MAD is based on the assumption of a normal data distribution of the core (inner 50 percent) of the data. Thus the data distribution must first be checked for symmetry of the majority of the data before performing any calculations of the boundaries for extreme values, even when using robust estimators.

The estimates obtained for the upper and lower outlier boundaries (Table 7.3) are very different from the estimates obtained when using the classical parameters (Table 7.1) or the boxplot (Table 7.2). Differences between estimates using the original and the log- and back-transformed values are now much smaller. The reason is that the MAD is much less influenced by skewed data than the SD and even less than the hinge width (IQR) used in the boxplot. The "MEDIAN ± 2 · MAD" rule delivers the lowest threshold values of all techniques discussed so far. Reimann *et al.* (2005) have shown that the percentage of extreme values is usually overestimated when using this rule. Again, different estimates for the threshold will be obtained when location or size of the survey area change.

Table 7.3 Lower and upper limits for extreme values calculated using the "MEDIAN ± 2 · MAD"
rule for Cu in the Kola Project O-horizon data. Results are shown when using the non-transformed
and the log-transformed values for the calculation. The log-transformed values have been
back-transformed to the original data scale

	[MEDIAN − 2 · MAD]		[MEDIAN + 2 · MAD]	
	original	log-based	original	log-based
Cu	−0.6	2.8	20	33
Cu-Finland	3.2	4.0	10	11
Cu-Norway	2.7	3.9	13	15
Cu-Russia	−5.3	3.8	40	79

7.1.4 Percentiles

The uppermost 2 percent, 2.5 percent, or 5 percent of the data (the uppermost extreme values)
are sometimes arbitrarily defined as "outliers" for further inspection. This will result in the
same percentage of extreme values for all measurements. This approach is not necessarily valid,
because the real percentage of extreme values could be very different. In a data distribution
derived from natural processes, there may be no extreme values at all, or, in the case of multiple
natural background processes, there may appear to be outliers in the context of the main mass
of the data that are not in fact outliers in the context of the background process with the
highest levels. However, in practice the percentile approach delivers a number of samples
for further inspection that can easily be handled. In some cases environmental regulators have
used the 98th percentile of background data as a more sensible inspection level (threshold) than
values calculated by the "MEAN ± 2 · SD" rule (see, e.g., Ontario Ministry of Environment
and Energy, 1993). The remaining problem is that even percentiles will change with size or
location of the survey area.

Table 7.4 shows the 2nd and 98th percentiles for Cu in the Kola Project O-horizon data set
for all data and for the three countries. Note that percentiles are solely based on sorting of the
data, log- and back-transformation will not change the resulting percentiles (see Section 4.3).
The results for the lower boundary for extreme values are of course again within the data
range and are actually the highest estimates of all the techniques presented so far. Results for
the upper threshold are high when compared to the other techniques. The exception is for the

Table 7.4 2nd and 98th percentiles of Cu from the Kola Project O-horizon data set. To highlight
the inherent problem of working with spatial data the 2nd and 98th percentiles for the three
country subsets are also shown

	2nd percentile	98th percentile
Cu	4.7	248
Cu-Finland	4.4	13
Cu-Norway	4.7	62
Cu-Russia	6.9	478

data from Finland. These provide a relatively stable upper threshold for all techniques. This indicates that the distribution of Cu in the Finnish samples is quite symmetrical and does not contain more than two percent upper extreme values.

7.1.5   Can the range of background be calculated?

As demonstrated above, several methods are available for calculating the range of geochemical background. When applied, all result in quite different estimates for the same variable. When changing size or location of the survey area, again all estimates will change. Reimann *et al.* (2005) studied in detail the different methods and their statistical behaviour with simulated and real data. The conclusion was reached that, given the special properties of geochemical data and that in fact real data outliers and not "extreme values of the normal distribution" are being sought, the "MEDIAN ± 2 · MAD" rule and the boxplot (if a suitable data transform is used) will provide the most reliable calculated results.

It should be noted that in the literature on robust statistics many other approaches for outlier detection have been proposed (see, e.g., Huber, 1981; Rousseeuw and Leroy, 1987; Barnett and Lewis, 1994; Dutter *et al.*, 2003; Maronna *et al.*, 2006). There are also multivariate methods for outlier detection (e.g., Chapter 13) and mixture decomposition methods have also been proposed as a solution (see, e.g., Graf and Henning, 1952; Carral *et al.*, 1995; Neykov *et al.*, 2007). However, none of these methods is able to solve the problem of working with spatial data. Estimated values will always change depending on size and location of the survey area. Applied geochemical data are not the well-behaved data from classical investigations, where statistical parameters will improve with the number of samples; they may in fact change dramatically when the survey boundaries are changed.

Percentiles based on the empirical data have the advantage that they will identify the uppermost (lowermost) two percent (or five percent) of the data as "unusual" and without being based on any distributional model. They will result in the identification of a reasonable number of samples that will need further investigation. These samples may, or may not, be true outliers. However, in a highly mineralised or contaminated survey area there may in fact be far more than, for example, five percent outliers, e.g., Kola Project mosses or O-horizon soils.

Considering the above results, it is clearly inappropriate that the estimation of background for applied geochemical data should be solely based on such calculations when the values are to be used for regulatory purposes. However, the resulting values may still provide useful estimates for data comparison purposes.

7.2   Detecting outliers and extreme values in the ECDF- or CP-plot

In Section 3.4 it has been shown that plots of the distribution function are powerful tools to visualise deviations from expected data behaviour. Tennant and White (1959), Sinclair (1974, 1976), and others originally introduced the cumulative probability plot (CP-plot) to geochemists. Figure 7.1 shows an ECDF- and a CP-plot for Cu in the O-horizon data, the example used above to calculate outlier boundaries. The borders obtained with "MEAN ± 2 · SD" and "MEDIAN ± 2 · MAD" rules for the log-transformed and then back-transformed results for the whole data set are also indicated.

One of the main advantages of plots of the distribution function is that each single data value remains visible. The range covered by the data is clearly visible, and extreme values are detectable as single points. It is possible to directly count the number of extreme values

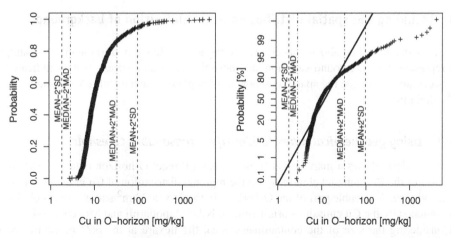

Figure 7.1   ECDF- (left) and CP-plot (right) for Cu from the Kola Project O-horizon data set. Boundaries for extreme values as identified with "MEAN $\pm$ 2 $\cdot$ SD" and "MEDIAN $\pm$ 2 $\cdot$ MAD" rules using the log-transformed and then back-transformed data are also shown; the 2nd and 98th percentile can be read directly from the plot

and observe their distance from the core (main mass) of the data. Identifying the threshold in ECDF- or CP-plots is, however, still not a trivial task. It is obvious that extreme values, if present, can be detected without any problem. However, several flexures can be identified in both plots. Which of the flexures is the "true" boundary between background and extreme values? In Figure 7.1, for example, it can be clearly seen in both versions of the diagram that the boundary dividing extreme values from the rest of the population is 1000 mg/kg. Experience with other data sets teaches that 1000 mg/kg Cu in the O-horizon is a highly unlikely value for an upper threshold; Cu values in such soils are usually much lower. When consulting a reference source collecting such values together from different surveys (see, e.g., Reimann and de Caritat, 1998), it can be clearly seen that median concentrations of Cu in natural soils are usually below 10 mg/kg.

Searching for breaks and inflection points is largely a graphical task undertaken by the investigator, and, as such, it is subjective and experience plays a major role. However, algorithms are available to partition linear (univariate) data, e.g., Garrett (1975) and Miesch (1981) who used linear clustering and gap test procedures, respectively. In cartography a procedure known as "natural breaks" that identifies gaps in ordered data to aid isopleth (contour interval) selection is available in some Geographical Information Systems (Slocum, 1999).

However, frequently the features that need to be investigated are subtle, and in the spirit of Exploratory Data Analysis (EDA) the trained eye is often the best tool. On closer inspection of Figure 7.1, it is evident that a subtle inflection exists at 13 mg/kg Cu, the 66th percentile, best seen in the ECDF-plot (left). About 34 percent of all samples are identified as extreme values if we accept this value as the threshold. This may again appear unreasonable at first glance and highlights the problem of working with plots of the distribution function – experience and additional information are needed to make best use of these plots. The challenge of objectively extracting a boundary for extreme values or the threshold from these plots is probably the reason that they are not more widely used.

7.3 Including the spatial distribution in the definition of background

The problems described above concerning finding a suitable method to identify the samples that represent the background variation and those that represent extreme values indicate that it is necessary to include the spatial component of applied geochemical (environmental) data in the definition.

7.3.1 Using geochemical maps to identify a reasonable threshold

Figure 7.2 (left) shows a map for Cu from the Kola Project O-horizon data set. The major influence of the Russian nickel industry on the regional distribution of Cu in the O-horizon is at once visible. A sizeable part of the O-horizon in the 188 000 km^2 survey area is obviously contaminated by the Cu emissions originating in Nikel/Zapoljarnij and Monchegorsk.

Considering the size of the contaminated area, the flexure at the 66th percentile in the ECDF-plot described above is in fact a likely candidate for an upper threshold value for natural background variation, dividing the contaminated from the uncontaminated land. A combination of ECDF- and CP-plots, a geochemical map, and some expert knowledge may thus all be needed to derive a prudent estimate of the upper limit of background variation. Relying on statistical calculations alone will most often result in inappropriate estimates. Regulators may then be interested in a map showing all locations where the Cu concentration in the O-horizon samples is larger than 13 mg/kg (Figure 7.2, right). The map demonstrates that the majority of Cu values that are higher than 13 mg/kg are probably directly related to emissions from the Russian nickel industry. However, there are in addition a number of geogenic outliers in the data set, e.g., two high values near Cape North in Norway, where a small intrusion of ultramafic, a relatively rare rock, occurs. It was thus possible to identify a reasonable, if not perfect, threshold for the upper boundary of natural background variation. This boundary, however, cannot be used to distinguish between man-made and natural high

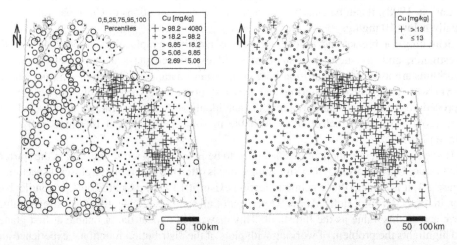

Figure 7.2 Regional distribution of Cu of the Kola Project O-horizon data (left) and regional distribution of samples with a Cu concentration >13 mg/kg in the O-horizon samples from the Kola Project area (right)

values of Cu in the O-horizon. For a decision about the likely cause of the high values the map is indispensable.

However, in terms of setting a regulatory value for the maximum allowable concentration of Cu in the O-horizon soils in the area, the derived threshold of 13 mg/kg may still not be the most appropriate value. Some information is still missing, i.e. what Cu concentration will be harmful to the environment? To make the issue even more complicated, the question needs to be extended: to which part of the environment – soil bacteria, some earth worm eating the soil, or human beings coming in contact with the soil – does this concentration represent a danger (Chapman and Wang, 2000)? It is obvious that these questions can only be answered via ecotoxicological investigations that should take the speciation (chemical form) of the element into consideration. Regulatory action levels or MAC-concentrations should rely on such ecotoxicologically-derived values rather than on statistically-derived values from regional geochemical data. Finally, if a comparison of the ecotoxicologically-derived value is to be made to regional geochemical data, for any conclusions to be valid the method of analysis used should estimate the amount of the element present in the sample in the form that the ecotoxicological study has shown to pose a risk.

7.3.2   The concentration-area plot

It has been demonstrated above that it is useful to consider the spatial scale of the distribution of the data in selecting break points or thresholds in plots of the distribution function. The concentration-area (CA-) plot achieves this by looking at the fractal structure of the data (Cheng *et al.*, 1994; Cheng and Agterberg, 1995). Many patterns in nature are self-replicating, or self-similar (Mandelbrot, 1982; Barnsley and Rising, 1993). Examples are the shape of fern leaves, a snow flake, or the dendritic patterns of manganese hydroxide coatings on the parting faces of rocks in desert environments. Geochemical patterns can also be investigated to determine if they are "fractal". Normally geochemical maps are thought of as two-dimensional, however, if they are spatially non-random they will have a fractal dimension between two and three. It is this partial dimension that contains spatial information on the structure of the data displayed in the two normal map dimensions. To use a CA-plot the data need to be relatively evenly spaced across the survey area. There should also be no major gaps, e.g., large areas where the sample medium is absent, in the data set. The Kola data meet this requirement and Figures 7.3 and 7.4 demonstrate the CA-plot procedure for the O-horizon soil data. First the data are spatially interpolated (see Section 5.6) to a fine (100 × 100) regular grid, increasing the number of data points so they can be used as a surrogate for measuring areas. Then the interpolated values outside the survey area boundary are trimmed away, so there is no interpolation to points outside the survey area. The original and interpolated data values together define the x-axis of the CP-plots displayed in the upper two plots of Figure 7.3. The number of values plotted is of course greatly increased by the spatial interpolation procedure. In the CA-plot (Figure 7.3, lower right) the y-axis shows the percentage of the interpolated values, plotted on a logarithmic scale, that are larger (or smaller) than each value plotted on a logarithmic scale on the x-axis. As the interpolated points are on a fine regular grid the number of points is a surrogate for the area of the interpolated map that is larger (or smaller) than the corresponding value on the x-axis. Thus the CA-plot displays the relationship between the percentage of the survey area that has a particular value and the actual value. Background levels will occur frequently and will represent the majority of the survey area, and low or high extreme value-containing areas will represent small percentages of the survey area. A single straight line indicates a single

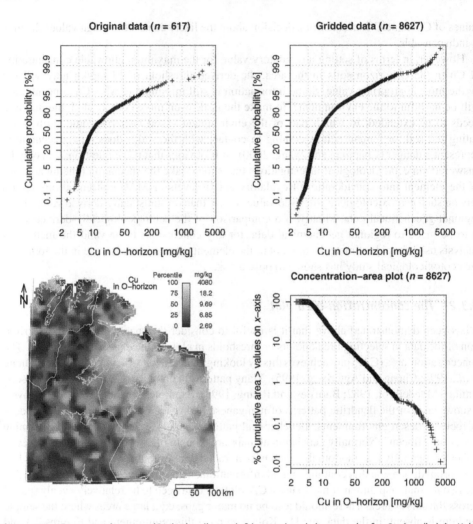

Figure 7.3 Concentration-area (CA-) plot (lower left) cumulated downwards, for Cu (mg/kg) in Kola O-horizon soil samples. The upper left plot is the CP-plot of the original data and the upper right plot is the CP-plot for the interpolated data. The lower left image is a grey-scale map of the interpolated data

fractal relationship controlling the data distribution; multiple straight lines indicate multiple fractal processes controlling the data distribution. The choice of interpolation algorithm and data transformation will influence the CA-plot. Firstly, the data distribution should be symmetrical so that extreme values do not over-influence the resulting interpolation. This is often achieved in applied geochemical data with a logarithmic transform. Secondly, in interpolation the data should not be excessively smoothed as that will "smear out" the very features being sought. To avoid this, simple triangulation is employed for the interpolation.

The CA-plot in Figure 7.3 (lower right) has been prepared by ordering the interpolated values from the maximum value downwards. As the areas are plotted on a logarithmic scale, this results in scale expansion and structural (fractal) detail being easily observable for high values, in this case related to the smelter facilities in the survey area. It is apparent that most

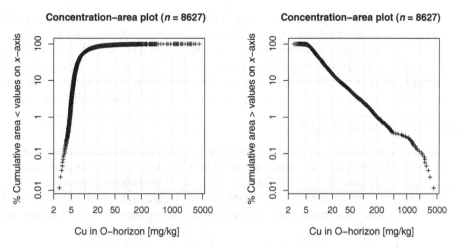

Figure 7.4 Concentration-area (CA-) plot, cumulated upwards (left) and downwards (right – see Figure 7.3), for Cu (mg/kg) in Kola O-horizon soil samples

of the data follow a single fractal relationship, with another relationship present at low levels and a more complex relationship above 500 mg/kg Cu near the smelters. Thus the distribution of the O-horizon Cu-data appears to be dominated by smelter-related processes. This does not help determine the lower limit of smelter influence; for this the CA-plot needs to focus on the lower end of the fractal distribution. This is achieved by ordering the interpolated values from lowest to highest, see Figure 7.4 (left). As distinct from the downwards cumulated display (Figure 7.4, right), it is immediately apparent that there are two major fractal processes present, one below about 10 mg/kg Cu, and one above.

The lower process is that controlling the regional background distribution of Cu in the O-horizon soil, and the upper process is related to the atmospheric dispersion of Cu from the smelter facilities. The fine detail present at the highest levels >500 mg/kg in Figure 7.4 (right) relates to very local processes close to the point sources, possibly fugitive releases of coarser-grained material incapable of long-range transport, or other local smelter-related activities.

It is interesting to note that the use of a simple triangulation interpolation procedure does not materially change the shape of the data distribution (compare the two upper plots in Figure 7.3). A comparison of the CA-plot (lower right) with the CP-plot (upper left) of the data is most informative. In Figure 7.3 not a lot of new information is learnt: the gaps in both plots are similar; however, Figure 7.4 (left) provides real new insight into the importance of the flexure in the CP-plot at around 13 mg/kg Cu (see previous discussion in Section 7.3.1).

Previously in this chapter, threshold estimates for Cu (mg/kg) in Finnish O-horizon soils of 12 ("MEAN $\pm$ 2 $\cdot$ SD" original scale), 13 ("MEAN $\pm$ 2 $\cdot$ SD" log-scale), 12 (boxplot upper fence original scale), 13 (boxplot upper fence log-scale), 10 ("MEDIAN $\pm$ 2 $\cdot$ MAD" original scale), 11 ("MEDIAN $\pm$ 2 $\cdot$ MAD" log-scale), and 13 (98th percentile) have all been made (see Tables 7.1 to 7.4). The upwards cumulated CA-plot confirms in a most convincing way that a useful threshold for Cu in O-horizon soils is in the 10 to 13 mg/kg range. This information would be most useful for developing a map of Cu in O-horizon soils where a symbol change or isoline at somewhere between 10 and 13 mg/kg would indicate the limit of measurable influence (in the O-horizon Cu data) of the smelter facilities (compare Figure 7.2, right). What is apparent is that in the context of the Kola Project the data from Finland provide a good estimate for the natural background range of the whole survey area. This also demonstrates how far one might have to go from a major point source of anthropogenic contamination to find a suitable area in which to establish a natural background range. Fortunately, northern Finland is generally geologically pedologically and ecologically similar to the Kola Peninsula; if this was not the case, as might be in other studies, finding a suitable "background" area could pose real challenges.

7.3.3 Spatial trend analysis

An alternative approach to visualising "background" and "contamination/anomaly" around a clearly defined anthropogenic source such as Monchegorsk is to draw transects across the expected source and study the distribution of the elements with distance from the source (see, e.g., Reimann *et al.*, 1997b; McMartin *et al.*, 1999; Bonham-Carter *et al.*, 2006; see Section 6.4). Figure 7.5 shows four such profiles for the O-horizon transecting Monchegorsk in an east–west

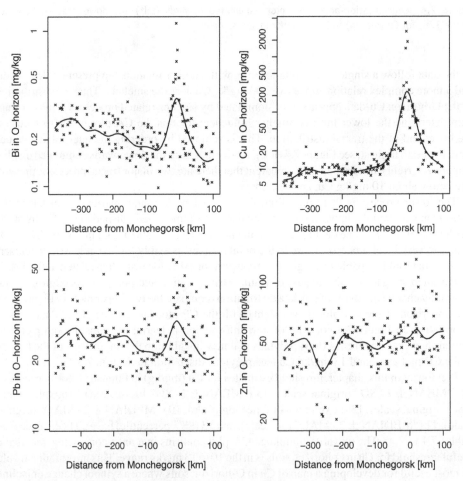

Figure 7.5 East-west geochemical transect through Monchegorsk for Bi, Cu, Pb, and Zn as determined in the Kola Project O-horizon samples

direction. For most of the elements the influence of Monchegorsk is clearly visible – as well as the distance to the point where this influence disappears into the background variation – often at less than 50 km (e.g., Bi, Figure 7.5, upper left) and up to 150 km (e.g., Cu, Figure 7.5, upper right) from source. The little anomaly at 300 km has a geogenic origin (Reimann and Melezhik, 2001). The diagram for Cu shows at once that the background is reached at around 10 mg/kg and that all samples with a value above 25 mg/kg are most likely influenced by the anthropogenic activities in the survey area.

For Pb the profile shows that the Pb emissions from Monchegorsk would not be easily discerned within the background variation were it not plotted as a profile (Figure 7.5, lower left). Apatity, some 30 km east of Monchegorsk, appears as a second source of Pb. Near the western project boundary (−300 km on the x-axis) the profile intersects a geogenic Pb anomaly related to natural hydrothermal alteration (Reimann and Melezhik, 2001), reaching levels almost as high as observed near Monchegorsk (Figure 7.5, lower left). The profile for Zn shows that no recognisable Zn emissions can be detected from Monchegorsk, the highest levels occur near Apatity (Figure 7.5, lower right). Such simple profiles, employing EDA principles and the user's optical processing capabilities, are a better indication of the influence of contamination than any statistically derived background variation estimates, TOP/BOT-ratios or EFs (see Sections 7.4.1 and 7.4.2). The same effects will be visible in properly-constructed geochemical maps (see Chapter 5), however the identification of the point where the input from contamination can no longer be separated from natural variation is easier on a transect. Again we derive a value of 10–12 mg/kg Cu.

7.3.4 Multiple background populations in one data set

The Cu distribution in Figure 7.2 is dominated by the massive input from the Russian nickel industry. In many cases, however, the regional distribution of an element will not be determined by one single process; several sources for unusually high (or low) element concentrations may be present. Figure 7.6 shows two examples, a map of Na in the O-horizon and a map of La

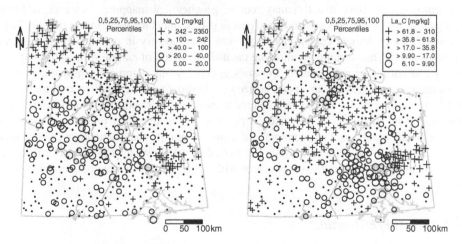

Figure 7.6   Regional distribution of Na (left) in the Kola Project O-horizon data set and of La (right) in the Kola Project C-horizon data set

in the C-horizon. Two major sources for high Na-concentrations in the O-horizon exist in the survey area: the steady input of marine aerosols along the coast and the dust related to mining and transport of the alkaline rocks near Apatity. Knowledge about these sources and their localisation allows their differentiation on the map (Figure 7.6, left). In the C-horizon, bedrock geology will usually dominate the observed elemental distribution patterns. The map of La shows that different rock units display very different element concentrations, in the example the Caledonian sediments along the Norwegian and parts of the Russian coast, the granulite belt in Finland and Russia, and the alkaline intrusions near Apatity are all marked by exceptionally high values of La (Figure 7.6, right, compare to Figure 1.2, the geological map of the survey area). Thus the different lithologies as identified in the geological map (Figure 1.2) will all have their own, and often quite different, background populations. In consequence, extreme values in soils collected on top of one of these lithologies will not be extreme values in soils collected on top of another lithology. For O-horizon samples something similar can be expected when comparing samples from different vegetation zones (different ecosystems).

It is thus an intrinsic problem when working with spatial data that no univariate statistical method will permit distinguishing between the different processes resulting in unusually high (or low) values. The estimation of background range and threshold are probably not the most important task, but rather the identification of the factors (or processes) causing "unusual" data behaviour. This requires geochemical mapping (Chapter 5) using appropriate techniques to visualise the data structure rather than to just highlight "extreme values" (see Chapter 5) and/or multivariate statistics (see Chapters 11 and following, especially Chapter 13 on multivariate outlier detection). Identifying these different processes in regional geochemical data should be an interactive and iterative process using a variety of graphical and data subsetting techniques rather than classical statistical calculations and tests.

7.4 Methods to distinguish geogenic from anthropogenic element sources

A typical task in environmental investigations is to distinguish data related to geogenic, or better natural, element sources from data influenced by anthropogenic sources. It has been demonstrated above that this may require geochemical mapping. However, in many investigations too few samples are taken to permit geochemical mapping over a usefully large area, or no clearly defined contamination sources (like industry in the Kola Project survey area) are present. Thus techniques allowing the differentiation of natural from anthropogenic element concentrations are required. For soils and sediments a comparison of the element concentrations measured in an upper horizon, believed to be exposed to contamination, against the concentrations of the same element measured at depth, the TOP/BOT-ratio, is often used. If no such direct comparison is possible (for example for precipitation samples or when using plants as "biomonitors" of contamination), the calculation of so-called "enrichment factors" is often employed. The Kola data set allows testing the performance of these techniques on a regional scale in an area with clearly defined major contamination point sources.

7.4.1 The TOP/BOT-ratio

When investigating sediment cores or soil samples, a popular technique to identify anthropogenic element additions is to calculate the ratio between a sample collected at depth (BOT), and thus most likely not influenced by anthropogenic activities, and a sample collected at the surface (TOP), and thus very susceptible to the atmospheric input of anthropogenic contamination.

This approach has serious shortcomings; it requires that the material collected at TOP be directly comparable to the material collected at BOT level and that no other processes than anthropogenic contamination can influence the observed element concentration in the TOP sample. At first glance there may appear to be no problem when collecting soils. However, an organic O-horizon soil (the TOP-layer for the Kola Project samples) has – other than the sample location – very little in common with the minerogenic C-horizon soil as collected at depth. The O-horizon reflects the biogeochemical cycles at the earth surface, while the C-horizon reflects mineralogical changes during rock weathering and, in the Kola area, glaciogenic processes. It is actually very uncommon that a TOP sample is 1:1 comparable to a bottom sample and the consequences are rarely considered when TOP/BOT-ratios are calculated, especially in all the cases where they appear to show an anticipated result (for example a massive anthropogenic contamination of the TOP-layer). For the Kola Project it is possible to directly compare B- and C-horizon soils, both are at least minerogenic soils, but even in this case soil-forming processes that will change the chemical composition have to be considered before any anthropogenic impact can be determined.

The results of the Kola Project permit the study of the effects of calculating such a TOP/BOT-ratio on a regional scale – for elements that are massively emitted in the survey area as well as for elements that are not part of the emissions spectrum of Kola industry (Figure 7.7).

At first glance the O-/C-horizon ratio maps appear to show pollution near industry quite nicely. However, comparisons with the maps for the raw data shown below in Figure 7.7 demonstrate that nothing has been gained. Contamination from the Cu/Ni industry is actually better depicted in the original distribution map of Cu in the O-horizon than in the ratio map, where some low values appear in the clearly contaminated part of the Russian survey area. For Pb, calculation of the O-/C-horizon-ratio results in the disappearance of the effect of emissions from the coal fired power plant in Apatity, which are visible in the raw data map. The reason is that there appear some rather high Pb-values in the C-horizon of that area due to the special lithology (alkaline rocks) present at Apatity. Lead is enriched in the O-horizon throughout the area without any spatial relation to the possible Pb-contamination sources within the area. This is due to the difference in the sample materials, minerogenic (C-horizon) versus organic (O-horizon). An uncritical interpretation of the O-/C-horizon-ratio without considering the spatial distribution may easily lead to the wrong inference that Pb is highly enriched in the upper soil layers due to anthropogenic activities.

7.4.2 Enrichment factors (EFs)

The formula to calculate EFs can be generalised from Chester and Stoner (1973) or Zoller et al. (1974) to be:

$$EF_i^{crust}(X1, X2) = \frac{\frac{X1_i}{crust\ X1}}{\frac{X2_i}{crust\ X2}} = \frac{\frac{X1_i}{X2_i}}{\frac{crust\ X1}{crust\ X2}} = \frac{crust\ X2}{crust\ X1} \cdot \frac{X1_i}{X2_i} \qquad for\ i = 1, \ldots, n$$

where "X1" is the element under consideration, the index i indicates the i-th concentration out of n samples (usually in mass/mass units, such as mg/kg), "X2" is the chosen reference

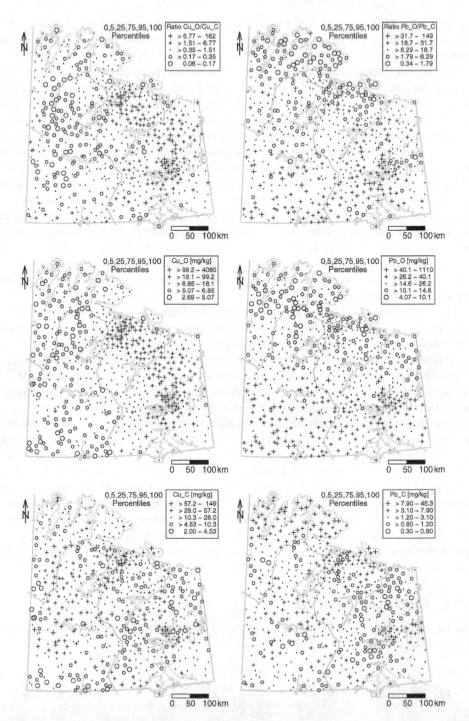

Figure 7.7 Maps of the distribution of the TOP/BOT-ratio (O-/C-horizon) for the elements Cu and Pb (upper row). The original distributions of the elements in the O- (middle row) and C-horizons (lower row) are shown in the accompanying maps, Cu (left) and Pb (right)

element (see below). "crust" indicates which medium the concentration (or enrichment) refers to. Instead of "crust" a local average background is occasionally used, e.g., the C-horizon at the sample point. In that case the above formula changes to:

$$EF_i^{local}(X1, X2) = \frac{\frac{X1_i}{X1\ Chor_i}}{\frac{X2_i}{X2\ Chor_i}} = \frac{\frac{X1_i}{X2_i}}{\frac{X1\ Chor_i}{X2\ Chor_i}} = \frac{X2\ Chor_i}{X1\ Chor_i} \cdot \frac{X1_i}{X2_i} \qquad \text{for } i = 1, \ldots, n.$$

Instead of the constant crustal ratio of the two elements, a variable local ratio is then used for the calculation of the EF.

Table 7.5 shows the average upper crustal concentrations for a number of elements used in the examples below.

It is assumed that the X1/X2-ratio stays constant and similar to the crustal (or other reference material) value during natural processes, but will change if an element is added via anthropogenic input, while the other element is not affected by the same process. If these assumptions hold, this double ratio could be used to differentiate between geogenic and anthropogenic element sources, independent of the measured concentration. This would allow detection of very low levels of contamination in a sample and even direct comparison between different sample materials, independent of concentration. The idea, as such, is thus elegant, and if it works would provide a very useful technique in environmental studies.

Reimann and de Caritat (2000, 2005) have recently shown that though EFs can be used as a means of comparison, they will not provide the sought differentiation between anthropogenic and geogenic element sources. One of the main reasons that these ratios do not work is that the biogeochemical cycling of elements is ignored. The biosphere and organic sample materials dominate at the earth surface and not "average crust" or any other minerogenic/geogenic material. It is not justified to ratio "crust" or other "minerogenic" materials against material belonging to another compartment in the biogeochemical cycle. Additional shortcomings of the procedure are detailed in the above two publications.

Data from the Kola Project afford the opportunity to study how EFs change on a regional scale in the presence of major metal contamination sources, the Russian nickel industry near Nikel, Zapoljarnij, and Monchegorsk in the western half of the Kola Peninsula. Figure 7.8 (upper left) shows the regional distribution of Ni in the O-horizon of the $188\,000\,km^2$ survey area. The location of the nickel industry is clearly marked by an extensive anomaly covering the Russian part of the survey area and small areas in northern Finland and Norway.

Table 7.5 Average concentration of selected elements in the upper crust. Values from Wedepohl (1995) and for S from Taylor and McLennan (1995)

			upper crust				
	mg/kg		mg/kg		mg/kg		mg/kg
Ag	0.055	Co	11.6	Ni	18.6	Se	0.083
Al	77440	Cr	35	Pb	17	Tl	0.75
As	2	Cu	14.3	S	300	V	53
Bi	0.123	Fe	30890	Sb	0.31	Zn	52
Cd	0.102	Mn	527	Sc	7		

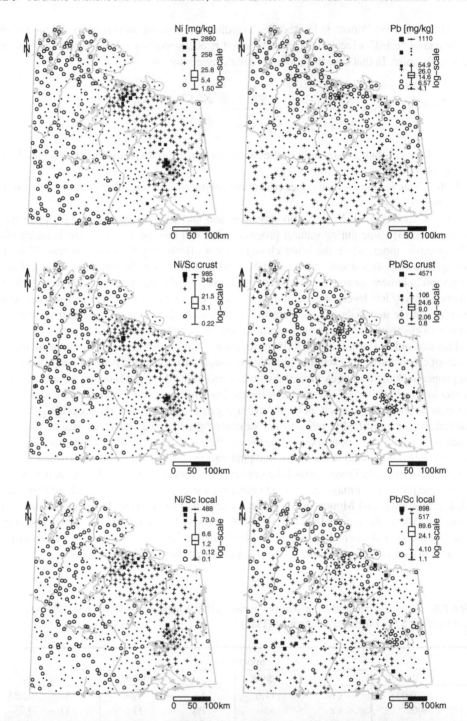

Figure 7.8 Regional distribution of Ni (upper left) and Pb (upper right) in the O-horizon; regional distribution of the EFs calculated relative to the Ni/Sc- (middle left) and Pb/Sc-ratio (middle right) in the earth crust and regional distribution of the EFs calculated relative to the "local background" as provided by the Ni/Sc- (lower left) and Pb/Sc-ratio in the C-horizon at each sample site

Figures 7.8 (middle left) and 7.8 (lower left) show the distribution of the EFs calculated for Ni using the crustal ratio and the local background (represented by the C-horizon at the same sampling site, rather than a local average), respectively. The EFs for Ni relative to the Sc in the Earth's crust are very high (up to 985), indicating the major input of Ni but not Sc from industry. However, the resulting map (Figure 7.8, middle left) demonstrates that by using EFs instead of the raw data, the map resolution clearly deteriorates (e.g., loss of the smooth radial decrease away from point-sources evident in Figure 7.8, upper left). The area indicated as influenced by contamination actually decreases and several low EF values occur proximal to the industrial centres at sites clearly within the contamination halo (Reimann *et al.*, 1998a).

The situation does not improve when using the local background (Ni/Sc-ratio in the C-horizon) instead of the crustal Ni/Sc-ratio (Figure 7.8, lower left). Although industry is marked by high EFs, there are also several sites within the contamination halo where the EFs exhibit low values, and the map is "noisy". Thus, EFs are not a sensitive indicator of anthropogenic contamination and, in this example at least, result in information loss rather than gain.

As a further example from the Kola Project, data for Pb distribution in the O-horizon (Figure 7.8, upper right) and EFs calculated relative to crustal (Figure 7.8, middle right) and local background (Figure 7.8, lower right) concentrations are shown. Cu-Ni smelters are generally not likely to be a major source of Pb emissions due to the characteristics of the ores they are processing. However, AMAP (1998) attributed 800 t/yr of Pb emissions to the Kola smelters. Boyd *et al.* (1998) demonstrated that the maximum amount of Pb entering the smelters with the ore feed is 13 t/yr. Some additional Pb may enter the Monchegorsk smelter with scrap metal that is occasionally recycled there, however, in any case the smelter is not a major Pb emission source and the AMAP estimate is far in excess of what actually enters the smelter. The map of the regional distribution of Pb in the O-horizon (Figure 7.8, upper right) is dominated by: (1) an anomaly near the northern coast of Norway (caused by the presence of small Pb-Zn vein occurrences in this area; Reimann and Melezhik, 2001); and (2) a prominent southerly increase in Pb concentrations. Monchegorsk and Apatity are situated at the border between low Pb values in the north and higher Pb values in the south, and it would be very interesting to be able to discriminate whether these sites are influenced more by natural phenomena or have been subjected to some level of contamination.

When EFs are calculated relative to the Pb/Sc-ratio in crust and mapped (Figure 7.8, middle right), the clear pattern displayed in the raw data map disappears. Monchegorsk is now marked as a low, although minor Pb-emission does occur there. Several high values occur at sporadic locations in the most pristine northernmost parts of the survey area. When using the local background (Figure 7.8, lower right), the real anthropogenic Pb-sources almost disappear from the map while a lot of spuriously high EFs appear scattered in the central part of the survey area. These sites might actually be of interest to exploration geochemists and certainly do not indicate anthropogenic Pb-contamination.

Terrestrial moss is considered a better sample medium than the O-horizon to indicate airborne Pb contamination. Figure 7.9 shows the regional distribution of Pb in moss in the Kola area. The three main anomalies mark the three major industrial and population centres in the survey area: Murmansk, Monchegorsk/Apatity, and Kandalaksha. Some additional high Pb values with no obvious industrial source are observed in northern Norway and along the southern project border in Finland.

When calculating EFs using the Pb/Al-ratio relative to the local C-horizon (Sc was not analysed in the moss), these industrial and population centres are, with the exception of one single point near Murmansk, marked by the lowest values on the map (Figure 7.9, upper right).

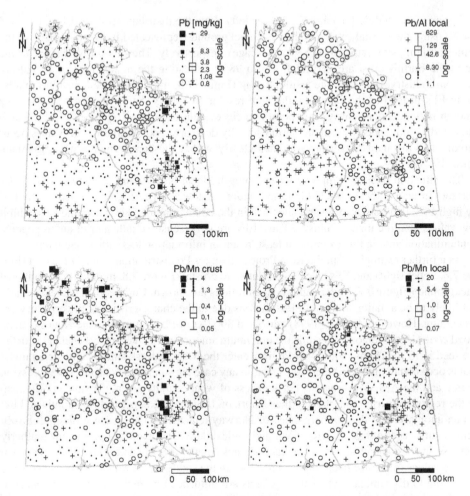

Figure 7.9 Regional distribution of Pb in moss (upper left); regional distribution of the EFs calculated relative to the Pb/Al-ratio in the local C-horizon (upper right); regional distribution of the EFs calculated relative to the crustal Pb/Mn-ratio (lower left) and regional distribution of the EFs calculated relative to the "local background" as provided by the Pb/Mn-ratio in the C-horizon at each sample site (lower right)

Instead spuriously high Pb EF values occur throughout many of the most pristine parts of the survey area.

Given that neither the average earth's crust nor rocks occur at the surface in most parts of the world, but in fact organic-containing soils and plants, one could question the general validity of using lithogenic (or crustal) element levels as reference values. It might actually be more realistic to use a biogenic element. One such element, which is also sometimes used as a "conservative lithogenic" reference element, is Mn (Loska *et al.*, 1997). The map in Figure 7.9 (lower left) was constructed using Mn instead of Al as the reference element for calculating Pb EFs in moss. This map shows the highest EFs where they are supposed to be: near industry (including Nikel and Zapoljarnij) and near population centres (Murmansk and Kandalaksha).

However, many high values occur also along the northern coast of Norway, and thus these EFs might again not differentiate between geo- (or bio-)genic element sources and contamination but show some other effect(s).

In fact, the regional distribution displayed in this map (Figure 7.9, lower left) is strongly influenced by vegetation zones. Low bio-productivity in the harsh northern coastal climate results in low Mn concentrations, in turn causing the observed higher EF pattern. Vegetation is also destroyed due to pollution near to contamination sources, and thus the clear anomaly around the industrial centres is caused by a combination of low Mn and slightly enriched Pb values. In general, it must be noted that the EFs calculated relative to Mn show rather low values (<4). Figure 7.9 (lower right) shows the EFs when using the local Mn values in the C-horizon. Here only two high EF values near the Monchegorsk refinery and one high value near the Nikel smelter remain visible on the map. This map may actually provide a realistic picture of sites with possible Pb contamination from industry in the area. In this case the EF map may actually be an improvement over the original Pb-distribution map (Figure 7.9, upper left).

When calculating EFs for all the typical contamination-related elements discussed in the current environmental literature for the Kola data set, a very interesting pattern emerges (Figure 7.10), given the known major emissions of Cu, Ni, Co, V and Cr in the survey area (Reimann *et al.*, 1998a). Many elements that are barely emitted at all by any source in the survey area show by far the highest EFs: Ag, Cd, Bi, Pb, Zn, Sb and As! Of the known emitted pollutants, both Cu and Ni show a median EF value well below ten (which is often arbitrarily used as the cut-off value for clearly indicating pollution) and some rather high individual EF values, which are clearly related to anthropogenic sources. Co, V and Cr show median EF values close to unity. Yet, the Kola smelters are among the largest Ni-Cu-Co emitters on a global scale (AMAP, 1998). The presence of the smelters is actually indicated by a large number

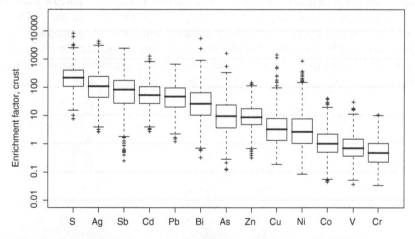

Figure 7.10   EFs for the Kola O-horizon samples calculated relative to the C-horizon using Sc as the reference element. Several of the elements that are major components of the smelter emissions (Cu, Ni, Co, V, Cr) show low EFs, while other elements that are hardly emitted at all within the survey area (e.g., Ag, Cd, Bi and Pb) show high EFs. The strong EF for S is due to the fact that S is a major plant nutrient

of upper outliers with high EF values for Cu, Ni and Co. This diagram thus indicates that a process other than contamination must cause the high EFs at the surface for elements like Se, Ag, Cd, Bi, and Pb in the Kola area.

The results presented here demonstrate that EFs may reveal patterns and can be an elegant method to directly compare elements measured in different sample materials independent of unit and range. They are, however, certainly no "proof" of anthropogenic contamination and should never be used in an attempt to differentiate between geogenic and anthropogenic elemental sources.

7.4.3 Mineralogical versus chemical methods

Although not the topic of this book, it should be mentioned that there are methods that may actually allow a clear differentiation between anthropogenic and natural sources. Even in difficult situations with several overlapping contamination sources, an apportionment of anthropogenic emissions to the likely cause may be possible. Chemical elements as determined in the different terrestrial sample materials are usually bound in certain minerals. Identification and analysis of these minerals may often provide a characteristic fingerprint of a single contamination source (Neinaweie and Pirkl, 1996). Gregurek *et al.* (1998a, 1999b) have, for example, used mineralogical investigations on the filter residues of snow samples to identify characteristic mineral assemblages for each of the Kola smelters. It was even possible to determine where in the metallurgical process the elements causing high concentrations in the snow samples were lost. These methods require, however, much more work than a chemical analysis.

7.5 Summary

Calculations of background and threshold should not be based on statistics alone. A sound knowledge of the regional distribution of the variables under study is required – and expertise about geochemical processes. Graphical data analysis, a combination of the ECDF- and/or CP-plot and suitable maps is the best approach towards identifying background and threshold. Many ratios that have been suggested in literature, including the popular Enrichment Factors (EFs) are problematic in their interpretation and may be misleading rather than providing useful information.

8

Comparing Data in Tables and Graphics

Data analysis often starts with some kind of comparison of data. It may be desired to compare data collected in one area with data collected from another. It may be necessary to compare project data for a certain sample material with "world average data" for that same material. It may be important to compare the data obtained from a certain area five years ago with data recently obtained from the same area. Typical questions include the following: "are there differences", "which factors can cause such differences", and last but not least a "statistical" question: "how significant are the differences"? This latter question is addressed in Chapter 9.

Often much can be learned about the processes that cause a certain data behaviour if it is possible to create subsets that differ in a known property from the data. In the case of the "complete" Kola data set there exist, for example, four different sample materials, each carefully selected to represent a certain part of the ecosystem. Moss is primarily reflecting the atmospheric input of elements in the survey area. The O-horizon reflects the interplay between atmosphere, biosphere and the pedosphere (see Chapter 1). The B-horizon is influenced by many soil-forming processes and is a representative of the pedosphere. The C-horizon is weathered soil parent material and its composition gives a representation of the bedrock in the survey area, the lithosphere (for a more complete discussion see Chapter 1). The project was designed to be able to directly compare the chemical composition of these four materials in tables and graphics. Such comparisons facilitate a better understanding of the cycling and fate of elements in and between the different compartments of the ecosystem.

8.1 Comparing data in tables

Because it is not usually possible to compare large data sets data item by data item, it is necessary to employ some kind of data summary. Most "classical" methods for comparing data populations are built around the MEAN and the SD, and thus the "location" or "central value" and "spread" of each variable. Most data tables summarising a data set report at least these two parameters, hopefully together with the data set size. It can also be informative to report the minimum (MIN) and maximum (MAX) values observed and certain percentiles of the data set (see summary table in Section 4.6). It has been demonstrated that MEAN and

Statistical Data Analysis Explained   Clemens Reimann, Peter Filzmoser, Robert G. Garrett, Rudolf Dutter
© 2008 John Wiley & Sons, Ltd.

SD are usually not good descriptors for applied geochemical and environmental data. Thus a better choice for reporting in a table should be MEDIAN and MAD; MEAN and SD can be provided in addition because so many scientists still use these classical measures and they might want to make data set comparisons. A large difference between the MEDIAN and MEAN (environmental data will usually show a higher MEAN than MEDIAN indicating a right skew) will show that the data are strongly skewed, and/or there are some extreme values, biasing the estimates of the MEAN and SD. In the simplest case of a table comparing data averages for a project, sample material is compared with averages of the same material collected elsewhere.

Table 8.1 shows an example of such a table using a selection of the elements from the Kola Project C-horizon data (<2 mm fraction, aqua regia extraction). These Kola Project data are compared to the MEDIAN of soil samples: collected in England & Wales (<2 mm fraction, "sub-soil", aqua regia extraction – McGrath and Loveland, 1992); from the Baltic Soil Survey (BSS), covering ten northern European countries (<2 mm fraction, B-/C-horizon, aqua regia extraction – Reimann *et al.*, 2003); and to the world average for soils (as provided in Reimann and de Caritat, 1998). The latter value is not based on measurements of soil collected all over the world but is an "estimate" of the most likely average of total concentration of these elements in soils based on limited data from soil surveys from different parts of the world. The table allows the reader to judge whether the project data are within a likely range and to note some major differences between the data sets.

Table 8.1 also demonstrates one of the problems with the approach; there are often elements where information is missing in one or another data set. Looking at Table 8.1 it is apparent that the world average values for soil are often much higher than the values found in any of the other data sets. This is in part caused by the fact that the world soil values are provided for "total" concentrations of the elements and not for an aqua regia extraction. The very high value for S in the world soils indicates that they represent agricultural soils, taken at the earth surface, while all other projects collected soils at a certain depth. The table also indicates that the As and Pb concentrations in the C-horizon soils from the Kola Project area are unusually low. A first reaction would be to check the quality control results for these two elements (see Section 18.2) to be sure that the data are correct. Once this is established one could start to consider why these elements might show such low concentrations in the survey area.

Table 8.1 MEDIAN (mg/kg) of some selected elements from the Kola C-horizon data set in comparison to soil average values as cited in literature

	Unit	Kola C-horizon MEDIAN	England & Wales MEDIAN	BSS MEDIAN	World soil AVERAGE
Ag	mg/kg	0.008	NA	NA	0.07
As	mg/kg	0.5	NA	2	5
Bi	mg/kg	0.03	NA	0.05	0.3
Cd	mg/kg	0.02	0.7	0.06	0.3
Co	mg/kg	7	9.8	5	10
Cu	mg/kg	16	18	7	25
Ni	mg/kg	19	23	9	20
Pb	mg/kg	1.6	40	5	17
S	mg/kg	30	NA	62	800
Th	mg/kg	6	NA	5.4	9.4

NA: not available.

Table 8.2 Comparison of some selected elements in the Kola Project O- and C-horizon data sets

	Number of samples		MEAN mg/kg		MEDIAN mg/kg		SD mg/kg		MAD mg/kg	
	O-hor	C-hor	O-hor	C-hor	O-hor	C-hor	O-hor	C-hor	O-hor	C-hor
Ag	617	606	0.283	0.011	0.2	0.008	0.325	0.011	0.160	0.0044
As	617	606	1.60	1.25	1.16	0.5	2.49	2.35	0.460	0.445
Bi	617	606	0.186	0.049	0.159	0.026	0.113	0.164	0.076	0.021
Cd	617	606	0.327	0.029	0.303	0.024	0.150	0.020	0.114	0.010
Co	617	606	3.1	8.2	1.6	7	7.1	5.03	1.11	3.71
Cu	617	606	44	22	9.7	16.2	246	18	5.15	10.8
Ni	617	606	51	23	9.2	18.7	199	21	7.74	11.6
Pb	617	606	24	2.75	18.8	1.6	49	3.3	7.41	0.741
S	617	606	1550	41	1530	30	334	43	297	17.8
Th	617	606	0.571	7.9	0.345	6.5	0.93	6.2	0.254	3.71

A serious shortcoming of just comparing MEDIAN values is that information about the spread of the data and the number of samples that support the reported MEDIAN value is missing. However, in the published literature this information is often not provided. For the Kola data the original data sets are available. For comparing the element concentrations in the collected sample materials it is thus possible to provide the information that is needed for a statistically more appropriate first comparison. Table 8.2 compares MEAN, MEDIAN, SD and MAD for the O- and C-horizon samples. Immediately such a small table covering just 10 elements (out of 50 that could be compared) gets quite complex and requires careful attention (and good editing and presentation of the table) in order to extract the required information.

In Table 8.2 it is apparent that for some elements there are large differences between MEAN and MEDIAN (even more so between SD and MAD). The elements Co, Cu and Ni in the O-horizon are prominent examples. All three are emitted by the Russian nickel industry, and the existence of a large number of upper outliers in the data sets causes extremely right-skewed data distributions. MEAN and SD should thus not be used for these data. Table 8.2 also shows that it is difficult to format such a table in a way that the inherent information can be retrieved at one glance.

In Table 8.3 the values for MEAN and SD are removed and instead the robust coefficient of variation (CVR) is added to get a measure of spread that is independent of the unit and range of the data. To easier "see" the differences between the two sample materials and the elements, the ratio between the MEDIANS for O- and C-horizon samples was calculated. It becomes apparent that there are large differences in these ratios between the different elements. To further improve the "readability" of such a table it is also possible to sort the variables according to one of these ratios.

In Table 8.4 the elements are no longer sorted in alphabetical order but according to the ratio of the MEDIANS for the O- and C-horizons. Here it is possible to immediately detect which elements, sulphur and silver, are most enriched in the O-horizon when compared to the C-horizon. This is because S is a major plant nutrient and thus it is found at high concentrations in the organic layer. The next element that is highly enriched in the O-horizon is Ag – as already observed in Figure 7.2. It is an interesting observation that all elements with a ratio >1 do not belong to the most important metals emitted by industry – Cu and Ni, which both have still

Table 8.3 Data from Table 8.2, MEAN and SD removed, CVR added, and the ratio of MEDIAN, MAD and CVR O-/C-horizon provided

	Number of samples		MEDIAN mg/kg			MAD mg/kg			CVR		
	O-hor	C-hor	O-hor	C-hor	O/C-rat	O-hor	C-hor	O/C-rat	O-hor	C-hor	O/C-rat
Ag	617	606	0.2	0.008	25	0.16	0.0044	36	0.80	0.56	1.44
As	617	606	1.16	0.5	2.32	0.46	0.445	1.03	0.40	0.89	0.45
Bi	617	606	0.159	0.026	6.12	0.076	0.021	3.64	0.48	0.80	0.60
Cd	617	606	0.303	0.024	12.6	0.114	0.010	11	0.38	0.43	0.87
Co	617	606	1.57	7	0.22	1.11	3.71	0.30	0.71	0.53	1.34
Cu	617	606	9.69	16.2	0.60	5.15	10.8	0.48	0.53	0.67	0.79
Ni	617	606	9.18	18.7	0.49	7.74	11.6	0.67	0.84	0.62	1.36
Pb	617	606	18.8	1.6	11.75	7.41	0.741	10	0.39	0.46	0.85
S	617	606	1530	30	51	297	18	17	0.19	0.59	0.33
Th	617	606	0.345	6.5	0.053	0.25	3.71	0.07	0.73	0.57	1.29

higher concentrations in the C-horizon. Thus there is every reason to suspect that a process other than anthropogenic contamination causes the enrichment of elements like Ag, Cd, Pb, Bi and As in the O-horizon (compare Goldschmidt, 1937; Reimann *et al.*, 2001a, 2007).

When looking at the additional information provided by MAD or CVR (Table 8.4), it is apparent that there are large differences in variation between the elements. For example S, with the highest MEDIAN-ratio, shows the lowest ratio for the CVR (Table 8.4). This is probably caused by the fact that S is a major plant nutrient and that plants keep uptake of this element regulated and in an optimal concentration range. It is also interesting to note the difference in the CVR between Cu and Ni, the major components in the emissions from the Russian nickel industry (Table 8.4). One important difference between the two elements is that Cu is a better-regulated micro-nutrient for the plants than Ni. Such tables, especially when sorted, can be used to help develop ideas about processes.

As long as only a few populations are to be compared, a table of summary statistics can be built in a word processor or a spreadsheet from the results displayed by the data analysis

Table 8.4 Data from Table 8.3 sorted according to the MEDIAN ratio O-/C-horizon

	Number of samples		MEDIAN mg/kg			MAD mg/kg			CVR		
	O-hor	C-hor	O-hor	C-hor	O/C-rat	O-hor	C-hor	O/C-rat	O-hor	C-hor	O/C-rat
S	617	606	1530	30	51	297	18	17	0.19	0.59	0.33
Ag	617	606	0.2	0.008	25	0.160	0.0044	36	0.80	0.56	1.44
Cd	617	606	0.303	0.024	13	0.114	0.010	11	0.38	0.43	0.87
Pb	617	606	18.8	1.6	12	7.4	0.741	10	0.39	0.46	0.85
Bi	617	606	0.159	0.026	6.1	0.076	0.021	3.64	0.48	0.80	0.60
As	617	606	1.16	0.5	2.3	0.460	0.445	1.03	0.40	0.89	0.45
Cu	617	606	9.69	16	0.598	5.1	11	0.476	0.53	0.67	0.79
Ni	617	606	9.18	19	0.492	7.7	12	0.669	0.84	0.62	1.36
Co	617	606	1.57	7	0.224	1.1	3.7	0.300	0.71	0.53	1.34
Th	617	606	0.345	6.5	0.053	0.254	3.7	0.068	0.73	0.57	1.29

software. It may even be possible to "cut and paste" the results (DAS+R is able to write these results directly into a ".csv" file). Once in a tabular form, results from other investigations that may be relevant can be added. However, the point is soon reached where such "summary tables" contain too much data to really grasp their content. For example, when looking at the complete Kola data set there is information for four materials and not only two as shown in the above tables (Tables 8.2–8.4). In addition this information exists for many more elements than just the selected ten shown in these tables. Thus the point where it becomes very tedious to work with tables is soon reached. It is at this point that graphical methods come to the fore.

8.2 Graphical comparison of the data distributions of several data sets

Figure 8.1 displays density traces of four elements in all four Kola Project sample materials (C-, B- and O-horizon, and moss) to compare the distribution of the elements in the different

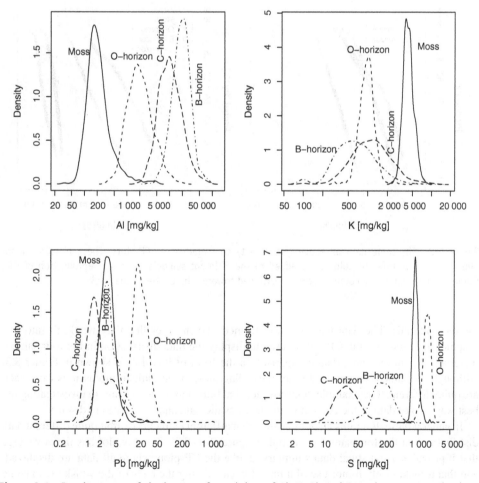

Figure 8.1 Density traces of the log-transformed data of Al, K, Pb and S used to compare the data distribution of these elements in the four main sample materials collected for the Kola Project

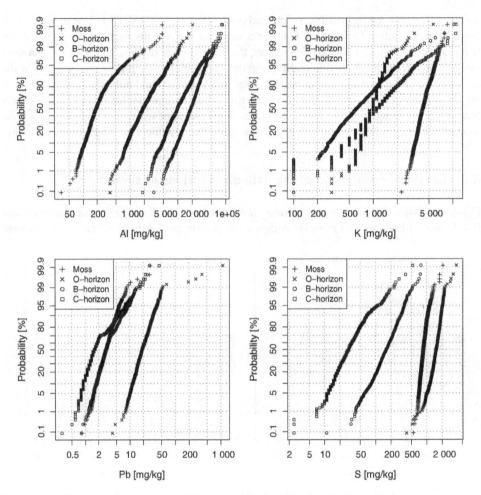

Figure 8.2 The same data as above (Figure 8.1) compared using CP-plots. These plots are more impressive when different colours are added to the different symbols. A colour reproduction of this figure can be seen in the colour section, positioned towards the centre of the book

sample materials. The data were log-transformed first in order that they would fit into one graphic. In Figure 8.2 the CP-plot is used to display the data distribution in the four sample materials. Again the same data comparison in the form of boxplots is shown in Figure 8.3. Although the same data are used in all these diagrams, the graphical impressions from each are quite different. The example demonstrates that time may be well spent in investigating the best way to plot data in the context of the features the data analyst wishes to discuss.

In general these diagrams all demonstrate that different elements exhibit quite different data distributions in the four sample materials (Figures 8.1–8.3). The boxplot has the advantage that it provides a statistical data summary, unlike the CP-plots where all data are displayed, and that it makes only minor use of a model in calculating the ends of the whiskers. In some instances it can be advantageous to see all data points as in the CP-plot and not just the form of the distribution as provided by the density trace.

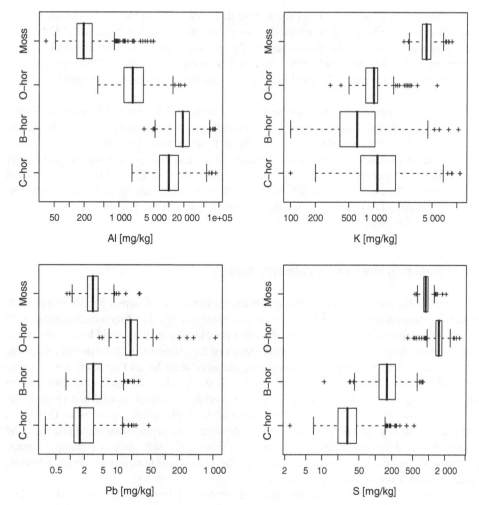

Figure 8.3 The same data as above (Figures 8.1 and 8.2) compared using Tukey boxplots with logarithmic scaling

Aluminium shows much higher concentrations in the B- and C-horizons than in moss or O-horizon samples. Each material has its own concentration range.

The existence of a large number of upper extreme values in moss is most visible in the boxplot comparison (Figure 8.3) but is also indicated in the break in the CP-plot at about 90 per cent (Figure 8.2). These high values in moss are due to the input of minerogenic dust, rich in Al, in the vicinity of the Russian industrial sites. Potassium exhibits a very different picture. Here moss displays by far the highest values and at the same time the lowest variation. This is obvious in all three plots (Figures 8.1–8.3). The reason being that K is one of the most important plant nutrients and the plants ensure K levels are maintained within a narrow range (spread) to support their existence. Sulphur is another important plant nutrient and again is highly enriched in moss, and even more so in the O-horizon soils. In addition, the spread is exceptionally low in these two materials due to their homeostatic control by plants. The density traces (Figure 8.1) may be the most impressive displays. Without the low variation and the comparison with K one could be tempted to assign the high S values in moss and O-horizon to the SO_2 emissions from the Kola smelters. Pb exhibits a special behaviour with its strong enrichment in the O-horizon, while its spread is low (with the exception of four outliers best seen in the CP-plot (Figure 8.2). The low spread is a strong indication that a natural process is causing this enrichment.

The boxplot, especially the notched boxplot (see Sections 3.5.4 and 9.4.1), is ideally suited for comparing different data sets, variables or data subsets, because many boxes can be plotted side by side and the graphic still remains readable (up to some number determined by the display format, portrait or landscape, and size). Notches may be added to the boxplots graphically displaying the 95 per cent confidence bounds on their MEDIANS, which permits a visual informal test of the equality of the MEDIANS to be undertaken (see Section 9.4.1). However, for this informal test to be valid the spreads for the data, as indicated by the span of the boxes, should be roughly equal (see Section 9.4.1).

8.3 Comparing the spatial data structure

In addition to the data distribution, the spatial data structure (see Chapter 5) will also provide important information about the factors or processes influencing the observed concentration of the elements under study. Maps may seem to be a poor choice as they would be uncomparable because of the very different concentration ranges of the elements. If, for example, maps are constructed for the four sample materials using the same scale for all four maps, the resulting maps will more or less show the same features as the graphics for their data distributions (see above, Figures 8.1–8.3), namely in which material the element is enriched or depleted. Figure 8.4 shows this for the element Bi. Almost all the high values appear in the O-horizon map because Bi is generally enriched in the O-horizon. The good intention of choosing the same scale for all four materials does not lead to the desired result: directly comparable maps displaying the spatial data structure permitting the drawing of conclusions about important processes within and between the four layers.

In Chapter 5 it was demonstrated that percentiles (or boxplot classes) used for class selection for regional mapping will display the spatial data structure in the resulting map. Such maps are directly comparable, independent of the concentration range, and the factors causing the data distribution can be immediately recognised. Figure 8.5 shows the four maps for Bi, where each map is constructed using the percentile values for the material displayed, but the same percentile boundaries, i.e. 0–2, >2–25, >25–50, >50–75, >75–98, >98–100 per cent. Although the class boundaries are now different from map to map the resulting presentations of the spatial data structure in the maps are directly comparable.

It can now be seen that the distribution of Bi in moss is dominated by emissions from the Ni smelters, especially around Monchegorsk (Figure 8.5, upper left). The map for the O-horizon looks surprisingly different. The impact of the emissions from industry is no longer dominant. The main factor visible in the map is a strong north-east to south-west gradient in the Bi concentrations (Figure 8.5). Two possible explanations for this pattern exist. The gradient could be caused by long-range atmospheric transport of Bi from central European sources that ceased to emit in the mid 1990s (otherwise the same gradient would be visible in the moss map). This explanation is unlikely in the light of the limited impact

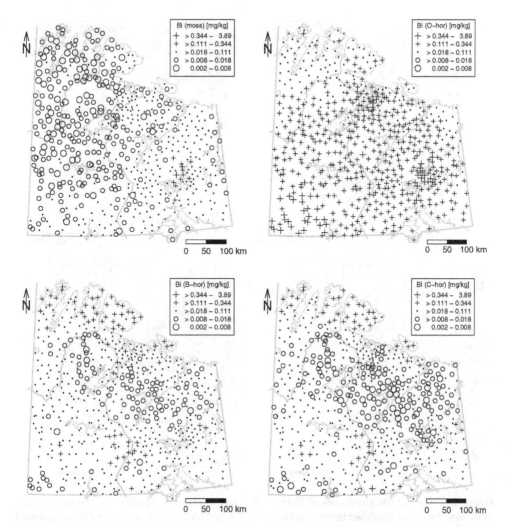

Figure 8.4 Maps of the distribution of Bi in all four sample materials collected for the Kola Project. The scale for all four maps is based on the 0, 5, 10, 25, 75, 98, 100-percentiles of all four data sets combined

of Bi on the O-horizon around Monchegorsk as an existing long-time Bi emitter right in the survey area. However, the gradient by and large follows the vegetation zones in the area from subarctic tundra in the north to boreal forests in the south (note that Figure 6.7, lower left, showed the same gradient in the form of a spatial trend diagram). It is thus very likely that Bi concentrations as observed in the O-horizon in the survey area are directly related to bio-productivity.

The maps of the B- and C-horizon show completely different regional distributions to those depicted in moss and O-horizon (Figure 8.5). These maps reflect geology (e.g., the Caledonian sediments along the Norwegian coast with some of the highest Bi concentrations). It is immediately clear that the factors causing the distribution of Bi in the O-horizon on one hand, and the B- and C-horizons on the other, in the survey area are unrelated.

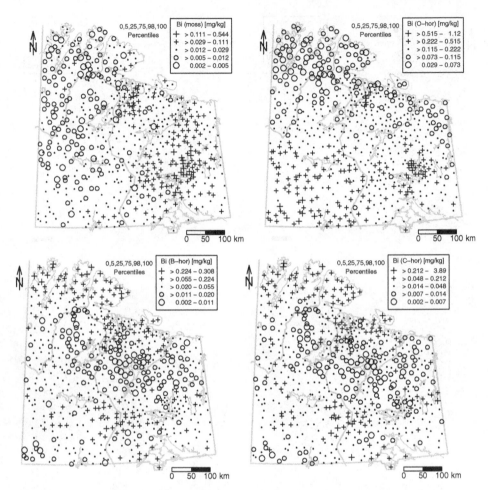

Figure 8.5 Maps of the distribution of Bi in the four sample materials collected for the Kola Project. Individual classes, as identified in the map legends, but based on the same percentiles, have been used for mapping each material

When choosing arbitrary (e.g., nice looking) class boundaries, the direct comparability between the maps may well be lost and the different factors causing the regional distribution of Bi in the survey area might be overlooked.

8.4   Subset creation – a mighty tool in graphical data analysis

The comparison of the four sample materials analysed in the Kola Project is an obvious choice. What can be done, however, if there is only one data set? Is it still possible to learn something via subsetting/grouping the data? Each data set can usually be further subdivided into subsets/groups. For example the Kola Project covered the territory of three different states (Finland, Norway and Russia). Already for national reasons it could be important to compare element levels and variation in the sample materials from the different countries. In addition, the Russian part of the survey area is heavily industrialised, while the Norwegian and Finnish parts are almost pristine (Norway being more influenced by the Russian industry than Finland). Thus, it could be informative to compare the behaviour of the elements in each of the materials for the three country data subsets to learn something about the impact of contamination. Other important factors in the survey area include the existence of three different vegetation zones, distance to the coast with a steady input of several elements via marine aerosols, distance to the different industries in the area and major differences in geology and topography. All these could be used to construct useful data subsets/groups (if the data file includes such information). It could also aid interpretation to use certain variables (e.g., pH) to construct data subsets – sometimes trends become more recognisable when large groups of samples are compared rather than single points in a scatterplot.

It is necessary to include information like "country of origin", "vegetation zone" or "geology" in the data file at the time of its construction to be able to define such subsets/groups that may become important later on during data analysis. Thus, it is time well spent to think in terms of "data analysis" already during the project design stages to ensure that data, especially field data, required for subsetting/grouping are captured during the execution of the survey. In effect, auxiliary data is being selected for inclusion on the supposition that it could influence the geochemistry of the study area, and its inclusion in the data file will permit some preliminary investigation and testing of the suppositions, i.e. hypotheses.

Figure 8.6 (left) shows that Cu concentrations and variation in C-horizon soils are quite comparable between the three countries. The samples from Russia actually have the lowest MEDIAN. For the O-horizon the picture is quite different (Figure 8.6, right). Here Cu data for Finland exhibit the lowest MEDIAN and the lowest spread, while Russia is at the other extreme. Both the MEDIAN and spread are much higher in Russia than in either of the other two countries, and additionally a large number of upper outliers are visible in the plot (Figure 8.6, right). This change is due to the effect of the Russian Cu-Ni industry on the O-horizon samples. Norway falls in an intermediate position in this plot as expected from the fact that many of the Norwegian sample sites are close to the Russian border and thus the Norwegian subset of samples is more affected by the emissions than the samples from Finland.

When directly comparing O- and C-horizon results for the whole Kola data set, it appears that these show little relation (e.g., Figure 8.5). Comparing the distribution of the variables in a number of subsets may provide a different impression and facilitate data interpretation. Figure 8.7 (lower half) is a boxplot comparison of Al_2O_3 and K_2O in C-horizon soils as collected in areas of several different bedrock lithologies (Caledonian sediments (lithologies 9 and 10), Palaeozoic basalts (lithologies 51 and 52), alkaline rocks (lithologies 81 and 82) and granites (lithology 7) see Figure 1.2). The boxplots illuminate pronounced differences in the chemical composition of the C-horizon soils collected in areas of these lithologies. The samples influenced by the alkaline rocks have the highest concentrations of both elements (Figure 8.7). In the O-horizon the Al distribution patterns are still similar to those exhibited in the C-horizon (Figure 8.7, upper left). For K, however, which is an important plant nutrient, the clear influence of the different bedrock types as seen in the C-horizon is greatly reduced in the O-horizon (Figure 8.7, upper right).

Note that when using different units or expressing the element contents in different ways, i.e. as oxides for major components, the direct comparability of data location and spread is lost (Figure 8.7). Transforming all variables to the same unit has thus important advantages during graphical data analysis.

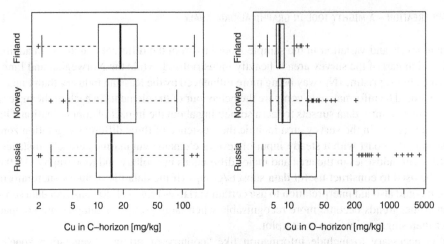

Figure 8.6 Tukey boxplot (logarithmic scale) comparison of Cu concentrations in the Kola O- and C-horizon soil data for the three countries where samples were collected

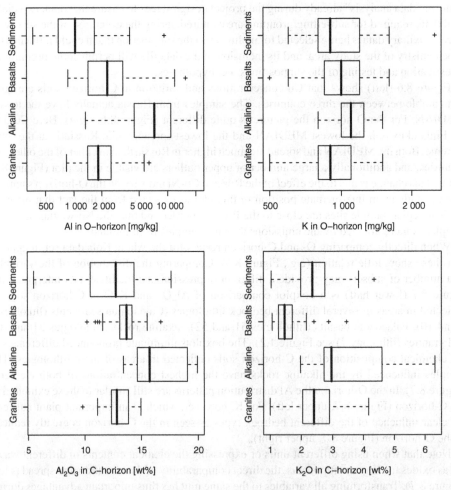

Figure 8.7 Tukey boxplot comparison of Al and K in the O-horizon (upper row – log-scale) and Al_2O_3 and K_2O in C-horizon samples (lower row) as collected above four abundant bedrock lithologies

Comparing data behaviour in a number of carefully defined data subsets in a variety of graphics (instead of boxplots, density traces or various CDF-plots could be used) is a very powerful tool in exploratory data analysis. However, it must be kept in mind that more than one factor or process may hide behind the selected subsets (for example, subsets for the vegetation zones are clearly linked to "distance from coast" and vice versa), and obscure the true causation or even provide graphical evidence for an erroneous conclusion. In this context the user needs to be aware of the possibility for "lurking variables", that is an apparent correlation with a factor where no true correlation exists, but both the factor and data are responding similarly to a third variable not directly present in the display. The graphics may thus not "prove" a process but can, in a truly "exploratory" sense be used to investigate and test hypotheses in an informal, but informative, way.

8.5   Data subsets in scatterplots

Scatterplots become more powerful if more information than just the analytical results of the two variables can be encoded. For example, it is possible to identify certain data subsets via different symbols and/or colours. In Figure 8.8 this is done using the country of origin of each sample to define the plotting symbols. Compared to Figure 6.1, where no subsets are encoded, the resulting diagram immediately reveals some additional interesting facts. The samples with the highest values of Cu and Ni in moss are all from Russia. This provides evidence that these high values are related to the Russian nickel industry as similar industrial activity does not occur in the two other countries. Two different trends present at high data values are apparent (Figure 8.8). In the "upper trend" (more Ni than Cu) there are not only samples from Russia but also some from Norway, while all the Finnish samples remain hidden in the main body of data (Figure 8.8). Knowing how close Nikel and Zapoljarnij are to the Russian–Norwegian border, one can speculate that the upper trend indicates deposition of emissions from Nikel/Zapoljarnij, while the lower trend likely represents the emissions from Monchegorsk.

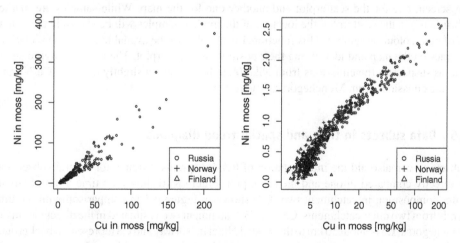

Figure 8.8 Scatterplot of Ni versus Cu in moss (compare Figure 6.1). The country of origin of the samples is encoded into the symbols. A colour reproduction of this figure can be seen in the colour section, positioned towards the centre of the book

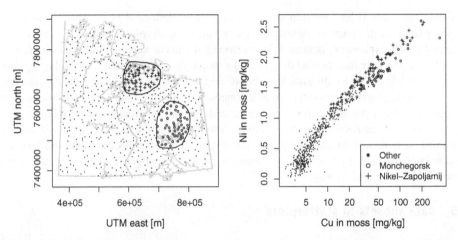

Figure 8.9 Map identifying the location of the subsets (left) and a scatterplot of Cu versus Ni (right), using the same symbols for the subsets. A colour reproduction of this figure can be seen in the colour section, positioned towards the centre of the book

This is an interesting finding and the hypothesis merits further investigation. The interactive definition of data subsets on the screen permits the definition of new data subsets directly from the map on the screen (Figure 8.9), one for "Nikel & Zapoljarnij", one for "Monchegorsk" and another one for "all other samples". These then can be encoded in the scatterplot. Another possibility is that if this "factor" had been thought of at the project planning and database building stages, the required information would have been already encoded, e.g., "samples collected close to Monchegorsk", "samples collected close to Nikel & Zapoljarnij" and "background" samples. In that case it is only necessary to select these subsets and plot them with different symbols in the scatterplot. Advanced software systems (e.g., GGOBI – see http://www.ggobi.org/) may actually permit the opening of two windows on the screen, one for the scatterplot and another one for the map. While samples are marked ("brushed") in the scatterplot, the location of the marked samples will be shown in the map in a different colour or symbol. This functionality should also be available in reverse – marking samples on the map and identifying their position in a scatterplot. The resulting figure shows that, as suspected, the emissions from Nikel/Zapoljarnij have a slightly different Cu/Ni-ratio than the emissions from Monchegorsk (Figure 8.9).

8.6   Data subsets in time and spatial trend diagrams

Subsetting can also aid the interpretation of trend diagrams if trends for several subsets can be directly compared in one and the same plot. Figure 8.10 shows the time trend of stream water composition in catchment two (C2) shown in Figure 6.5 in comparison with the time trends from two other catchments, C5 and C8. Catchment two is situated in the direct vicinity of Monchegorsk, C5 is about 30 km to the west of Nikel in Norway and C8 represents a background catchment at the western project boundary in Finland, far removed from the influence of the smelter emissions. Trends for electrical conductivity (EC) are the same in all three catchments, the highest values for EC are observed in C2, possibly due to enhanced bedrock weathering as

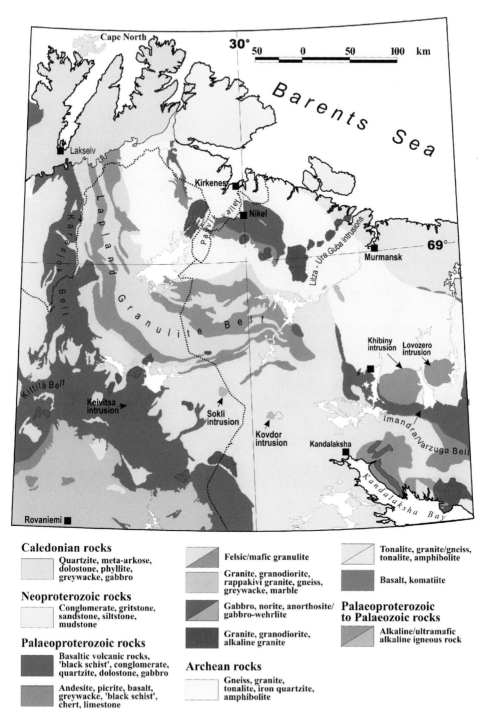

Figure 1.2 Geological map of the Kola Project survey area (modified from Reimann et al., 1998a)

Caledonian rocks

Quartzite, meta-arkose, dolostone, phyllite, greywacke, gabbro

Neoproterozoic rocks

Conglomerate, gritstone, sandstone, siltstone, mudstone

Palaeoproterozoic rocks

Basaltic volcanic rocks, 'black schist', conglomerate, quartzite, dolostone, gabbro

Andesite, picrite, basalt, greywacke, 'black schist', chert, limestone

Felsic/mafic granulite

Granite, granodiorite, rappakivi granite, gneiss, greywacke, marble

Gabbro, norite, anorthosite/ gabbro-wehrlite

Granite, granodiorite, alkaline granite

Archean rocks

Gneiss, granite, tonalite, iron quartzite, amphibolite

Tonalite, granite/gneiss, tonalite, amphibolite

Basalt, komatiite

Palaeoproterozoic to Palaeozoic rocks

Alkaline/ultramafic alkaline igneous rock

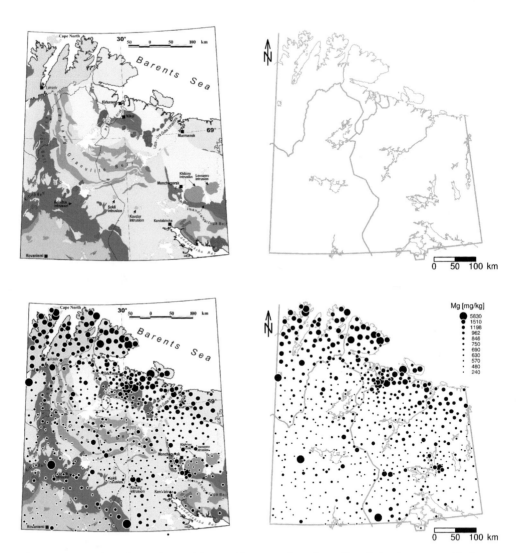

Figure 5.1 Geological map (upper left, see Figure 1.2 for a legend) and simple topographical map (upper right) of the Kola Project area used as base maps for geochemical mapping (lower maps). Lower maps: distribution of Mg in O-horizon soils added to base maps.

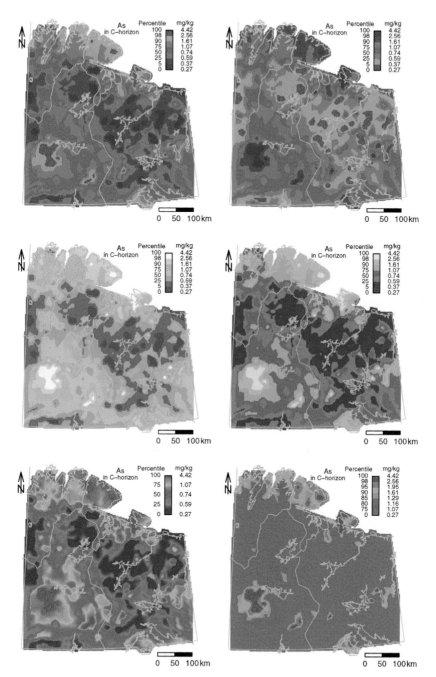

Figure 5.12 Colour smoothed surface maps for As in Kola C-horizon soils. Different colour scales are used. Upper left, percentile scale using rainbow colours; upper right, inverse rainbow colours; middle left, terrain colours; middle right, topographic colours; lower left, continuous percentile scale with rainbow colours; and lower right, truncated percentile scale

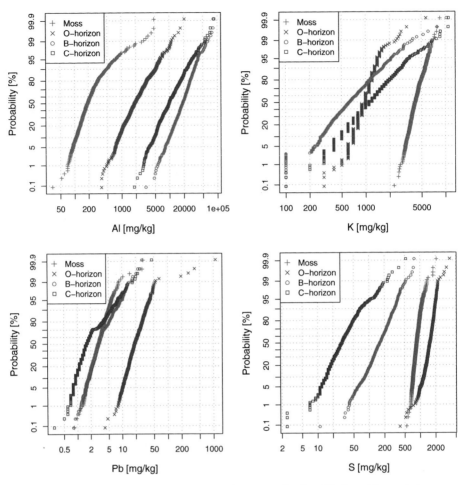

Figure 8.2 The same data as above (Figure 8.1) compared using CP-plots. These plots are more impressive when different colours are added to the different symbols

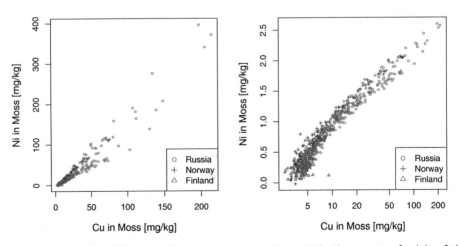

Figure 8.8 Scatterplot of Ni versus Cu in moss (compare Figure 6.1). The country of origin of the samples is encoded into the symbols

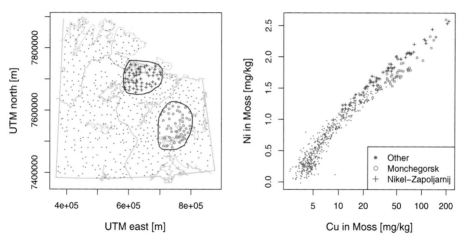

Figure 8.9 Map identifying the location of the subsets (left) and a scatterplot of Cu versus Ni (right), using the same symbols for the subsets

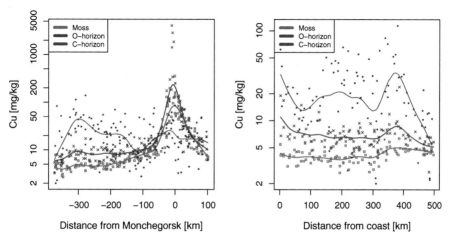

Figure 8.11 Cu in moss and the O- and C-horizon soils along the east–west transect through Monchegorsk (compare Figure 6.6) and along the north–south transect at the western project boundary (compare Figure 6.7)

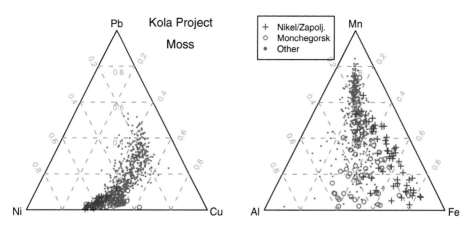

Figure 8.12 Two ternary diagrams plotted with data from the Kola moss data set

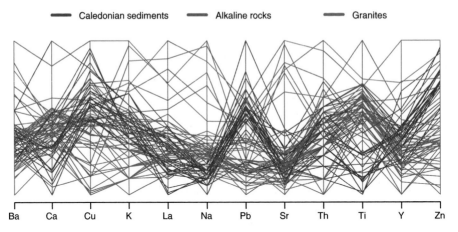

Figure 12.9 Parallel coordinates plot using three lithology-related data subsets (lithology 9, Caledonian sediments; lithology 82, alkaline rocks; and lithology 7, granites – see geological map Figure 1.2) from the Kola C-horizon soil data set

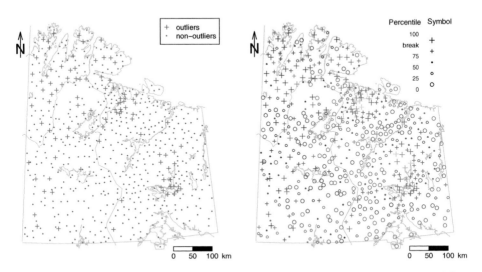

Figure 13.5 Maps identifying multivariate ($p = 7$) outliers in the Kola Project moss data. The left map shows the location of all samples identified as multivariate outliers, the right map includes information on the relative distance to the multivariate data centre (different symbols) and the average element concentration at the sample site (colour, here grey-scale, intensity)

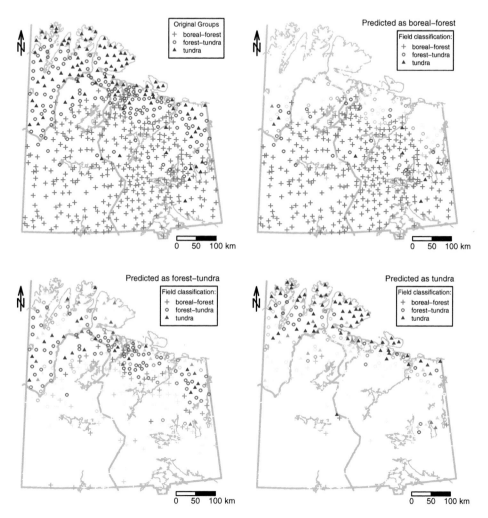

Figure 17.3 Map showing the distribution of the vegetation zones in the Kola Project survey area (upper left) and maps of the predicted distribution of the three vegetation zones in the survey area using LDA with cross-validation. The grey scale corresponds to the probability of assigning a sample to a specific group (boreal-forest, upper right; forest-tundra, lower left; tundra, lower right)

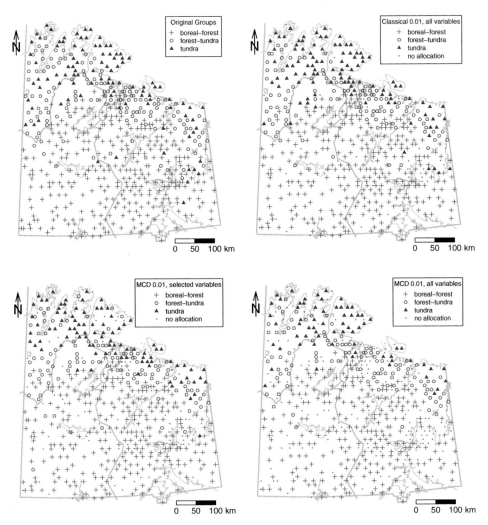

Figure 17.6 Map showing the distribution of the vegetation zones in the Kola Project survey area (upper left) and maps of the predicted distribution of the three vegetation zones in the survey area using allocation. Classical non-robust estimates were used in the upper right map, robust estimates (MCD) were used for the map below (lower right), and in the lower left map the allocation was based on a subset of 11 elements using robust estimates

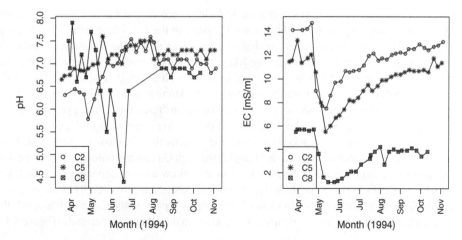

Figure 8.10 Time trends in groundwater pH and electrical conductivity (EC) as observed in three different catchments (C2, C5 and C8) (compare with Figure 6.5)

a result of smelter emissions (Figure 8.10). In contrast the pH trends are quite different between the catchments, surprisingly C8 shows the strongest "acid peak" at snow melt (Figure 8.10). This is caused by lithology; basic rocks (i.e. those high in Ca and Mg) dominate in C2 and C5 and buffer the acid peak (there is actually no acid peak observed in C5, just lower pH values in the first months of the year when compared to the second half); acid rocks (i.e. those low in Ca and Mg) dominate in C8, which is thus very vulnerable to acidification (for a detailed interpretation see de Caritat et al., 1996b).

In Figure 8.11 spatial trends for three different sample materials along the north–south (left) and the east–west (right) transects presented in Chapter 6 are compared. Both plots show that

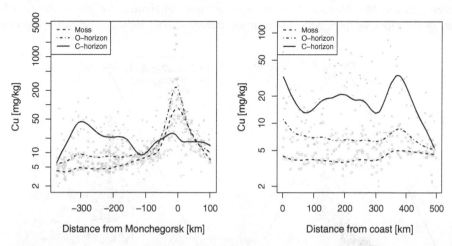

Figure 8.11 Cu in moss and the O- and C-horizon soils along the east–west transect through Monchegorsk (compare Figure 6.6) and along the north–south transect at the western project boundary (compare Figure 6.7). A colour reproduction of this figure can be seen in the colour section, positioned towards the centre of the book

the Cu concentrations in the C-horizon have little influence on the Cu concentration in the O-horizon and moss (Figure 8.11). The minor Cu peak in the C-horizon near Monchegorsk (Figure 8.11, left) could be an indication that contamination due to the emissions has already reached the C-horizon. It is, however, more likely that the peak indicates the presence of the basic rocks that occur near Monchegorsk (including a Cu-Ni deposit). The high values of Cu in the western half of the transect are due to a large greenstone belt containing rocks relatively rich in Cu and Ni. The plot also shows that in the O-horizon "threshold", the upper limit of the background range and the point where input from the smelter can no longer be distinguished from background variation is reached closer to the sources than in moss (Figure 8.11, left). Along the north–south transect (Figure 8.11, right) the high Cu concentrations in the C-horizon are the most interesting feature of the plot. Copper may show a slight dependency on distance to coastline in the O-horizon (slightly decreasing concentrations from 0–300 km, despite a peak in the C-horizon and no trend in moss). The strong Cu peak in the C-horizon towards the southern end of the transect finds only a weak reflection in the other two materials (Figure 8.11, right).

8.7   Data subsets in ternary plots

In environmental sciences these plots can be especially informative if different subsets of data are encoded in different symbols and/or colours.

The example plot is taken from Figure 6.15 and shows the proportions of Ni, Cu and Pb, and Al, Fe and Mn in moss from the Kola Project. However, the two industrial sites and "all other samples" are now marked by different symbols. In the Ni-Cu-Pb plot (Figure 8.12, left) the dominance of Ni and Cu in the smelter emissions is obvious – typically Cu-Ni ores do not contain Pb. It is apparent also that the Monchegorsk emissions tend to be Cu-rich relative to those from Nikel/Zapoljarnij, while the Nikel/Zapoljarnij emissions tend to be more Ni-rich (compare with Figure 8.9).

The Al-Fe-Mn plot (Figure 8.12, right) shows that the moss samples near the industrial sites contain less Mn. Manganese is an important nutrient element for plants, the poor Mn status of the moss samples collected near the industrial sites indicates probable vegetation

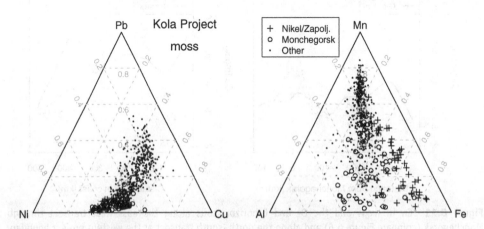

Figure 8.12   Two ternary diagrams plotted with data from the Kola moss data set. A colour reproduction of this figure can be seen in the colour section, positioned towards the centre of the book

stress through SO_2 emissions. In addition, fugitive dust-rich samples will be drawn towards the Al–Fe line of the plot. A further interesting difference between the Monchegorsk and Nikel/Zapoljarnij samples is highlighted. The former are more Al-rich, while the samples collected near Nikel/Zapoljarnij are more Fe-rich (Figure 8.12, right). Within this plot there is another group of samples that show unusually high proportions of Al. Just as in the scatterplot, it would be desirable to identify such samples and to define subsets within the plot to be able to identify their origin and the process(es) influencing them.

Ideally it should be possible to open a second window with a map, brush the samples in the ternary plot and see their location in the map. In the example plot (Figure 8.12, right) the majority of the relatively Al-Mn enriched samples come from the surroundings of the huge open pit apatite (calcium phosphate) mine near Apatity, and from the highlands of the Finnmark plateau in Norway where vegetation is sparse. The relative increase in Al is an indication of the input of terrigenous dust to the moss. Such a "brushing" feature is at present not available in DAS+R, thus it is necessary to use the interactive subset creation tool to mark all unusual

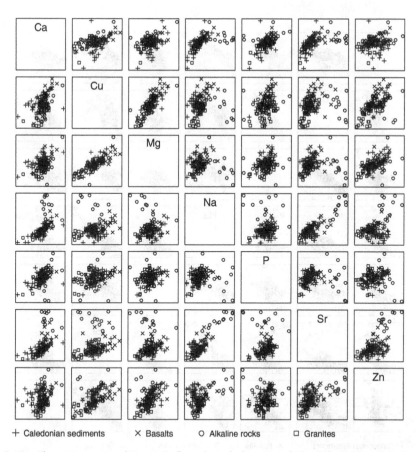

+ Caledonian sediments   × Basalts   O Alkaline rocks   □ Granites

Figure 8.13 The same scatterplot matrix for selected elements of the Kola Project C-horizon data set as shown in Figure 6.14, but with additional information about the underlying lithology (compare Figure 1.2) encoded in the plotting symbols

groups of samples in the diagram and define sample subsets for each of these. The distribution of these subsets can be subsequently studied in a map.

8.8   Data subsets in the scatterplot matrix

The real power of the scatterplot matrix (Section 6.7) becomes visible when more information is encoded in the plot. The careful selection of subsets and plotting these subsets in a scatterplot matrix via different symbols or colour will often provide an extremely informative display (Figure 8.13).

To enhance the visibility of different data behaviour in different subsets, regression lines for the different subsets can be added to the plot (Figure 8.14). In this plot the two halves of the plot may reveal quite different features, because the resulting regression line depends on the choice of the *y*-variable (see Section 6.2). The dependence of the line in the plot on the choice of the *y*-variable could also be overcome by using orthogonal or total least squares

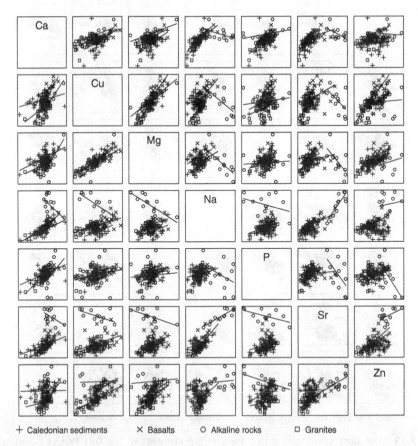

+ Caledonian sediments   × Basalts   ○ Alkaline rocks   □ Granites

Figure 8.14   The same scatterplot matrix as above (Figure 8.13); regression lines for each subset are added to highlight the different elemental relationships in the different data subsets

regression (Van Huffel and Vandewalle, 1991; see Chapter 16). However, these methods are at present not implemented in R for the scatterplot matrix. Furthermore showing both lines provides additional information about the relationships between a pair of variables.

8.9   Data subsets in maps

When preparing geochemical maps, the definition of percentile-based class boundaries and thus the appearance of the resulting map will crucially depend on size and location of the area mapped. Once a certain factor causing a special feature in a map is recognised, it may be interesting to remove the data related to this feature and prepare a new map. The new map will usually show far greater detail in the remaining data. The Cu map for the O-horizon data

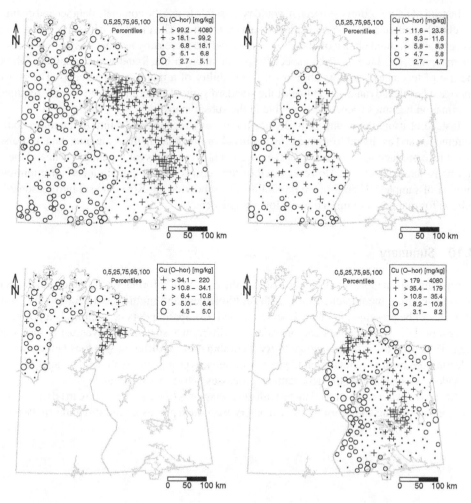

Figure 8.15   Cu in the O-horizon soils of the Kola Project samples. Total data, upper left; Finland only, upper right; Norway only, lower left; and Russia only, lower right

of the Kola Project data set can be used as an example. When mapping the whole data set, the resulting map is completely dominated by the input of the Cu emissions from the Russian Cu-Ni industry (Figure 8.15). Data for samples from Finland and Norway plot almost exclusively in the lowermost two classes. It is thus attractive to plot a map for each country data subset. The resulting maps provide important information that was not visible when plotting the whole data set. For example it appears that quite a number of sample sites along the eastern project boundary in Finland are clearly influenced by the Russian emissions. In addition a number of Cu anomalies become visible that were hidden when mapping the complete data set (Figure 8.15, compare upper left and right). When mapping Norway separately, it is immediately visible how far into Norway the influence of the Nikel emissions reach (Figure 8.15, lower left). When mapping the Russian data separately (Figure 8.15 lower right), the resulting map provides a much better impression of the fast decline of the Cu values with distance from the industrial sites. This fast decline was also clearly visible in the spatial trend plots (Figures 6.6, 6.9 and 6.10).

Instead of using country subsets, it would be possible to use the complete data set and change the class boundaries in order to display the above features. The problem with this approach is, however, the optimal choice of class boundaries and the increasing number of symbols needed for mapping. The latter will be especially problematic in black and white mapping, where the use of too many symbols obscures the readability of a map (see Chapter 5). The above approach has the advantage that one of the standard symbol sets can be used and the required information becomes immediately visible in the subset maps.

Instead of using the countries to define subsets, any recognised feature in the maps could be removed and excluded from mapping to provide more detail in the remaining data. It is also possible to produce separate maps for the different bedrock types (or vegetation zones) if these can be isolated as subsets. However, each of the subsets should contain a certain minimum number of samples. Therefore the possibilities to subdivide the Kola data set into reasonable subsets for mapping are limited by its low sample density.

8.10 Summary

Comparing data in tables has limitations due to the amount of information the tables can include before becoming incomprehensible. With multi-element and multi-media data sets the limit of what can be presented in tables is soon reached. Comparing data in graphics is a powerful EDA tool. Different graphics can provide very different visualisations even using the same data. Thus, it is often an advantage to try a combination of different graphics before finally selecting the most convincing plot. The use of subsets/groups in graphical data analysis can provide great insight into the geochemical processes active in a study area.

An interactive combination of maps with other diagrams (e.g., scatterplots) can be a powerful tool for data mining and be used to reveal otherwise hidden processes and structures in the data.

9

Comparing Data Using Statistical Tests

In Chapter 3 simple graphical methods for studying data distributions have been described. Histograms, density traces, and diverse plots of the distribution function (ECDF-, QQ-, CP-, and PP-plots – Section 3.4) can all give an indication as to whether data follow a normal or a log-normal data distribution. It has been demonstrated that knowledge about the data distribution is important when it is required to estimate the central tendency and spread of the data (Chapter 4).

In the QQ- or CP-plot a straight line indicates a normal or a lognormal distribution depending on the data transformation. Because the data are a random sample of an underlying distribution (the hypothetical distribution used to scale the y-axis of the CP-plot) deviations from the straight line are to be expected. These deviations should decrease the larger the data set becomes if the data really follow the hypothetical distribution, and if these random deviations are sufficiently small the data may really follow the hypothetical distribution. In the CP-plot for Co (Figure 9.1, left) some deviations from a straight line can be observed, especially at the lower and upper ends of the distribution. The comparison of the empirical and hypothetical density trace (Figure 9.1, right) also shows slight deviations from the model. The question is, whether these deviations are pure chance (lognormal distribution), or whether they are systematic (no underlying lognormal distribution with the assumed parameters for mean (central value) and variance (spread)).

Confidence intervals around the straight line in the CP- (or QQ-) plot can be used to help decide whether the empirical data distribution follows the hypothetical distribution (see Figure 3.9). Often confidence intervals and a statistical test are equivalent, and some statisticians consider the use of confidence intervals to be more informative. The more formal alternative to check if data are drawn from a particular distribution is, however, a statistical test.

For statistical testing the values of the empirical, collected, data are always compared to an underlying hypothetical data distribution, or to parameters of this distribution (see, e.g., Lehmann, 1959; Kanji, 1999). Typical parameters to test are central value and spread. Most statistical tests assume that the data are a random sample drawn from a normal distribution. If a test for the mean is carried out the hypothesis refers to the central value of the underlying data distribution and not to the arithmetic mean of the data at hand. The question is not whether the arithmetic mean of a data set is equal to a certain value. This value will always change when another random sample is taken. Therefore it is only informative to test whether the value is representative of the central value of the underlying data distribution of all possible samples.

Statistical Data Analysis Explained   Clemens Reimann, Peter Filzmoser, Robert G. Garrett, Rudolf Dutter
© 2008 John Wiley & Sons, Ltd.

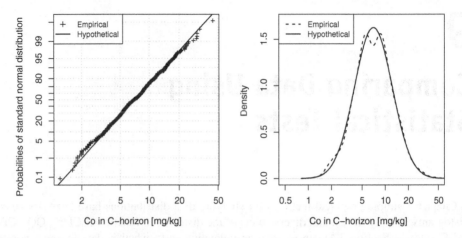

Figure 9.1 CP-plot for Co in the Kola C-horizon soil data showing the empirical distribution and the straight line of the underlying hypothetical normal distribution (left). Density trace of the original data and the hypothetical normal distribution in log-scale (right)

A statistical test always follows a strict scheme (see, e.g., Lehmann, 1959):

(1) Formulation of a *hypothesis*, a statement about the behaviour of the underlying data distribution. The hypothesis must contain two parts, a statement that is supposed to be true (null hypothesis) and a statement formulating the contrary (alternative hypothesis).
(2) Choice of the *significance level* α for the test. The significance level identifies the probability that the null hypothesis, although correct, is wrongly rejected. Usually $\alpha = 0.05$ (i.e. $\alpha = 5$ per cent) is chosen.
(3) Choice of a *suitable test statistic*. Many tests require certain statistical assumptions about the data, e.g., an underlying normal distribution.
(4) Calculation of the *p-value* corresponding to the chosen test. The *p*-value is a probability. If the *p*-value is smaller than the chosen significance level α the null hypothesis of the test is rejected.
(5) Stating the *consequences* of the test.

In addition to the significance level, α, characterising the probability of rejecting a correct null hypothesis, there exists a second possible error probability β to accept a wrong null hypothesis. A decrease in α automatically increases β. For this reason α cannot be chosen arbitrarily small.

It is statistically not acceptable to perform multiple tests with slightly changed hypotheses until the result fits the expectation because this will change the significance level.

9.1 Tests for distribution (Kolmogorov–Smirnov and Shapiro–Wilk tests)

As noted above, statistical testing requires that a null hypothesis be formulated. An example of such a null hypothesis could be:

Null hypothesis: *the hypothetical distribution of the log-transformed Co data in the Kola C-horizon soil data is a normal distribution.*

In addition an alternative hypothesis has to be formulated, for example:

Alternative hypothesis: *the hypothetical distribution is different from the above stated distribution.*

Of course "real" data will never follow exactly the distribution stated in the null hypothesis (see Figure 9.1). Thus some random deviation from the hypothetical distribution and its parameters must be tolerated in practice. If this deviation is larger than a boundary defined by the chosen significance level of the test, the null hypothesis cannot be accepted. This means that the alternative hypothesis is accepted.

There are different possibilities for testing for univariate normality. Two widely used and popular tests are the Kolmogorov–Smirnov test (Smirnov, 1948; Afifi and Azen, 1979) and the Shapiro–Wilk test (Shapiro and Wilk, 1965). In general, the Shapiro–Wilk test is statistically preferable to the other available tests. All tests compare an independent identically distributed sample from an unknown univariate distribution, the data, with a reference distribution (in this case the lognormal distribution). The test result is a probability, the p-value, which can support a decision as to whether the null hypothesis is to be rejected. If the p-value is smaller than the pre-defined significance level, the null hypothesis has to be rejected. Usually the chosen significance level is 5 per cent, i.e. the p-value has to be smaller than 0.05 to reject the null hypothesis. A significance level of 5 per cent infers that in 95 per cent of all cases the correct decision will be taken. In general it is not wise to further reduce the significance level (e.g., to 1 per cent) because the probability of the error of accepting the wrong hypothesis will increase.

As an example, the log-transformed Co data are subjected to the Shapiro–Wilk test, and the test returns a p-value of 0.84. This p-value is much larger than the chosen significance level of 5 per cent (0.05), and thus it can be accepted that the Co data follow a lognormal distribution.

The Kolmogorov–Smirnov test compares the differences between the ECDF of the observed distribution of the log-transformed data with the theoretical lognormal distribution with the same mean and variance. For Co this test yields a p-value of 0.66. With the Shapiro–Wilk test yielding a p-value of 0.84, both tests indicate that the hypothesis that the Co data follow a lognormal distribution with mean and variance of the data can be accepted.

In contrast, for the log-transformed data for Au both tests deliver p-values close to zero (both $p = 10^{-16}$). This provides a quantitative confirmation of the impression obtained for the QQ-plot for Au with confidence intervals (see Figure 3.9) where systematic deviations from a straight line are visible. Thus both tests indicate that the Au data do not follow a lognormal distribution, and the null hypothesis has to be rejected at the stated significance level of 5 per cent.

9.1.1 *The Kola data set and the normal or lognormal distribution*

In the scientific literature it is often stated that geochemical data follow a lognormal distribution. This has been questioned by a number of scientists, e.g., more than 40 years ago (see, e.g., Aubrey, 1954, 1956; Chayes, 1954; Miller and Goldberg, 1955; Vistelius, 1960), or quite recently by Reimann and Filzmoser (2000). The Shapiro–Wilk test can be applied to the Kola data sets to test for normal data distributions of the untransformed and the log-transformed values. Table 9.1 provides the resulting p-values for the Shapiro–Wilk tests for the 24 variables where more than 95 per cent of all data in all four materials returned analytical results above the respective limits of detection.

Table 9.1 demonstrates that for the Kola data set the vast majority of variables in all four layers do not follow a normal or lognormal distribution. This is not really surprising. Environmental/geochemical data are spatially dependent, and spatially dependent data almost never show a normal or lognormal distribution. Furthermore, environmental/geochemical data are based on rather imprecise measurements (see Chapter 18). There are many sources of error involved in sampling, sample preparation, and analysis. Trace element analyses are often plagued by detection limit problems, i.e. a substantial number of samples are not characterised by a real measured value. In addition, the precision of the measurements changes with element concentration, i.e. values are less precise at very low and very high concentrations. The existence of data outliers – in most cases the existence of some samples with unusually high concentrations – is a very common characteristic of such data sets. They are thus strongly

Table 9.1 Resulting p-values of the Shapiro–Wilk (SW) test for normal (norm) and lognormal (lognorm) distribution for 24 variables in the four sample materials collected for the Kola Project

	Moss		O-horizon		B-horizon		C-horizon	
	SW norm	SW lognorm	SW norm	SW lognorm	SW norm	SW lognorm	SW norm	SW lognorm
Ag	0	0	0	0.048	0	0.001	0	0
Al	0	0	0	0.001	0	0.228	0	0.003
As	0	0	0	0	0	0	0	0
Ba	0	0	0	0.025	0	0.022	0	0.002
Bi	0	0	0	0.062	0	0	0	0
Ca	0	0	0	0	0	0	0	0
Cd	0	0	0	0.002	0	0	0	0
Co	0	0	0	0	0	0	0	0.840
Cr	0	0	0	0	0	0	0	0
Cu	0	0	0	0	0	0.840	0	0.257
Fe	0	0	0	0	0	0.004	0	0.269
K	0	0.001	0	0	0	0	0	0.001
Mg	0	0.006	0	0	0	0.028	0	0
Mn	0	0	0	0.066	0	0	0	0
Mo	0	0	0	0	0	0	0	0
Na	0	0	0	0	0	0	0	0
Ni	0	0	0	0	0	0.239	0	0.001
P	0	0	0	0	0	0.130	0	0
Pb	0	0	0	0	0	0	0	0
S	0	0	0	0	0	0	0	0
Sr	0	0	0	0	0	0	0	0
Th	0	0	0	0	0	0	0	0
V	0	0	0	0	0	0	0	0.841
Zn	0	0.587	0	0	0	0.446	0	0.012

skewed. Even worse, these outliers originate from (a) different population(s) than the main body of data. Not only do the outliers originate from another population, in the majority of cases there exist several different primary and secondary factors that influence the chemical composition of the samples at every sample site. Primary factors include geology, the existence of mineralisation and/or contamination sources in an area, vegetation zones and plant communities, and distance to coast and topography. Examples of secondary factors having influence on the chemical composition of a soil sample include pH, the amount of organic material, the amount and exact composition of Fe-Mn-oxyhydroxides, and the grain size distribution. These factors will change from site to site. The emerging empirical data distribution is thus a result of a mixture of several unspecified populations, and it should not be surprising that the data fail formal tests for normality. What is interesting is that data often best mimic a lognormal distribution, and for those interested in this observation Vistelius (1960) is the classic publication.

Cobalt as measured in the C-horizon of the Kola data set was one of the few elements where the tests indicate the data were drawn from a lognormal distribution (Section 9.1 and Table 9.1). Figure 9.2 shows the density trace for a number of lithology-related data subsets in C-horizon soil samples collected on top of the five main lithologies in the survey area (see geological map, Figure 1.2). The density trace of the Co distribution for all the selected subsets (Figure 9.2, left) clearly suggests a lognormal distribution. The Shapiro–Wilk test delivers a p-value of 0.88 (note, not 0.84 as in Table 9.1, as this test is for the combined data from only five lithologies). However, when the density traces of the different subsets are plotted separately (Figure 9.2, right) several of the selected subsets clearly deviate from a lognormal distribution. A highly significant lognormal distribution can thus still consist of groups with very different data structures.

Testing a wide variety of different regional geochemical data sets, Reimann and Filzmoser (2000) came to the conclusion that the data almost never follow a normal or lognormal distribution. In the majority of cases a data transformation (e.g., log, ln, logit, square root, range

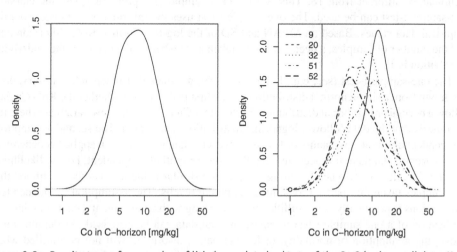

Figure 9.2 Density traces for a number of lithology related subsets of the Co C-horizon soil data, Kola Project C-horizon. Left: one density trace plotted using all selected samples; right: density traces for each of the selected subsets. Lithologies: 9, sedimentary rocks; 20, felsic gneisses; 32, mafic granulites; 51, andesites; 52, basalts (see Figure 1.2)

or Box–Cox – see Chapter 10) did not result in a normal distribution. The vast majority of classical statistical methods are based on the assumption that the data follow a normal distribution. When using them with non-normally distributed data, one should be very aware that this could give biased, or even erroneous, results. Data outliers do not influence, or have minimal influence, on robust methods, and non-parametric methods are not based on such strict model assumptions. These are thus preferable to the classical methods. In any case, a thorough graphical exploratory data analysis and documentation of geochemical and environmental data sets is an absolute necessity before moving on to more advanced statistical methods.

9.2   The one-sample t-test (test for the central value)

The name of the t-test refers to the Student t distribution of the test statistic (see Figure 4.1, middle left). The mean concentration of Cu in soils (<2 mm fraction, aqua regia extraction) from England and Wales was reported to be 18 mg/kg (see Table 7.1 – McGrath and Loveland, 1992). This is one of few published data sets where a substantial number of soil samples were analysed for Cu using the same extraction as employed for the Kola Project.

When calculating the mean concentration of the log-transformed Cu concentrations for the Kola Project C-horizon soils and transforming it back to the original data scale, the value obtained is 16.6 mg/kg. In Chapter 4 it was shown that there are several possibilities to estimate the central value of a data distribution and that each of the resulting values is usually slightly different. The estimate of 16.6 mg/kg for the project area also depends on the number of sample sites (statistically speaking, the sample size) and their geographic locations. The actual suite of survey samples was just one "realisation" of an infinite number of possibilities. Thus the task is to determine the centre (statistical location) of the underlying data distribution from which the statistical sample is drawn. To do this it is necessary to test whether the centre of the underlying distribution is 18 (the mean of Cu in the England and Wales data set) or significantly different from 18. This is a typical example of a problem where the classical one-sample t-test can be used. The one-sample t-test uses the information obtained from the empirical data values. Based on MEAN and SD of the log-transformed Kola C-horizon data and the number of samples, a decision can be taken as to whether the centre of the underlying distribution is 18.

The one-sample t-test (see, e.g., Lehmann, 1959) assumes that independent observations are drawn from a single normal distribution, so the first task is to test whether the Kola Cu data follow a normal or lognormal distribution. Fortunately Cu is one of the few variables where the hypothesis that the data follow a lognormal distribution is accepted. The second assumption, independent observations, cannot be tested. However, when working with applied geochemical and environmental data, it is accepted that the data are spatially dependent, i.e. the likelihood that the Cu values of two neighbouring samples are similar is higher than the likelihood that two samples taken far apart return similar analytical results. The assumption of independent observations is practically never fulfilled when working with environmental data, and thus the application of a one-sample t-test to environmental data can be criticised. As demonstrated above, the assumption of normality, or log-normality, also often will not be fulfilled when working with environmental data. Nevertheless, the one-sample t-test is still one of the most frequently used tests, and users should be aware that the spatial dependencies and any non-normality introduce bias into the one-sample t-test and that other tests (e.g., the Wilcoxon signed-rank test – see Section 9.3) are generally a better choice.

To carry out the one-sample t-test we formulate the following hypotheses:

Null hypothesis: *the central value of the underlying distribution of Cu in the C-horizon is equal to 18 mg/kg*, and

Alternative hypothesis (1): *the central value of the underlying distribution of Cu in the C-horizon is different from 18 mg/kg.*

Here two further alternative hypotheses are possible:

Alternative hypothesis (2): *the central value of the underlying distribution of Cu in the C-horizon is less than 18 mg/kg*, or

Alternative hypothesis (3): *the central value of the underlying distribution of Cu in the C-horizon is greater than 18 mg/kg.*

It is important to decide against which of these alternatives the test should be made. It is not permitted to carry out these tests one after the other, because the overall significance level of the tests will change with each test carried out. Because of this fact the alternative hypothesis should be chosen carefully. Alternative hypothesis (1) is the most general possibility and it is decided to use it for the test.

The MEAN for the log-transformed Kola C-horizon soil data is 2.81, equal to 16.6 mg/kg if back-transformed. The resulting p-value of the t-test is 0.008, which is clearly smaller than the significance level of 0.05. Thus the test result is highly significant and the central value of the Kola data is different from 18 mg/kg (2.89 in log-units). This depends largely on the high number of samples that represent the Kola C-horizon data set ($n = 606$). For a distribution with the same MEAN and SD the test would no longer be rejected if the sample size (n) were 332 or less.

Figure 9.3 (left) shows the density function of the assumed underlying normal distribution with centre 18 (right line) and the same spread as the empirical data, overplotted with the density

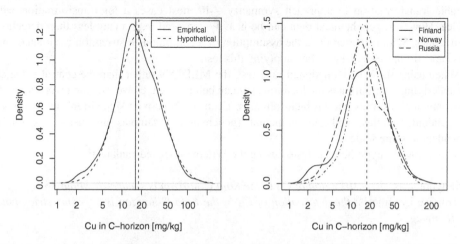

Figure 9.3 Comparison of the assumed hypothetical and the empirical distribution for Cu in the Kola C-horizon soils (left plot) and comparison of the empirical distribution of Cu in the soils from Finland, Norway and Russia (right plot). The vertical line indicates the test value 18 mg/kg (right plot)

trace of the empirical data with MEAN 16.6 mg/kg (left line). Note that the MEAN plots to the right of the maximum of the empirical distribution because even the log-transformed values still show a certain right-skewness.

Another question might be: are the means for Cu concentrations in the C-horizon soils from the three countries (Finland, Norway, and Russia) separately comparable to that for England and Wales? Again, it is first necessary to test whether the Cu data in each country subset follows a lognormal distribution. In Finland the results are surprising: the Shapiro–Wilk test yields a p-value of 0.047, and the Kolmogorov–Smirnov test a p-value of 0.60. This demonstrates nicely that with environmental data the limits of statistical testing are being stretched. The results of the test for normal distribution of the log-transformed data indicates that in the case of Cu in the C-horizon soils from Finland, both requirements for the one-sample t-test (normal distribution of the log-transformed data and independence of the samples) may be violated.

The p-value of the one-sample t-test for the values from Finland (back-transformed MEAN = 17.0 mg/kg) is 0.31, for Norway (19.0 mg/kg) it is 0.38 and for Russia (15.4 mg/kg) it is 0.0006. Thus the means of the underlying data distribution in Finland and Norway could be 18 mg/kg, while the mean Cu concentration of the underlying distribution in Russia is different from 18 mg/kg (compare Figure 9.3). Providing an explanation for the relatively low Cu concentrations in the C-horizon soils from Russia would be desirable, and an inspection of a geological map will show that different bedrock types than occur in Finland and Norway dominate in Russia.

9.3 Wilcoxon signed-rank test

When working with environmental data, a different test is preferable because the one-sample t-test can neither handle non-normally distributed nor dependent samples. The Wilcoxon signed-rank test (Wilcoxon, 1945; Hollander and Wolfe, 1973) is a non-parametric test that only assumes a continuous and symmetric distribution. It is thus only necessary to find a suitable transformation to approach symmetry – in most cases a log-transformation will suffice. However, geochemical data can be heavily censored, e.g., many less than detection limit values. With censored data, the assumption of a continuous distribution is violated and thus care still has to be taken when applying this test.

When using the Wilcoxon signed-rank test, the MEDIAN rather than the central value of the underlying distribution is the location estimate being tested. However, when working with environmental samples one problem remains. Even the Wilcoxon signed-rank test requires independent samples, and environmental (geochemical) samples are usually spatially dependent at some scale.

The following hypotheses is tested using the Wilcoxon signed-rank test:

Null hypothesis: *the distribution of Cu in the Kola C-horizon is symmetric about 18 mg/kg.*
Alternative hypothesis: *the distribution of Cu in the Kola C-horizon is not symmetric about 18 mg/kg.*

Carrying out the Wilcoxon signed-rank test, a p-value of 0.013 is obtained (compared to 0.008 from the one-sample t-test above). The test result thus does not change. Using the Wilcoxon signed-rank test for the three countries separately, the following p-values are obtained (in brackets: results of the one-sample t-test): Finland 0.56 (0.31), Norway 0.46 (0.38) and Russia 0.0003 (0.0006). Thus the non-parametric test leads to the same conclusions as the parametric test.

9.4   Comparing two central values of the distributions of independent data groups

9.4.1   The two-sample t-test

In the above example the central value of Cu in the Kola data has been compared against a single value, the mean of Cu in the England and Wales data set. It may also be of interest to directly compare the central values of the "country" subsets. It has been observed above that there are probably differences between the samples collected in Finland and Norway, and Russia. So a test to determine whether the central value for the combined data for Finland and Norway (group 1) is significantly different from the central value of the data for Russia (group 2) is required. This would be a typical example where the two-sample t-test (see, e.g., Lehmann, 1959) could be applied. The test requires that the data in both groups follow a normal distribution, are independent, and show equal variance (pooled two-sample t-test). If the assumption of equal variance is not fulfilled, a different version of the two-sample t-test called the separate t-test (see, e.g., Lehmann, 1959) can be applied.

The following hypotheses will be tested:

Null hypothesis: *the central value of Cu of the underlying distribution of group 1 (Finland and Norway) is equal to the central value of the underlying distribution of group 2 (Russia).*
Alternative hypothesis: *the central value of Cu of the underlying distribution of group 1 (Finland and Norway) is unequal to the central value of the underlying distribution of group 2 (Russia).*

The pooled two-sample t-test yields a *p*-value of 0.023. The separate two-sample t-test yields the same *p*-value. This test result implies that the null hypothesis of equal central values is rejected at the significance level of 5 per cent. Thus the mean Cu concentration of the underlying distribution of group 1 (Finland and Norway) is with 95 per cent confidence different from the central value of group 2 (Russia).

The notched boxplot (mentioned in Section 3.5.4) can be used as an alternative to check for equality of the MEDIANS of two (or more) data groups. The notches are placed at $1.58 \cdot \frac{HS}{\sqrt{n}}$ on either side of the MEDIAN. These notch positions, representing the 95 per cent confidence bounds on the MEDIAN, are based on a normal approximation. If the data spread, as indicated by the width of the boxes, is comparable between the data sets, non-overlapping notches provide strong evidence that the MEDIANS are significantly different at the 5 per cent significance level. Instead of using a normal approximation for computing the notches, it would be possible to use a non-parametric binomial approximation. However, when working with very small numbers of observations, the resulting confidence bounds can become very imprecise.

Figure 9.4 shows the Tukey boxplot of the log-transformed data of the two data groups Finland and Norway, and Russia. The notches of the two Tukey boxplots do not overlap. This provides clear evidence that the two location measures, MEDIANS, are different. As for the pooled t-test, equality of variances is important when using the notched boxplot for testing.

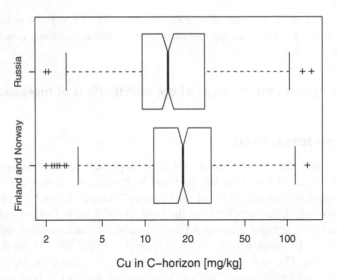

Figure 9.4 Notched Tukey boxplot comparison for Cu in the C-horizon soils from Finland and Norway versus Russia

Figure 9.4 graphically confirms that the variances in both groups are comparable: the lengths of the two boxes are almost the same.

9.4.2 The Wilcoxon rank sum test

The Wilcoxon rank sum test (equivalent to the Mann–Whitney U-test) is a non-parametric test for equal distribution (Mann and Whitney, 1947; Wilcoxon, 1945). Requirements are independent samples and a continuous data distribution. The Wilcoxon rank sum test tests whether the MEDIANS of two samples drawn from the same distribution are equal.

The two hypotheses are:

Null hypothesis: *the underlying distributions of variable Cu in group 1 (Finland and Norway) and group 2 (Russia) are equal.*

Alternative hypothesis: *the underlying distributions of the variable Cu of group 1 and group 2 are shifted (i.e. they have different MEDIANS).*

The resulting p-value for the Mann–Whitney test is 0.006. The conclusion is that the central location of Cu in group 1 (Finland and Norway) is different from the central location of Cu in group 2 (Russia). The Wilcoxon rank sum test should always be used instead of the two sample t-test when the hypothesis of normality is not justified (see, e.g., Hollander and Wolfe, 1973).

9.5 Comparing two central values of matched pairs of data

9.5.1 The paired t-test

Another comparison that may be tested is whether the measurements for a variable, e.g., Cu in the soil samples from the C- and B-horizons, are sufficiently similar to be considered equal.

In this instance the test involves dependent pairs of samples because the samples were taken at the same location. Another application of this test would be to investigate if analyses for the same samples by two different methods yield similar results. In the temporal domain, the question is whether the results of two sampling campaigns are equal, if the same area had been sampled at the same locations twice, e.g., in a monitoring program in 1995 and 2005. In all these instances the paired t-test (see, e.g., Lehmann, 1959) is required.

Requirements for the paired t-test are again that the data follow a normal distribution and that the samples are independent. The Cu values in the B-horizon follow a lognormal distribution. The following hypotheses will be tested:

Null hypothesis: *the distribution of differences between the pairs of C- and B-horizon samples has central value zero.*

Alternative hypothesis: *the distribution of differences between the pairs of C- and B-horizon samples has a central value different from zero.*

The resulting *p*-value of the test is practically zero (10^{-16}). Figure 9.5, left, shows a notched Tukey boxplot of the differences of the sample pairs. To accept the hypothesis that the MEDIAN of the differences between the sample pairs is zero, the notch would have to straddle zero. This is clearly not the case, supporting the result of the paired t-test. One might be tempted to use the notched Tukey boxplot to directly compare the results of the B- and C-horizon samples (Figure 9.5, right). However, for this to be valid the independence assumption for the two groups of samples has to be fulfilled. This is not the case as the soil horizon samples are collected at the same sites. Thus the graphic relevant to the two-sample t-test can easily be misleading, and a visualisation of the test must be carried out as shown in Figure 9.5, left.

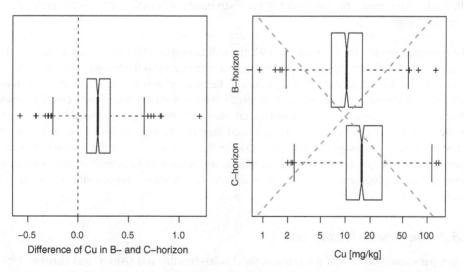

Figure 9.5 Notched Tukey boxplot of the differences of all sample pairs for Cu in the B- and C-horizon of the Kola Project data (left plot). Note that the notched boxplots (right) for independent samples should not be confused with the test for equality of the MEDIANS between dependent samples (left)

9.5.2 The Wilcoxon test

A non-parametric version of the paired t-test is the Wilcoxon test, also called Wilcoxon signed rank test for data pairs (Wilcoxon, 1945). Requirements are that the differences of the data pairs are independent samples and follow a continuous and symmetric distribution around the MEDIAN.

Null hypothesis: *the MEDIAN of the differences of the pairs of samples is zero.*
Alternative hypothesis: *the MEDIAN of the differences of the pairs of samples is different from zero.*

The resulting p-value of the Wilcoxon signed rank test for Cu in the B- and C-horizon of the Kola data set is again close to zero (10^{-16}). Thus the hypothesis of equal MEDIANS in both soil horizons has to be rejected. This test should be preferred if the data do not follow a normal or lognormal distribution.

9.6 Comparing the variance of two data sets

9.6.1 The F-test

In the above example the separate two-sample t-test had the additional requirement of equal variance. To test for equal variance, an F-test would be applied (see, e.g., Lehmann, 1959). Requirements for an F-test are again normal distribution and independent samples in both groups.

Null hypothesis: *the variances of the distributions of the variable Cu for group 1 (Finland and Norway) and group 2 (Russia) are equal.*
Alternative hypothesis: *the variances of the distributions of the variable Cu for group 1 and group 2 differ.*

The resulting p-value for the F-test is 0.94, the null hypothesis that the variances are equal is accepted. This confirms the graphical impression obtained from the boxplot (Figure 9.4).

The Kola data demonstrate that unusual factors or geochemical processes express themselves more often in an increase in spread than in a difference in the central location. Thus tests for variance are of a more general interest than the above example might imply. Changes in spread can also be easily detected graphically, e.g., in boxplots. However, the F-test is more rigorous because it accounts for the number of samples whereas boxplots always give, more or less, the same graphical impression independent of the number of samples (some programs allow one to scale the box widths proportional to the sample size).

9.6.2 The Ansari–Bradley test

A non-parametric version of the F-test is the Ansari–Bradley test (Ansari and Bradley, 1960) that requires independent samples (note that the alternative Mood test (Conover, 1998) in addition requires equal MEDIANS, which limits its use in applied geochemistry and the environmental sciences considerably). The p-value of the Ansari–Bradley test for the above example is 0.70, again leading to the conclusion that the spread for Cu in the two data subsets can be considered equal.

9.7 Comparing several central values

9.7.1 One-way analysis of variance (ANOVA)

It is often required to simultaneously compare the central values of several data groups, i.e. more than the two the t-test is appropriate for, and for this one-way analysis of variance (ANOVA) (see, e.g., Scheffé, 1959, 1999) is appropriate. Again the requirements are for independent observations, normally distributed data in each group, and equal variances for all groups. It should be noted that the requirements have increased and that these will be rarely met when working with applied geochemical and environmental data. It is a question as to whether classical, parametric ANOVA tests should be employed, or rather the non-parametric version of the test(s) should be used.

The comparison of the central values for Cu in the data for Finland, Norway and Russia provide an illustration. As has been noted in Section 9.2, the requirement for a normal distribution may not be met by the data for Finland. The test for equality of the variances (see below) indicates that the variances are comparable. When Finland, Norway and Russia are treated as three separate groups, ANOVA can be used to test for equality of the central values of the Cu concentrations in the three groups simultaneously.

Null hypothesis: *the central values of the groups are equal.*
Alternative hypothesis: *at least two groups have different central values.*

The *p*-value returned from the test is 0.12 and thus the null hypothesis cannot be rejected. Thus ANOVA suggests that the central values of the underlying distributions of the three groups are equal. This is certainly a surprise considering the results previously presented!

9.7.2 Kruskal–Wallis test

A non-parametric version of the above test is the Kruskal–Wallis test (Kruskal and Wallis, 1952; Hollander and Wolfe, 1973). It does not require that the data be normally distributed, but just that they follow a continuous distribution in each of the groups and, of course, that the samples are independent. The same hypotheses as above are tested, and this time the *p*-value is 0.013. Thus the null hypothesis is now rejected and one can assume that there is a difference between at least one pair of the three groups, which would confirm the previous results. This example highlights the problems of hypothesis testing for applied geochemical and environmental data that time and again violate at least one, and often several, of the statistical assumptions of the tests. Again, the non-parametric test is more reliable because it is dependent on fewer assumptions, and thus more trust should be placed on this result.

9.8 Comparing the variance of several data groups

9.8.1 Bartlett test

The Bartlett test (Bartlett, 1937) is for the equality of the variance of several data groups against the difference of variability in one of the possible group-pairs. The assumptions require that the data for each group be normally distributed and the samples independent.

Null hypothesis: *the group variances are equal.*
Alternative hypothesis: *at least two groups have different variances.*

For the example of Cu in the C-horizon soils from Finland, Norway, and Russia the resulting p-value is 0.08. Thus equality of the variances in the three data groups, which was one of the requirements for carrying out the above ANOVA, is actually just fulfilled. However, as noted above, the requirement for a normal distribution may be violated in the case of the Finnish data.

9.8.2   Levene test

Like the Bartlett test, the Levene test (Levene, 1960; Brown and Forsythe, 1974) also tests for equality of the variances of several data groups. It allows for certain deviations from normal distribution by replacing the MEAN with the MEDIAN in the formula for the test statistic.

Carrying out the Levene test for the above example delivers a p-value of 0.49. This result is far less marginal than the result of the Bartlett test. The null hypothesis that the group variances are equal cannot be rejected.

9.8.3   Fligner test

The non-parametric version of the Bartlett test is the Fligner test (Conover *et al.*, 1981). This test is most robust against departures from normality, and tests, as above, the null hypothesis of homogeneity (equality) of variances. The resulting p-value is 0.15 and we would again accept the null hypothesis.

It is noteworthy that this result would not necessarily be expected when studying the boxplots of the three national subsets (Figure 9.6). When studying these, the graphical impression is that

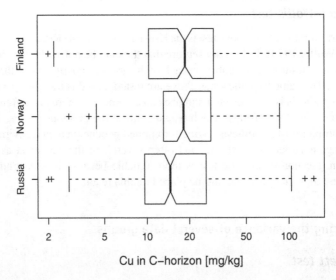

Figure 9.6   Notched Tukey boxplot comparison of the distribution of the log-transformed values of the variable Cu in the Kola C-horizon samples in the three country subsets

the spread of the Finnish and Russian samples may be comparable while the Norwegian samples exhibit less variability. This shows that the graphical impression alone can be misleading.

9.9 Comparing several central values of dependent groups

9.9.1 ANOVA with blocking (two-way)

It may be necessary to compare results from several data sets. For instance, in the Kola Project, data were collected from four layers at the same location. As another example, monitoring results for samples collected at different times, e.g., 1995, 2000, and 2005 at the same locations might be available. It is desired to determine whether the results for the different layers (surveys) are comparable. Here the results of the groups are clearly dependent because the samples originate from the same locations, thus one-way ANOVA cannot be used. The paired t-test for dependent data can only be used to compare two groups of data, but not simultaneously three, four, or even more groups. Thus an extension of ANOVA that accounts for a blocking effect (several different sample materials taken at one site or one material sampled several times at the same location) has to be employed. This type of ANOVA, often referred to as two-way (see, e.g., Scheffé, 1959, 1999), simultaneously tests two null hypotheses against two alternative hypotheses. ANOVA with blocking requires that the data in each group are normally distributed with equal variances. The tests for normality for the log-transformed Cu data show that the data from O-horizon and moss do not follow a normal distribution. There is in fact no variable in the Kola data that follows a normal or lognormal distribution in all four layers. However, the test will be continued in order to carry out a comparison with the non-parametric equivalent.

Null hypothesis (1): *the central values of the Cu concentrations of the four sample materials are equal.*

Alternative hypothesis (1): *there is a difference in the central values between at least two layers.*

Null hypothesis (2): *the average concentration of Cu in the four layers is the same at all sample sites.*

Alternative hypothesis (2): *the average Cu results differ in at least two locations.*

In both cases the p-values are approximately 0, and both null hypotheses have to be rejected. This implies that there are significant differences in the Cu concentrations observed between the layers, and that there are significant differences between the average Cu concentrations (averaged over all layers) at sites across the whole region.

Figure 9.7 shows a Tukey boxplot comparison of Cu in the four collected layers. It is clearly visible that the MEDIANS in the four layers differ substantially. The notches, however, may be misleading because of the lack of spatial independence (see Figure 9.5). In moss and O-horizon an unusual number of outliers are visible. This is caused by the additional input of Cu from the Russian nickel industry disturbing the natural distributions.

9.9.2 Friedman test

The Friedman test (Friedman, 1937; Hollander and Wolfe, 1973) is the non-parametric equivalent of (two-way) ANOVA with blocking. Here the only assumption is that the samples are

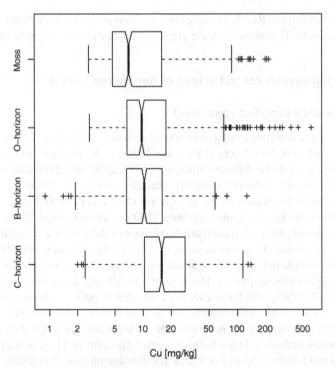

Figure 9.7 Notched Tukey boxplots for log-transformed Cu concentrations in C-, B-, and O-horizon soils and terrestrial moss samples from the Kola Project area

drawn from continuous distributions. In contrast to the above ANOVA with blocking only one null hypothesis is now tested:

Null hypothesis: *the MEDIAN values of the underlying distributions of the Cu concentrations in the four layers are equal.*

Alternative hypothesis: *the MEDIAN values of the underlying distributions of the Cu concentrations in the four layers are different.*

Again the *p*-value is almost zero ($p = 10^{-16}$), and the Friedman test thus confirms the results of the classical parametric ANOVA with blocking. Given the distribution of Cu in the O-horizon and moss, the Friedman test is the best choice for testing this hypothesis.

9.10 Summary

Statistical tests require the careful formulation of a null hypothesis and an alternative hypothesis. Many of the classical statistical tests build on quite strong statistical assumptions concerning the data and are as such not very well suited for environmental data. Despite the fact that the assumptions of normality and independence, and for some tests equality of variance, required for formal parametric hypothesis testing are not met, surprisingly the inferences drawn from the non-robust and robust procedures are often similar. This does not justify the continued use of the classical parametric procedures, though perhaps it helps to explain it along with a lack of availability of suitable software to undertake the more preferable non-parametric procedures. Results of statistical tests should thus be interpreted with great care.

Finally, EDA graphical procedures and formal statistical analyses complement each other. In many cases the inferences drawn by the practitioner may be similar, but in others illuminating differences are present that may stimulate further questions and analyses. Therefore, hypotheses should be investigated both graphically and formally, and each used to avoid misinterpretation of the other.

10
Improving Data Behaviour for Statistical Analysis: Ranking and Transformations

Sorting the data will often provide a better appreciation of the values of a variable. It is also possible to replace the data by the ranks of the sorted values and continue statistical analysis with these ranks.

Data transformations, which usually consist of relatively simple mathematical operations, are frequently required. There are several reasons for carrying out transformations. One reason may be that it is required that the data approach a normal distribution. Another reason could be that some statistical methods require comparability (equality) of the variance, or comparability of both mean and variance. For graphical representations it can also be useful to transform variables so that they span the same range of data. Transformations can condition the data so that they more closely meet these requirements.

Sometimes the deviations from the average concentration of an element may be of interest. In that case it can be advantageous to move the origin (0) to the centre of the data. This can be achieved by subtracting the location (MEAN, MEDIAN, etc.) from each single data value. Because applied geochemical and environmental data often contain outliers, preferably one should use the MEDIAN as the estimate of central location as it is unchanged by extreme values.

When working with at least two variables in a statistical analysis, it will often be advantageous for all variables to have the same weight (have the same scale). A typical example in applied geochemistry is the enormous difference in concentration between major and trace elements (often five orders of magnitude). For example, when plotting a ternary diagram it may be necessary to multiply (or divide) one (or two) of the variables by a certain factor so that the three variables come within the same data range.

Because of the frequent use of transformations, it should be easy to execute the most common data transformations within a data analysis system. Variables or files that contain transformed data need to be clearly marked as such. In some data analysis systems there may be a real danger that the original data file can be destroyed in saving a data file with transformed variables.

Commonly, the data are first transformed (e.g., to approach a normal distribution), then, if required, centred and then scaled.

Statistical Data Analysis Explained   Clemens Reimann, Peter Filzmoser, Robert G. Garrett, Rudolf Dutter
© 2008 John Wiley & Sons, Ltd.

10.1 Ranking/sorting

Ranking assigns each data value a number according to its rank in the data set, starting from the minimum to the maximum value. The rank equals the position of the value if the data set were sorted in increasing order. When values are identical, they are all given a rank equal to the average of the ranks of the equal values. A disadvantage of ranking is that the detailed information of the original data is lost. However, ranking will remove skewness and the influence of outliers from the data. In some classical statistical analyses, such as analysis of variance (ANOVA), data that do not fulfil the assumptions of normality and homogeneity of variance may be replaced by their ranks and the analysis successfully undertaken (Conover and Iman, 1982; Conover, 1998). Ranking the data often allows better insight into the dependencies between the variables (e.g., correlation – see Chapter 11).

Sorting of single variables can also be helpful to detect unexpected structures within a variable. A typical example would be a sorted data table, prepared in a spreadsheet program (*NOTE*, with some software, sorting involves the serious danger of destroying a data file because the relation to the identifier (and other data/information) can be lost; the "original" data file must always be securely stored, backed up, before starting to sort). Extensive editing of a spreadsheet may be required to prepare a sorted data table for all variables, where the relation to the ID (and/or further required information) is retained (see Figure 10.1).

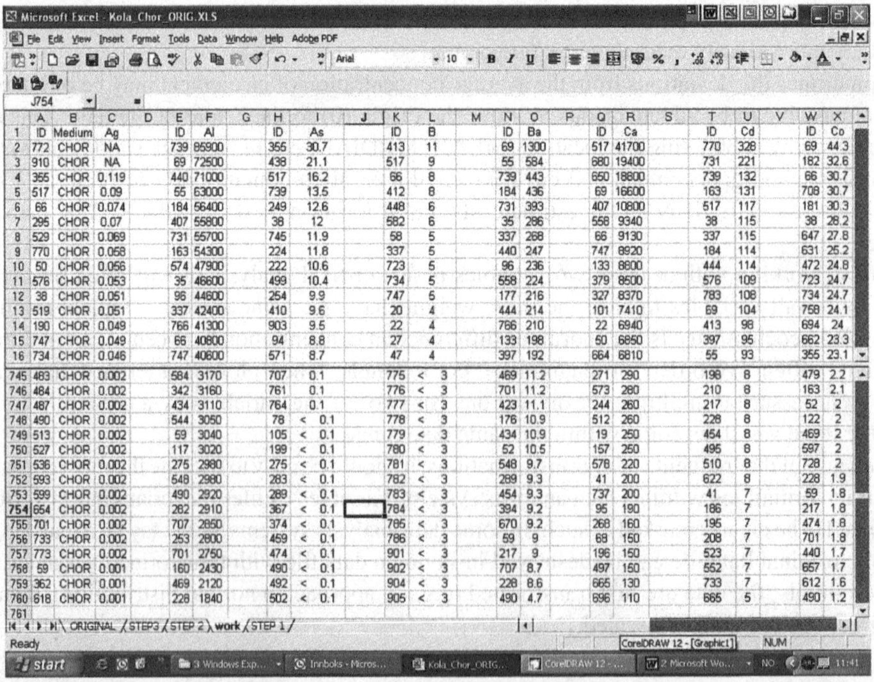

	A	B	C	E	F	H	I	K	L	N	O	Q	R	T	U	W	X
1	ID	Medium	Ag	ID	Al	ID	As	ID	B	ID	Ba	ID	Ca	ID	Cd	ID	Co
2	772	CHOR	NA	739	85900	355	30.7	413	11	69	1300	517	41700	770	328	69	44.3
3	910	CHOR	NA	69	72500	438	21.1	517	9	55	584	680	19400	731	221	182	32.6
4	355	CHOR	0.119	440	71000	517	16.2	66	8	739	443	650	18600	739	132	66	30.7
5	517	CHOR	0.09	55	63000	739	13.5	412	8	184	436	69	16600	163	131	708	30.7
6	66	CHOR	0.074	184	56400	249	12.6	448	6	731	393	407	10800	517	117	181	30.3
7	295	CHOR	0.07	407	55800	38	12	582	6	35	286	558	9340	38	115	38	28.2
8	529	CHOR	0.069	731	55700	745	11.9	58	5	337	268	66	9130	337	115	647	27.8
9	770	CHOR	0.058	163	54300	224	11.8	337	5	440	247	747	8920	184	114	631	25.2
10	50	CHOR	0.056	574	47900	222	10.6	723	5	96	236	133	8800	444	114	472	24.8
11	576	CHOR	0.053	35	46600	499	10.4	734	5	558	224	379	8500	576	109	723	24.7
12	38	CHOR	0.051	96	44600	254	9.9	747	5	177	216	327	8370	783	108	734	24.7
13	519	CHOR	0.051	337	42400	410	9.6	20	4	444	214	101	7410	69	104	758	24.1
14	190	CHOR	0.049	766	41300	903	9.5	22	4	766	210	22	6940	413	98	694	24
15	747	CHOR	0.049	66	40800	94	8.8	27	4	133	198	50	6850	397	95	662	23.3
16	734	CHOR	0.046	747	40600	571	8.7	47	4	397	192	664	6810	55	93	355	23.1
745	483	CHOR	0.002	584	3170	707	0.1	775	< 3	469	11.2	271	290	198	8	479	2.2
746	484	CHOR	0.002	342	3160	761	0.1	776	< 3	701	11.2	573	280	210	8	163	2.1
747	487	CHOR	0.002	434	3110	764	0.1	777	< 3	423	11.1	244	260	217	8	52	2
748	490	CHOR	0.002	544	3050	78	< 0.1	778	< 3	176	10.9	512	260	228	8	122	2
749	513	CHOR	0.002	59	3040	105	< 0.1	779	< 3	434	10.9	19	250	454	8	469	2
750	527	CHOR	0.002	117	3020	199	< 0.1	780	< 3	52	10.5	157	250	495	8	597	2
751	536	CHOR	0.002	275	2980	275	< 0.1	781	< 3	548	9.7	578	220	510	8	728	2
752	593	CHOR	0.002	548	2980	283	< 0.1	782	< 3	289	9.3	41	200	622	8	228	1.9
753	599	CHOR	0.002	490	2920	289	< 0.1	783	< 3	454	9.3	737	200	41	7	59	1.8
754	654	CHOR	0.002	282	2910	367	< 0.1	784	< 3	394	9.2	95	190	186	7	217	1.8
755	701	CHOR	0.002	707	2850	374	< 0.1	785	< 3	670	9.2	268	160	195	7	474	1.8
756	733	CHOR	0.002	253	2800	459	< 0.1	786	< 3	59	9	68	150	208	7	701	1.8
757	773	CHOR	0.002	701	2750	474	< 0.1	901	< 3	217	9	196	150	523	7	440	1.7
758	59	CHOR	0.001	160	2430	490	< 0.1	902	< 3	707	8.7	497	150	552	7	657	1.7
759	362	CHOR	0.001	469	2120	492	< 0.1	904	< 3	228	8.6	665	130	733	7	612	1.6
760	618	CHOR	0.001	228	1840	502	< 0.1	905	< 3	490	4.7	696	110	665	5	490	1.2
761																	

Figure 10.1 Screenshot of the Kola Project C-horizon soil data resorted variable by variable according to decreasing analytical results. Samples with "special signs" like "NA" or "<" are sorted to the top or bottom of the array when using Excel™

In a sorted table it is possible to observe at once how many samples (if any) are below the detection limit, e.g., no samples for Al, 13 samples for As, and 686 samples for B (Figure 10.1). It is also visible whether there are unusual numbers of samples returning the same result, whether rounding of results by the laboratory (e.g., to integer numbers like 1, 2, 3, ... or even 10, 20, 30, ...) has caused artificial structures in the data. For an example, see variable Al where all values above 10 000 are rounded to the nearest "100", and the data above 1000 to the nearest "10". Any unusual "gaps" with no analytical results (e.g., two samples marked NA for the variable Ag) are visible. It can be detected how far removed from the main set of data the high (or low) values are. The number of unusually high (or low) values is visible at one glance (see for example the maximum value for Ca) (Figure 10.1). Some "unlikely" values, such as the very high results for Cd, will also become far more obvious than in an unsorted table (Figure 10.1). Checking with the original data file as received from the laboratory will, in this case, reveal that Cd was reported in "μg/kg" and that it was forgotten to recalculate these values to "mg/kg", so those Cd values still need to be divided by 1000.

10.2   Non-linear transformations

10.2.1   Square root transformation

For count data, e.g., numbers of mineral grains or geochemical data where the concentration of an element is a direct function of the number of rare mineral grains present, it is often possible to approach normal distribution via a simple square root transformation. Instead of the original data, the distribution of the square root of the result for each sample is studied and used for further data analysis. This reflects the fact that count data follow a Poisson distribution (Bartlett, 1947; Krumbein and Graybill, 1965; Weissberg, 1980) and that the square root transform is the generally accepted procedure for normalising such data.

10.2.2   Power transformation

Instead of applying the square root transformation, which is a special case of the power transformation $(x^{1/2})$, a transformation to any other power can sometimes be useful to approach a normal distribution. Even negative powers can be used, e.g., the inverse of the square root transformation $(x^{-1/2})$. In the special case of a power of zero the power transformation is defined as $\log(x)$. Power transformations require that the data are positive, which is the usual case in environmental sciences and applied geochemistry. If negative data are present, these must first be brought into a positive range via addition of a suitable constant. The choice of an appropriate power for the transformation is the objective of the Box–Cox procedure (see Section 10.2.4).

Figure 10.2 shows as an example the original, strongly right-skewed distribution for Cr from the Kola C-horizon data set, and the effect of power transformation with powers of $+\frac{1}{4}$ and $-\frac{1}{4}$ on the distribution. Both transformations result in more symmetrical distributions, however, some skewness still exists.

10.2.3   Log(arithmic)-transformation

In geochemistry a log-transformation is often used to approach a normal distribution. A log-transformation will reduce the very high values and spread out the small data values and

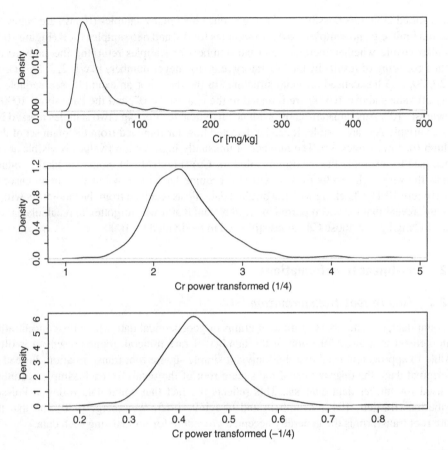

Figure 10.2   Density trace for Cr of the Kola C-horizon soil data set. Upper diagram: original data; middle diagram: power transformed data with power $+\frac{1}{4}$; lower diagram, power transformed data with power $-\frac{1}{4}$

is thus well suited for right-skewed distributions. In practice there are two commonly used log-transformations, the transformation to the natural logarithm (ln) and the transformation to base 10 logarithms (log10). Some statistical data inspection techniques (e.g., the CP-plot) benefit from a logarithmic transformation. If the objective is to check whether data points follow a straight line, either a ln or a log10 transformation will suffice for a CP-plot. Otherwise, when working with environmental data, a transformation to the base 10 logarithms may be the better choice, because it is considerably easier to relate to the original data, as it is common knowledge that 0-1-2-3 on the log10-scale corresponds to 1-10-100-1000 in the original scale.

Data transformations are also often carried out to make different variables more comparable. For example, when plotting Cr versus Cu for the Kola C-horizon soil data, the majority of points fall in the lower left corner of a plotting area that is dominated by a few extreme values. Plotting the log-transformed values, the data points are spread over the whole plotting area, and the structure of the two-dimensional data can be studied in far greater detail (see Figure 10.3).

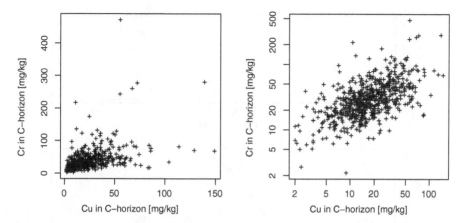

Figure 10.3 Scatterplot for Cu and Cr from the Kola Project C-horizon soil data set. Left diagram: original data scale; right diagram: log-scaled data

10.2.4 Box–Cox transformation

The algorithm for the Box–Cox transformation (Box and Cox, 1964) estimates the power that is most likely to result in a normal distribution when applied to the given data set. This is an extremely flexible transformation, and as noted above, includes a logarithmic transformation for the special case of the estimated power being zero. In reality the power may not transform the data to perfect normality, however, the estimate is the power that brings the data closest to normality. If the parameter is close to some specific value, for example -1, 0, $\frac{1}{3}$, 0.5, it is often sufficient to use a reciprocal (-1), logarithmic (0), cube root $(\frac{1}{3})$, or square root (0.5) transform, respectively. This is most appropriate when there is evidence that the underlying physical or chemical processes controlling the data are best modelled by a specific distribution normalised by that power.

As an example, the variable Fe_XRF from the Kola C-horizon soil data set can be used. The Box–Cox transformation estimates a power of 0.31. There is the choice of using a log-transform or a power transformation. Using a log-transformation, a Shapiro–Wilk test for normality (see Chapter 9) returns a p-value of $4 \cdot 10^{-6}$, and thus the hypothesis that the data follow a normal distribution is rejected. Applying a power transformation with the value of $\frac{1}{3}$ to the data, the Shapiro–Wilk test delivers a p-value of 0.057 and the hypothesis that the data follow a normal distribution cannot be rejected. Figure 10.4 shows this example in the form of density traces. The direct relation to the unit of the measurements is lost under power transformation. This can be overcome by an appropriate transformation of the scale of the x-axis (not routinely provided in DAS+R, but possible in R).

10.2.5 Logit transformation

To approach a normal distribution for proportional data, the logit transformation (Berkson, 1944) performs better than the log-transformation in many instances. It is particularly appropriate for proportions or probabilities as it opens up the data to an unbounded scale, though zero is not permitted. To carry out a logit transformation, if the data are not already

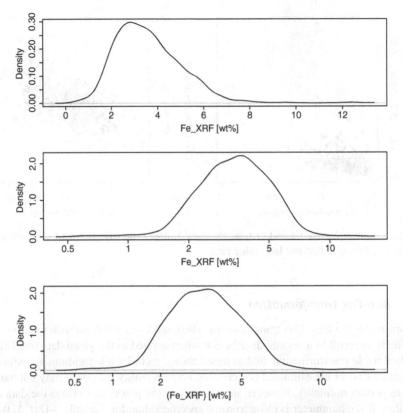

Figure 10.4 Density trace for Fe_XRF from the Kola C-horizon data set. Upper diagram: original data; middle diagram: log-transformed data; lower diagram: power transformed data with power 0.31

on a 0–1 scale, the data need to be transformed such that they fall into the range between 0 and 1. For data given in wt% they thus need to be divided by 100; data given in mg/kg need to be divided by 1 000 000 and data given in μg/kg need to be divided by 10^9. The data are then divided by the inverse proportion of the data and log-transformed (i.e. if the value is 0.1 the transform is $\ln(0.1/(1 - 0.1))$; if the value is 0.3 the transform is $\ln(0.3/(1 - 0.3))$. Because the relation to the original data is lost, this transformation is not favoured for applied geochemical and environmental data. In addition, when working with skewed data (as is often the case in environmental sciences), the calculation of the inverse proportion will increase the skewness of the distribution, which is not a desirable result. The logit transformation will perform much better than the log-transformation when dealing with uniformly distributed data, or data that are expressed as proportions or probabilities.

10.3 Linear transformations

10.3.1 Addition/subtraction

For centring the data, or more generally moving the origin for the data, it can be useful to add or subtract a constant to/from all values. Figure 10.5 shows the original right-skewed

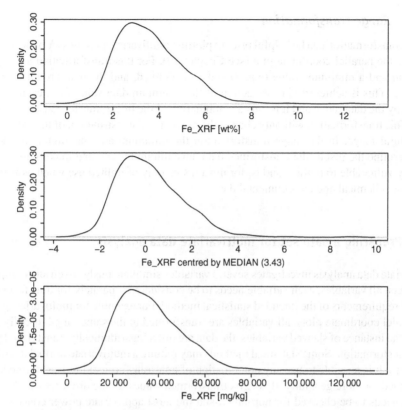

Figure 10.5 Density trace for Fe_XRF from the Kola C-horizon soil data. Upper diagram: original data; middle diagram: the MEDIAN is subtracted from all samples to move the centre of the data to zero; lower diagram: all values are multiplied by 10 000 to change the unit from wt% to mg/kg

distribution of the variable Fe_XRF from the Kola C-horizon soil data. When subtracting the MEDIAN (3.43) from all samples, the centre of the distribution is shifted to zero but the shape of the distribution does not change (Figure 10.5, middle). Centring of data is required for many multivariate methods.

10.3.2 Multiplication/division

To change the data range, e.g., for scaling, it can be useful to multiply/divide all values by a constant. This transformation needs to be applied quite often to one, or two, of the selected variables when plotting ternary diagrams. Figure 10.5, lower diagram, shows that when multiplying the values with 10 000 (to shift the unit from wt% to mg/kg), the scaling of the diagram changes, but the distribution remains the same.

In another instance of scaling, the constant could be a measure of data spread (dispersion), e.g., standard deviation (SD) or median absolute deviation (MAD) (see Section 10.4.2).

10.3.3   Range transformation

A range transformation can be helpful prior to plotting multivariate graphics. A typical example would be the parallel coordinate plot (see Chapter 13). For this transformation a range via a maximum and a minimum value (e.g., 0 and 1) is defined, and the data are all forced into this range. This is achieved via subtraction of the minimum data value from each value and division by the data range and it is thus a combination of the linear transformations mentioned above. This transformation will only change the range, not the distribution of the values within their natural range. In the range transformation the minimum and maximum values of the data determine the result (the transformed data have minimum $= 0$ and maximum $= 1$). It is thus very vulnerable to outliers and is, for this reason, only of limited use when working with applied geochemical and environmental data.

10.4   Preparing a data set for multivariate data analysis

Multivariate data analysis investigates several variables simultaneously. To ensure comparability between all variables, each variable needs to be considered separately for transformation to fulfil the requirements of the intended statistical method. For example for multivariate graphics like parallel coordinate plots, all variables are transformed to the same range (usually zero to one). In the instance of skewed variables, the data are often logarithmically transformed prior to range transformation. Some statistical methods may require a multivariate normal distribution. In this case it is possible to use any combination of non-transformed, log-transformed, power transformed, or more generally, Box–Cox transformed values for the different variables. Each variable needs to be checked for normality, and the most appropriate power constant should be chosen.

Many multivariate methods require that the data have a common origin and that the spread of each variable is the same. This is achieved via centring and scaling the data.

10.4.1   Centring

For some multivariate data analysis techniques (e.g., Principal Components Analysis – PCA, Factor Analysis – FA, see Chapter 14), it is helpful if the origin of the coordinate system of the data (not the spatial coordinates) is moved to the centre of the multivariate data cloud. This is done via calculating the average (MEAN, MEDIAN or whatever is appropriate for the data at hand) of each transformed variable and subtracting it from each single value. Centring is thus a simple linear transformation (subtraction).

10.4.2   Scaling

Even after transformation and centring the data may still not be directly comparable because the spread of the variables selected for analysis is different. For example, the Kola data set consists of a combination of major elements (e.g., Al, Ca, Fe, K, Na, Mg, Si) and trace elements (e.g., Ag, As, Co, Cr, Cu) that have very different concentration ranges. Major elements occur in concentrations of 10 000 to several 100 000 mg/kg. Trace elements occur at concentrations far below 1000 mg/kg. Major elements would completely dominate the results of some multivariate techniques (e.g., PCA, FA, cluster analysis) if no allowance was made. The variables thus need to be brought to a comparable scale.

This could be done via a range transformation, however, as mentioned in Section 10.3.3, the result of a range transformation depends critically on the minimum and maximum concentration of each variable, and is thus extremely vulnerable to data outliers. The SD of a variable is a measure of its scale (spread or dispersion). Thus a better way to scale the data will be to divide all values of each variable following its centring by its SD (or, preferably, the robust measure MAD, as outliers are common in environmental data). Therefore, scaling most frequently involves applying a simple linear transformation (multiplication or division). Centring and scaling data is often referred to as data standardisation.

This can be easily demonstrated through plotting the density traces of two variables that have different concentration ranges, e.g., Fe_XRF and Cr from the Kola C-horizon soil data (Figure 10.6). If the samples are only log-transformed a symmetrical distribution is achieved, but the centres differ by orders of magnitude (Figure 10.6, upper plot). Centring the data still leaves the width and height of the two distributions uncomparable, the scaling is clearly

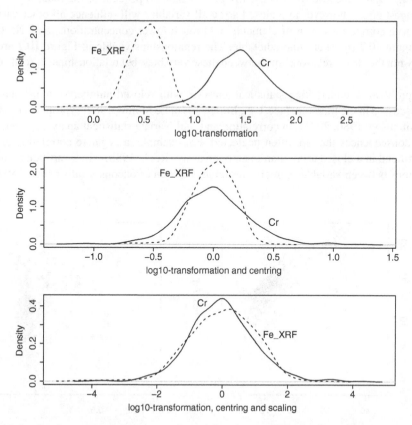

Figure 10.6 Density traces of the log-transformed data for Cr and Fe_XRF from the Kola C-horizon soil data (top), the log-transformed and centred data (middle) and the log-transformed, centred and scaled data (bottom)

different (Figure 10.6, middle plot). After scaling both variables, the density traces are far more comparable (Figure 10.6, lower plot).

10.5   Transformations for closed number systems

A closed array or closed number system is a data set where the individual variables are not independent of each other but are related by, for example, being expressed as a percentage (or parts per million – ppm (mg/kg)). They sum up to a constant, e.g., 100 per cent or 1. To understand the problem of closed data, it just has to be remembered how percentages are calculated. They are ratios that contain all variables that are investigated in their denominator. Thus, variables expressed as percentage data are not free to vary independently.

This can be easily seen in Figure 10.7 by plotting SiO_2, which is often the major component in a rock or soil analysis, against other major or minor components – the apparent strong negative relationships are most likely caused by the simple fact that as the major component (SiO_2) increases, all other components will decrease because the sum is fixed (100 per cent). Such negative relationships will result in a forced negative correlation (see, e.g., Rock, 1988). Because SiO_2 is usually greater than 50 per cent of the total, it will show the strongest effect, however in a closed array all variables will influence all other variables and the true correlations even of elements with much lower concentrations, e.g., Na_2O and K_2O (Figure 10.7 right) are unpredictable. The relationships visible in Figure 10.7 are thus probably not the "true" relationships between these variables, but relationships "forced" due to closure.

The problem of undertaking statistical analyses with "closed number systems" has been much discussed in the literature (see, e.g., Butler, 1976; Le Maitre, 1982; Woronow and Butler, 1986; Aitchison, 1986, 2003). In correlation and multivariate statistical analyses, it has quite serious consequences that are often neglected. For example, a negative correlation is more often artificial in a closed array than in an open array (Rock, 1988). High "artificial" internal correlations between variables can give unstable and even erroneous results (Rock, 1988).

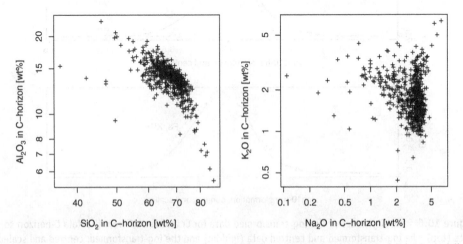

Figure 10.7 Scatterplots of SiO_2 versus Al_2O_3, and Na_2O versus K_2O for Kola Project C-horizon soils

To open the data and destroy the effects of closure, various forms of data transformation are available. The pioneer work on addressing the closure issue was undertaken by Aitchison in the 1980s (Aitchison, 1986, 2003), and additive logratio and centred logratio transformations derive from that work. Egozcue *et al.* (2003) introduced the isometric logratio transformation which has good mathematical and geometric properties.

10.5.1 Additive logratio transformation

The additive logratio transformation is a simple and straight-forward procedure. One of the variables in the data set is "sacrificed" to open the data array. The transformation itself consists simply in dividing each analytical result of each sample by the analytical result of the variable that is to be "sacrificed" (e.g., TiO_2). The logarithm of all results is then used for further work. Common choices in geochemistry for selection of this variable have been the minor and trace elements Ti (TiO_2) or Zr, based on the fact that in some circumstances during various geochemical processes these elements are believed (careful, in many cases in environmental applications this is not a scientific fact but wishful thinking) to be conserved; it is hoped that they are neither lost nor accumulated. However, often the minor and trace elements are being measured closer to their analytical detection limits, and thus include relatively large analytical variability, i.e. imprecision. Furthermore, at those lower levels the data are often influenced by discretisation, i.e. rounding to a relatively few actual values (see sorted table in Figure 10.1). So the petrologically best choice may not be the best computationally. It can be advantageous to experiment with a number of possibilities for this variable.

Igneous petrologists have long recognised the closure problem in bivariate data plotting. Simple element or oxide versus element or oxide plots for major and minor element data are known as Harker diagrams. Pearce (1968) introduced the element-ratio plot, where the oxides or elements to be plotted were first divided by a common third element or oxide. The parallel to additive logratioing is immediately apparent. Figure 10.8 displays the Pearce-element-ratios for SiO_2 vs. Al_2O_3 and Na_2O vs. K_2O for Kola Project C-horizon soils using TiO_2 as the divisor with logarithmic scaling; these are the same data as displayed in the traditional scatterplots

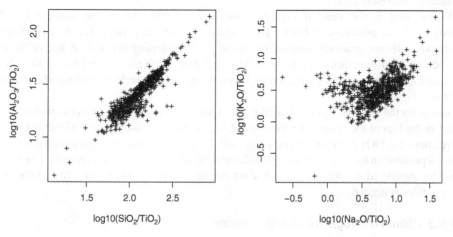

Figure 10.8 Scatterplots of SiO_2 versus Al_2O_3, and Na_2O versus K_2O after additive logratio transformation using TiO_2 as divisor

(Harker diagrams) in Figure 10.7. The difference between the relationships displayed in Figure 10.7 and 10.8 should convince any reader that the closure effect can be a serious problem when working with geochemical (compositional) data. A serious problem with the additive logratio transformation is that a different choice of the ratioing variable can result in quite different diagrams.

It is now clear that there is in fact a positive relationship between SiO_2 and Al_2O_3 (Figure 10.8, left plot) and not the negative relationship indicated in Figure 10.7. This is not surprising to a mineralogist. Most common rock-forming minerals and their weathering products contain both Si and Al. Quartz is the only abundant common rock-forming mineral that contains only SiO_2.

The Pearce plot for Na_2O vs. K_2O (Figure 10.8, right) reveals that two relationships are present: a dominant one where K_2O increases with Na_2O and a minor one where increasing K_2O occurs with decreasing Na_2O. The first is typical of the major relationship present for most rocks where Na and K increase as Ca and Mg decrease. The latter minor relationship is related to the rare alkaline intrusive rocks present in the Kola area where K increases relative to Na.

A disadvantage of the method is that the direct relation to the original data and the unit is lost under this transformation. However, as the objective is to correctly visualise and quantify the bivariate and multivariate relationships this is not a great hindrance. Furthermore, the ratioing variable is lost for further data analysis.

10.5.2   Centred logratio transformation

An alternative way of opening the data is by dividing each value of a variable for an individual by the geometric mean G of all the variables for that individual and then taking logarithms. It is critical in this transformation that all the variables should be expressed in the same measurement unit. In practice it is easiest to log-transform the data first and then subtract the mean of the logarithms of the data for each individual (row of the data matrix). This procedure has the advantage that a divisor does not have to be selected. However, it has the disadvantage that the resulting covariance matrix (see Chapter 11) cannot be inverted as it is numerically singular, which thus makes it impossible to undertake some multivariate statistical analyses with centred logratio transformed data.

Figure 10.9 is the centred logratio transformed equivalent of the additive logratio transformed data presented in Figure 10.8. As can be seen the patterns for the Kola Project C-horizon soils are generally similar, graphically demonstrating the equivalence of the two procedures. The dispersion in the SiO_2 vs. Al_2O_3 relationship is greater, and the Na_2O vs. K_2O relationship due to underlying common rock forming mineral and feldspar relationships is even clearer.

Just as for the additive logratio transformation, the centred logratio transformation will also result in the loss of the relation to the original data and unit, however, no variable is lost, the correlations for TiO_2 can thus be studied as well. Because the centred logratio transformation is not dependent on the results of a single selected variable, but uses the average of all variables, it can be considered the preferable method for opening a data set whenever a direct relation to the variables is needed.

10.5.3   Isometric logratio transformation

The isometric logratio transformation (Egozcue *et al.*, 2003) overcomes the singularity problem of the centred logratio transformation. Thus it is now possible to calculate the covariance

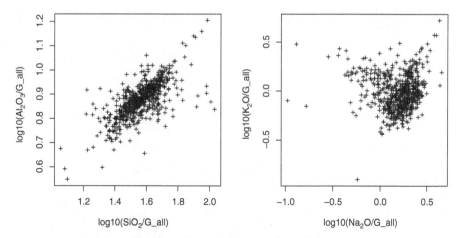

Figure 10.9 Scatterplots of SiO_2 versus Al_2O_3, and Na_2O versus K_2O after a centred logratio transformation. G_all is the geometric mean of all variables for each sample

matrix and its inverse. Moreover it has preferable geometric properties (Egozcue *et al.*, 2003). However, during this transformation the dimension of the data set is reduced by one and the direct relation to the original variables is completely lost, although it is possible to back-transform the data. Due to this fact the above examples can no longer be plotted and used. Advanced multivariate techniques can be applied to the transformed data and the results can be back-transformed to the original data space, allowing interpretation.

10.6 Summary

Data transformations are frequently used in graphical and more formal statistical data analyses. They will reduce the influence of some unusual observations and draw attention to the main body of data or result in improved comparability between the variables. For environmental data, which are characterised by the existence of data outliers and most often right-skewed data distributions, experience shows that a log-transformation is in the majority of cases advantageous for data analysis. It is always prudent to check the results of a data transformation graphically. Furthermore, care should be taken in subsequent data analyses to ensure that a transformed variable is clearly indicated as such.

The majority of advanced statistical methods will not function well with compositional, i.e. closed, data. Note that closure effects can seriously affect even simple scatterplots. Notwithstanding this, experience has shown that if the data have generally similar levels, i.e. no one or few variables have wide data ranges across the composition, the traditional statistical approaches will usually provide reasonable results. Whenever needed, additive logratio, centred logratio or isometric logratio transformations can be used to open such closed data sets, and it is informative to investigate the difference that such transformations can make. These transformations also have some serious shortcomings: the relation to the data unit or the original data is lost, results may depend on the variable chosen to calculate the ratios (additive logratio transformation), or the resulting covariance matrix (see Chapter 11) cannot be inverted as it is numerically singular (centred logratio transformation).

11

Correlation

Correlation analysis estimates the extent of the relationship between any pair of variables. The covariance is a measure of this relationship and depends on the variability of each of the two variables. Because covariances can take any number, only the sign (+ or −) is informative; the strength of the relation between the two variables, however, cannot be interpreted. To obtain standardised numbers, the correlation coefficients, it is necessary to eliminate the dependency on the variability of each variable.

Three widely-used methods to calculate a correlation coefficient are the Pearson, Spearman and Kendall correlation (Galton, 1890; Spearman, 1904; Kendal, 1938). These methods result in a number between −1 and +1 that expresses how closely the two variables are related, ±1 shows a perfect 1:1 relationship (positive or negative) and 0 indicates that no systematic relationship exists between the two variables. A correlation of ±0.5 usually indicates a significant relationship, but this depends on the number of samples. With increasing sample numbers, decreasing absolute values of the correlation coefficients become significant. Relationships between two variables are best visualised in a scatterplot (Section 6.1). However, when working with many variables, scatterplots occupy a lot of physical space. Reducing the information of multiple scatterplots to one number per plot may simplify studying the relationships (similarly to using location and spread measures to characterise the data distribution). If correlation analysis results are presented without a name for the method, usually the Pearson correlation coefficient has been computed.

Some other methods will be presented below because applied geochemical and environmental data are often plagued with outliers. The Pearson method is very sensitive to data outliers as it is dependent on estimates of the variance and covariance, and the sensitivity of variance to outliers has been discussed in Chapter 4. Correlation coefficients are one starting point for multivariate data analysis. A reliable set of correlation coefficients or covariances are of utmost importance for all correlation-based multivariate data analysis methods, such as Principal Component Analysis and Factor Analysis (Chapter 14) or Discriminant Analysis (Chapter 17).

Correlation analysis and correlation based multivariate methods are very popular in environmental sciences and geochemistry. Great care is, however, necessary when attempting any correlation analysis with compositional data (Aitchison, 1986, pp. 48–50). The interpretation of correlation coefficients may be misleading, even when appearing quite convincing. While opening the data provides a solution, the resulting coefficients do not reflect the correlations

Statistical Data Analysis Explained   Clemens Reimann, Peter Filzmoser, Robert G. Garrett, Rudolf Dutter
© 2008 John Wiley & Sons, Ltd.

of the original variables. Careful attention needs to be paid to these issues when correlation analysis is used with compositional data.

11.1 Pearson correlation

The "classical" Pearson method (Galton, 1889, 1890) is ideal if the data follow a bivariate normal distribution (the resulting density function has the form of a three-dimensional bell), an unlikely reality with environmental or geochemical data. A correlation coefficient of ± 1 indicates a perfect linear relationship between the two variables, when $+1$ the variables vary sympathetically, when -1 the relationship is antipathetic, i.e. as one variable increases the other decreases. The Pearson method is very vulnerable to data outliers of any kind, not only very high (or low) values but also deviations from the main structure of the data (see Figure 11.1). Thus, when using the Pearson method, geochemical data will usually need to be transformed to approach the required bivariate normal distribution. Using the Pearson method therefore requires a careful study of the univariate distribution of each of the variables prior to estimating the correlation coefficient. When working with geochemical data, it may be wise to routinely investigate a log-transformation of the data prior to calculating Pearson correlation coefficients.

The scatterplot in Figure 11.1 shows the relationship between Be and Sr in Kola Project C-horizon soils. Because some outliers dominate the plot, the main body of data is almost invisible in the lower left corner. The outliers are caused by the existence of a particular bedrock (alkaline intrusives) in the survey area, which exhibit high concentrations of both Be and Sr. It is impossible to see any relationships in the main body of data. The Pearson correlation coefficient responds to the strong linear relationship imposed by the few outliers, resulting in a very high coefficient of 0.9. After log-transformation, a considerable number of outliers still exist resulting in a linear relationship (Figure 11.1 right), while the main body of data, whose structure is now visible, shows at best a weak relation between the two variables. It is possible to see in Figure 11.1 (right) that there may be two relationships present. The dominant one running diagonally across the plot, and a second one with a lower slope at levels below 1 mg/kg Be. The Pearson correlation coefficient is now 0.66 and still strongly influenced by the outliers.

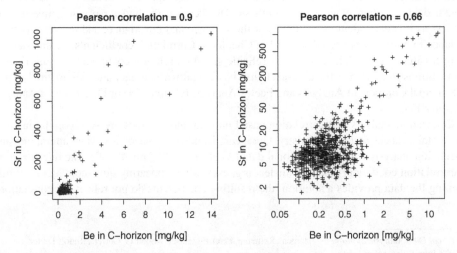

Figure 11.1 Scatterplot and Pearson correlation coefficient between Be and Sr, Kola Project C-horizon soil data. Left plot: untransformed data; right plot: log-transformed data

11.2 Spearman rank correlation

The Spearman rank method (Spearman, 1904) provides a non-parametric (distribution free) measure of correlation between two variables. It does not call for a linear relation, but requires that the association is monotonically (steadily) either increasing or decreasing. Searching for a monotonic relationship is far more general, and less restrictive, than searching for a linear one. However, in the (rare) case of a bivariate or multivariate normal distribution the Pearson method performs better because it is more precise. Again a correlation coefficient of ± 1 indicates a perfect monotonic relationship. In the Spearman coefficient it is the ranks of the sorted values that determine the result, not the actual data values. Thus the data are first ranked (sorted), and the Pearson correlation of the ranks of the data is then computed. This is the reason that the Spearman rank correlation coefficient is relatively robust against data outliers (see discussion in Section 10.1). One of the important advantages of the Spearman rank correlation is that results will be the same for the original data or any linear transformation, as the transformation does not disrupt the order of the data-values from lowest to highest. Thus, log-transformation is unnecessary when estimating the Spearman rank correlation.

Using the example of the correlation of Be with Sr (Figure 11.1) the Spearman rank correlation coefficient is 0.37, a much lower value than the Pearson correlation coefficient for the log-transformed data of 0.66. Figure 11.2, where the ranks of Be are plotted against the ranks of Sr, demonstrates why the resulting correlation coefficient is lower than the Pearson correlation of the original data. The Spearman rank correlation coefficient is far less influenced by the outliers and thus probably provides a more realistic estimate of the correlation in the main body of data. The outliers are, however, still visible in the upper right corner of the plot (Figure 11.2, left) because they have the highest rank in both variables. The Spearman rank correlation is thus to a certain extent still influenced by outliers, however, their "leverage" is reduced by using ranks rather than absolute values. The plot of the ranks of La versus Ce in the C-horizon provides a graphical impression of a high Spearman rank correlation. Note again that one of the main advantages of the Spearman rank method is that the data do not need to be transformed.

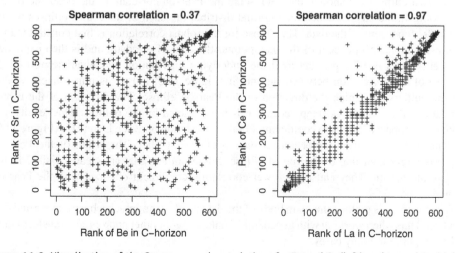

Figure 11.2 Visualisation of the Spearman rank correlations for Be and Sr (left) and La and Ce (right)

11.3   Kendall-tau correlation

The Kendall-tau method (Kendal, 1938) is quite similar to the Spearman rank method. It also measures the extent of monotonically increasing or decreasing relationships between the pairs of variables. However, it uses a different method of calculation (looking at the sign of the slope of the line connecting each existing pair of points, summing the signs, and dividing the result by the number of pairs). The method is relatively robust against data outliers – as long as the sign of the slope does not change the result will stay the same. Thus Kendall-tau is widely independent of the actual values of the data, and a linear transformation will not alter the estimated correlation coefficient. For the example of the correlation of Be and Sr the Kendall-tau correlation coefficient is 0.27, an even lower value than the Spearman rank correlation of 0.37. The Kendall-tau correlation cannot be visualised in an easy graphic. To plot the Kendall-tau correlation would require connecting all possible pairs of data points by lines and to study their slopes. For the complete C-horizon data set this would require plotting 183 315 lines.

11.4   Robust correlation coefficients

Although Spearman rank and Kendall-tau methods are relatively insensitive to data outliers, they are still not robust and can be easily influenced by unusual data. In contrast, robust methods focus on the main body of data and are not influenced by deviating data points. A robust correlation can be constructed via a robust covariance estimator. One possibility is to take the Minimum Covariance Determinant (MCD) estimator (Rousseeuw and Van Driessen, 1999) which searches for the most compact ellipse containing, e.g., 50 per cent of the data points. The percentage can be selected between 50 and 100 per cent – at 100 per cent the robustness of the estimator is, of course, lost. The points outside of the ellipse can be located anywhere in the data space without influencing the robust correlation coefficient for the data points included in the most compact, e.g., 50 per cent, ellipse. Both direction and shape of the ellipse are an indication of the correlation.

Figure 11.3 demonstrates the idea using the Kola Project C-horizon soil data used above. Two ellipses are shown, one represents the Pearson correlation, the other the robust correlation. In the case of a bivariate normal distribution, each of the drawn ellipses would contain 97.5 per cent of the data. The ellipse for the robust correlation is first constructed for the most compact 50 per cent of the data points (the core of the data) and is then expanded to cover an area of 97.5 per cent as if the underlying distribution was bivariate normal. In the case of the correlation between Be and Sr (Figure 11.3, left), it is immediately visible that the outliers determine the direction of the Pearson ellipse. The direction of the robust correlation ellipse is not as steep because it is not attracted by the outliers. This results in a much lower robust correlation coefficient. For the correlation between La and Ce (Figure 11.3, right plot) the two ellipses plot almost on top of one another. The width of the Pearson ellipse is larger than that of the robust ellipse. Some outliers that are clearly visible in the plot are the reason. They cause the lower correlation coefficient obtained with the Pearson method.

When using this method the main body of the data needs to approach a bivariate normal distribution. Thus when working with geochemical data, a log- or other appropriate transformation will be required in many cases.

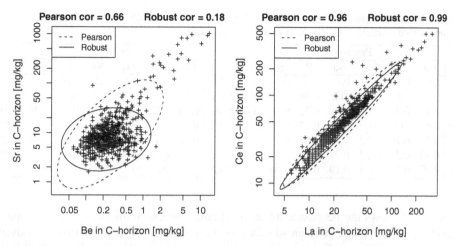

Figure 11.3 Visualisation of the Pearson correlation and the robust correlation of Be versus Sr and La versus Ce

11.5   When is a correlation coefficient significant?

Correlation coefficients vary in the range −1 to 0 to +1. When studying a correlation table, one of the first questions is "which of these coefficients indicates a clear relation between two variables?" Statisticians tend to formulate such a question differently. They will formulate a null hypothesis and an alternative hypothesis (see Chapter 9):

Null hypothesis: *the correlation coefficient is zero.*
Alternative hypothesis: *the correlation coefficient is different from zero.*

Different tests have to be used for the different methods (see, e.g., Hollander and Wolfe, 1973). The number of observations and the correlation coefficient both play an important role in the tests. The larger the sample size, n, the lower the correlation coefficient needs to be to accept the hypothesis that a pair of variables are uncorrelated at a certain significance level (e.g., 0.05 or 0.01). For the above examples (Be versus Sr, and La versus Ce) all tests result in very small p-values (10^{-16}), i.e. all the estimated correlation coefficients are significantly different from zero. When working with correlations, it is usually of more interest to look for strong relationships (e.g., correlation > 0.7 or < -0.7) than to test against zero correlation and to estimate the significance of the relationship. The scatterplot matrix as shown in Figure 6.14 is a far better tool to visualise close relationships between any two pairs of variables.

11.6   Working with many variables

When working with a multivariate data set, it is necessary to study the relationship between all pairs of variables simultaneously. The results are usually displayed in tabular form, the correlation matrix.

Table 11.1 Pearson (left) and Spearman rank (right) correlation matrix for selected log-transformed elements from the Kola C-horizon soil data set

	Pearson								Spearman						
	Ca	Cu	Mg	Na	P	Sr	Zn		Ca	Cu	Mg	Na	P	Sr	Zn
Ca	1	0.26	0.31	0.65	0.61	0.52	0.07	Ca	1	0.29	0.26	0.79	0.62	0.7	0.08
Cu	0.26	1	0.75	0.2	0.19	0.23	0.63	Cu	0.29	1	0.77	0.26	0.18	0.35	0.66
Mg	0.31	0.75	1	0.09	0.22	0.22	0.66	Mg	0.26	0.77	1	0.16	0.21	0.39	0.71
Na	0.65	0.2	0.09	1	0.25	0.77	0.19	Na	0.79	0.26	0.16	1	0.38	0.63	0.05
P	0.61	0.19	0.22	0.25	1	0.27	0.16	P	0.62	0.18	0.21	0.38	1	0.47	0.2
Sr	0.52	0.23	0.22	0.77	0.27	1	0.48	Sr	0.7	0.35	0.39	0.63	0.47	1	0.41
Zn	0.07	0.63	0.66	0.19	0.16	0.48	1	Zn	0.08	0.66	0.71	0.05	0.2	0.41	1

Table 11.1 shows the Pearson and Spearman rank correlation matrices for a selection of elements from the Kola C-horizon soil data set. Results obtained from the two methods are sometimes quite different. Because the correlation between variable A and variable B is the same as the correlation of variable B and variable A, only one half of the complete matrix needs to be shown. As an option, the correlation matrix can be plotted with additional symbols indicating the degree of significance (the p-value) of the correlation for each pair of variables, e.g., *** for very high significance ($p < 0.001$), ** for high significance ($0.001 \leq p < 0.1$), * for significance ($0.1 \leq p < 0.05$), · for weak significance ($0.05 \leq p < 0.10$), and no symbol

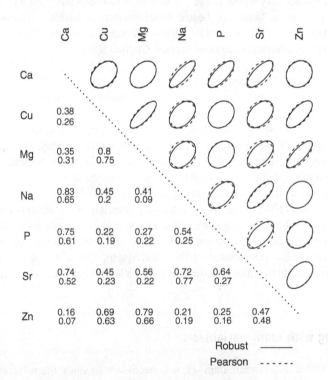

Figure 11.4 Correlation coefficients and correlation ellipses. Upper number: robust (MCD-based) correlation; lower number: Pearson correlation

Table 11.2 Pearson (left) and Spearman rank (right) correlation matrices for selected log-transformed elements from the Kola C-horizon soil data set. The degree of significance of the correlation of each pair of elements is indicated by stars, *** for very high significance ($p < 0.001$), ** for high significance ($0.001 \leq p < 0.1$), * for significance ($0.1 \leq p < 0.05$), · for weak significance ($0.05 \leq p < 0.10$), and no symbol for no significance ($p \geq 0.10$)

			Pearson									Spearman				
	Ca	Cu	Mg	Na	P	Sr	Zn		Ca	Cu	Mg	Na	P	Sr	Zn
Ca	1	***	***	***	***	***	·	Ca	1	***	***	***	***	***	*
Cu	0.26	1	***	***	***	***	***	Cu	0.29	1	***	***	***	***	***
Mg	0.31	0.75	1	*	***	***	***	Mg	0.26	0.77	1	***	***	***	***
Na	0.65	0.2	0.09	1	***	***	***	Na	0.79	0.26	0.16	1	***	***	
P	0.61	0.19	0.22	0.25	1	***	***	P	0.62	0.18	0.21	0.38	1	***	***
Sr	0.52	0.23	0.22	0.77	0.27	1	***	Sr	0.7	0.35	0.39	0.63	0.47	1	***
Zn	0.07	0.63	0.66	0.19	0.16	0.48	1	Zn	0.08	0.66	0.71	0.05	0.2	0.41	1

for no significance ($p \geq 0.10$). Table 11.2 shows this version of the Pearson and Spearman rank correlation matrix for the elements in Table 11.1. Many of the pairs in Table 11.2 show a highly significant correlation.

When calculating the robust MCD-based correlation coefficients, the method considers all selected variables at the same time – thus the correlation coefficients are computed in the multivariate space. The MCD-based method provides true multivariate correlation coefficients and is robust against multivariate data outliers that do not need to be extreme values in any of the variables. Table 11.3 shows the robust correlation matrix. At present there exists no significance test for this kind of robust correlation measure.

Figure 11.4 compares the Pearson and the robust MCD-based correlation matrices. The lower half provides a comparison of the coefficients derived by both methods. The upper half shows the ellipses corresponding to the two methods (see Figure 11.3). The ellipses highlight at one glance for which pairs of elements differences can be expected (e.g., P versus Sr).

11.7 Correlation analysis and inhomogeneous data

It has been discussed (e.g., Chapter 1) that for applied geochemical and environmental data, a variety of different factors and processes will influence the analytical result of each variable at each sample site. In Figure 11.1 the samples collected in proximity to the alkaline rocks

Table 11.3 Robust (MCD-based) correlation matrix for selected elements from the Kola C-horizon soil data set

	Ca	Cu	Mg	Na	P	Sr	Zn
Ca	1	0.38	0.35	0.83	0.75	0.74	0.16
Cu	0.38	1	0.8	0.45	0.22	0.45	0.69
Mg	0.35	0.8	1	0.41	0.27	0.56	0.79
Na	0.83	0.45	0.41	1	0.54	0.72	0.21
P	0.75	0.22	0.27	0.54	1	0.64	0.25
Sr	0.74	0.45	0.56	0.72	0.64	1	0.47
Zn	0.16	0.69	0.79	0.21	0.25	0.47	1

show both high Be and Sr concentrations. These few samples are outliers when looking at the complete data set and dominate the resulting correlation coefficients when using Pearson, Spearman rank, and even Kendall-tau methods. Before using correlation-based methods for multivariate data analysis it can be informative to compare the correlation structure for data subsets within a data set (e.g., the parent material lithologies encoded in the C-horizon data set) using correlation analysis.

In Figure 11.5 it is apparent that different subsets in the C-horizon soil data defined according to parent material lithology can yield very different correlation coefficients. Regional geochemical data sets are often extremely inhomogeneous. Usually they consist of a mixture of data from different populations representing different geological and geochemical processes. In different subsets antagonistic data structures may exist. In the case of the Kola C-horizon soil data there is some knowledge concerning the processes governing the data structure. For example, the geological map could be used to construct more homogeneous subsets according to the underlying soil parent material lithology. When this is done and the correlation matrix for each of these subsets is estimated, it becomes clear that almost all of these lithologically-based data subsets have quite different correlation structures (Figure 11.5).

This was demonstrated in Figure 8.14, showing a scatterplot matrix with regression lines for the subsets. It is always informative to look at a scatterplot matrix, plotting all variables against all variables before proceeding to the estimation of a matrix of coefficients. Undertaking correlation analysis with inhomogeneous data will almost invariably lead to unstable results. At an initial univariate data inspection, breaks in ECDF- or CP-plots present evidence for the presence of multiple data populations, and should be a warning that the data may confound

	Ca	Cu	Mg	Na	P	Sr
Cu	0.28					
	0.16					
	0.23					
	0.34					
Mg	0.21	0.76				
	0.2	0.77				
	0.19	0.82				
	0.49	0.72				
Na	0.73	0.24	0.08			
	0.63	0.09	0.08			
	0.68	0.03	−0.16			
	0.6	0.33	0.2			
P	0.67	0.23	0.24	0.32		
	0.61	0.06	0.18	0.14		
	0.57	0.22	0.26	0.35		
	0.58	0.23	0.22	0.26		
Sr	0.55	0.22	0.19	0.75	0.35	
	0.45	0.12	0.13	0.83	0.08	
	0.6	0.32	0.18	0.73	0.53	
	0.56	0.3	0.35	0.76	0.28	
Zn	0.06	0.62	0.69	0.17	0.23	0.48
	−0.06	0.52	0.56	0.27	−0.04	0.5
	−0.02	0.77	0.75	−0.14	0.19	0.28
	0.22	0.67	0.68	0.25	0.23	0.54

Legend box:
Caledonian sediments
Basalts
Alkaline rocks
Granites

Figure 11.5 Pearson correlation coefficients for the four different bedrock-based subsets of the Kola C-horizon soil data set. Caledonian sediments, lithologies 9,10; basalts, lithologies 51,52; alkaline rocks, lithologies 81,82; granites, lithology 7

a satisfactory correlation analysis. Such data behaviour can be expected in the vast majority of cases when working with regional geochemical or environmental data. It is doubtful that correlation analysis is the correct method with which to start the multivariate analysis of such data. A better procedure would be to first disaggregate the total data set into more structurally homogenous subsets, e.g., via a clustering procedure (see Chapter 15) or by the use of already existing knowledge (e.g., the geological map) and then to carry out a correlation analysis for each subset separately.

Correlation analysis is a very data-sensitive method, a fact that is often neglected. A careful univariate analysis of each variable entered into correlation analysis is a necessity.

11.8 Correlation results following additive logratio or centred logratio transformations

As discussed in Chapter 10, when working with geochemical data, one should be aware of the "closed array" problem (see, e.g., Aitchison, 1986, 2003). Major elements determined by XRF are an obvious example of closed array data, they sum up to 100 wt%. However, closure is an inherent problem for all compositional data; it does not matter whether the major components were determined or not, even if only trace elements are used it will in most cases be advantageous to "open" the data array. To carry out correlation analysis with such data

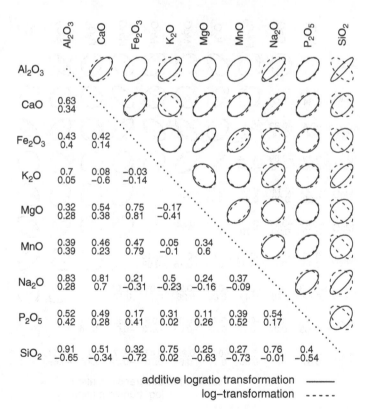

Figure 11.6 Correlation coefficients and correlation ellipses before and after additive logratio transformation. Upper number, additive logratio transformation; lower number, log-transformation

requires that the data are "opened" first, using either an additive logratio (Section 10.5.1), or a centred logratio transformation (Section 10.5.2). With the isometric logratio transformation the relation to the original variables would be lost, this transformation is thus not suitable for studying bivariate correlations.

Figure 11.6 displays the correlation matrix and ellipses for the major element data for the Kola Project C-horizon soils. Note that as TiO_2 has been used as the divisor for the additive logratio transformation, it is no longer in the standard major and minor element geochemical matrix. Moreover the correlation coefficients for the opened data are now related to the ratios to TiO_2 and no longer to the original variables.

The changes in the correlation coefficients when using the additive logratio transformation and the effects of removal of the closure effect are immediately apparent, especially in the correlation ellipses. These new correlation coefficients are far more interpretable in terms of soil and bedrock mineralogy. Of particular note are the complete changes in structure for CaO and K_2O, reflecting the relationship between increasing Ca-rich plagioclase at the expense of K-rich orthoclase; the Na_2O-K_2O relationship discussed earlier; and the many complete changes of relationships with SiO_2, the major compositional component in the soil samples.

A centred logratio transformation (Section 10.5.2) is an alternative method of opening the data. Here the correlations for TiO_2 can also be studied (Figure 11.7). However, again the correlation coefficients no longer relate to the original variables but to the ratios.

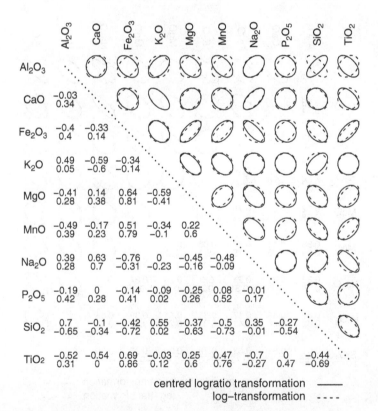

Figure 11.7 Correlation coefficients and correlation ellipses before and after a centred logratio transformation. Upper number, centred logratio transformation; lower number, log-transformation

Because a centred logratio transformation is not dependent on the results of just one other variable as the divisor but uses the geometric mean of all variables it should be the preferable method in many applications (Section 10.5.2). Ratios to one selected component can depend as much, or more, on the selection of that variable and the characteristics of the data for it, as on the remaining data. This could become an important issue when working with spatial data influenced by multiple factors and processes, and result in misinterpretations. The differences between Figures 11.6 and 11.7 are related to such features. The important relationship between the two major components of the C-horizon soils, SiO_2 and Al_2O_3, is that which shows the greatest change after opening the data. Relative to other elements the changes in their relationships with Al_2O_3 are the most apparent. This infers that Si dominantly dilutes the samples due to the abundance of free SiO_2, quartz, in the C-horizon soils and the important role of alumino-silicate weathering products in the soils as hosts for a wide range of elements.

11.9 Summary

The problems with correlation analysis of inhomogenous data were discussed in Section 11.7. An overall correlation coefficient for two variables may not be very informative if several populations are hidden in a data set (e.g., different underlying bedrock types, different vegetation zones, men and women, apples and pears).

When carrying out correlation analysis it must be kept in mind that a high correlation is by no means a proof of a relation between the two variables. There may exist a third, "lurking", variable that the other two variables are independently related to and that causes the high correlation between two otherwise unrelated variables.

Statistical tests for correlation usually test for zero correlation and not for "high" correlation thus even a "highly significant" correlation must by no means indicate a strong relation between the variables. It is also important to note that the significance depends on the number of samples; the more samples, the lower absolute correlation coefficients will be shown as "significant". Before any far-reaching conclusions are drawn from correlation coefficients it is thus advisable to check the correlations in a scatterplot.

Correlation should not be confused with regression (see Section 6.2 and Chapter 16). For correlation the sequence of the two variables plays no role for computing the correlation coefficient. In regression analysis completely different results can be expected when exchanging the two variables (see Figure 16.1), i.e. x for y, and y for x.

Correlation analysis and all correlation-based statistical methods will be seriously affected by the closure effect, always inherent in compositional data, and independent of whether the major components of a sample have been analysed or not.

12

Multivariate Graphics

Most of the graphics presented so far permit the display of up to two variables simultaneously. In a classical map (Chapter 5) it is only possible to visualise one variable at a time. In many instances it may, however, be desirable to study the behaviour of several variables at the same time. For this purpose multivariate graphics have been developed (see, e.g., Chambers *et al.*, 1983; Cleveland, 1994; Everitt and Dunn, 2001), and in this chapter a small selection of procedures that have been proven to be informative in geochemical applications are illustrated (for worked examples see also Garrett, 1983a; Howarth and Garrett, 1986). As with so many displays, if necessary, the original data should be first transformed in order to remove the influence of data outliers and achieve more symmetrical data distributions. The data should then be range transformed (0, 1) so that each variable has equal weight in the plot. In the examples following in this chapter the data have been both logarithmically and range transformed.

Multivariate graphics can be presented in a tabular display, where each observation is identified via its sample number. The resulting table can be used to visually classify the observations according to the appearance of the graphics. The graphics can also be placed at the spatially correct position on a map, the multivariate graphic simply replaces the symbols otherwise used for mapping. One disadvantage of most multivariate graphics is that they take more space than other (e.g., EDA) symbols. This reduces the number of them that can be shown in a map without increasing the map-size dramatically or bringing clutter to an unacceptable level. Small data sets with few variables and few samples are thus the main realm of multivariate graphics. An exception is the parallel coordinates plot (Section 12.6), where a large number of observations can be displayed. Chernoff's faces (Chernoff, 1973) can also accommodate a high number of variables but have not been presented here due to the perceptual biases that can be achieved by manipulating the association of the variables with specific features of the faces (Howarth and Garrett, 1986).

12.1 Profiles

For constructing profiles the selected variables are plotted along the x-axis. For one observation, sample, the log- and range transformed value of each variable is then shown along the y-axis, and a line is used to connect the values. The order of the variables is arbitrary, though the

Statistical Data Analysis Explained   Clemens Reimann, Peter Filzmoser, Robert G. Garrett, Rudolf Dutter
© 2008 John Wiley & Sons, Ltd.

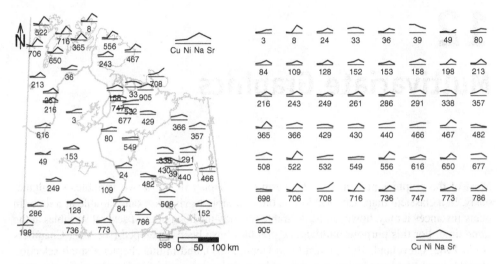

Figure 12.1 Profile plot for 49 selected samples from the Kola O-horizon soil data set. The variables Cu, Ni, Na and Sr were used to construct the profiles

user may wish to plot associated or contrasting variables adjacently for graphical effect. Thus connected lines, giving the impression of a profile, represent each sample.

Figure 12.1 shows profiles for 49 selected observations from the Kola O-horizon soil data set. The elements Cu, Ni, Na and Sr were chosen to construct the profiles. Co and Ni reflect contamination from the Russian nickel smelters, while Na and Sr reflect the input of marine aerosols along the coast of the Barents Sea, but also dust emissions from mining and processing the alkaline rocks near Apatity. The map in Figure 12.1 (left) shows clearly that the profiles take on characteristic forms at the coast, near the smelters, and near Apatity. The table (Figure 12.1, right) permits easy comparison of the profiles of all observations and the identification of unusual profiles at a glance (e.g., observations 39 and 708). The figure also demonstrates one of the key problems when using multivariate graphics. The symbols need a lot of plotting space and the point where the map becomes unreadable due to serious over-plotting is soon reached.

12.2   Stars

Instead of plotting the variables along the x-axis, it is also possible to plot the variables radially with equi-angular spacing, with each variable plotted as a line whose length is proportional to its value. The resulting graphic may be plotted in three styles: (1) as radii, (2) as polygons drawn around the ends of the radii, and (3) as polygons alone.

In Figure 12.2 the stars clearly show the location of the Monchegorsk smelter (Figure 12.2, left), where a large star is developed because the emissions from nickel smelting and mining and the nearby processing of the alkaline rocks at Apatity overlap (see observation 39 in the table, Figure 12.2, right). The special Na-Sr signature along the coast is visible (see observation 8 or 522 in the table), however, the profiles showed this feature considerably better. There are also some unusual stars at the south-western project boundary, which are dominated by high Bi and Rb (e.g., observations 128, 249).

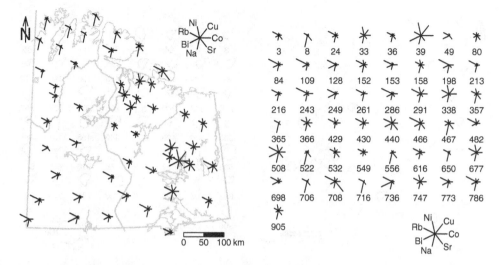

Figure 12.2 Stars in Style (1) constructed for 49 selected samples from the Kola O-horizon soil data set. The variables Bi, Co, Cu, Ni, Na, Rb and Sr were used to construct the symbols. For sample locations see Figure 12.1 (left)

Figure 12.3 shows the same selection of elements displayed as Style (2) stars. The visual impression changes quite dramatically because now it is not only the radii that indicate the importance of a certain element but an area (compare Figures 12.2 and 12.3). The information that can be extracted from the plot is of course the same.

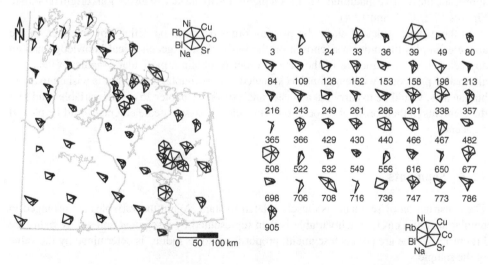

Figure 12.3 Stars in Style (2) constructed for 49 selected samples from the Kola O-horizon soil data set. The variables Bi, Co, Cu, Ni, Na, Rb and Sr were used to construct the symbols. For sample locations see Figure 12.1 (left)

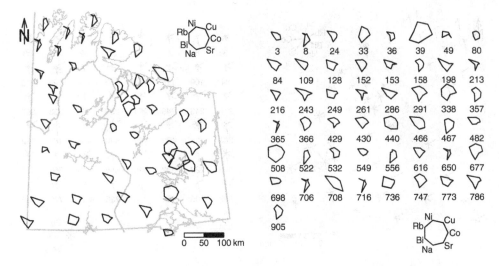

Figure 12.4 Stars in Style (3) constructed for 49 selected samples from the Kola O-horizon soil data set. The variables Bi, Co, Cu, Ni, Na, Rb and Sr were used to construct the polygons. For sample locations see Figure 12.1 (left)

If the radii are deleted from the symbol (Style (3)) it takes the appearance of a polygon (Figure 12.4).

Figure 12.4 again shows the same selection of elements as used for constructing the other two styles. The visual impression changes substantially once more. When Style (3) is compared to Style (1), Style (1) may be preferable because the origin of the star, the location of the observation, is more easily identified. Another issue with Style (3) is that it is difficult to appreciate the relative magnitudes of the elements due to the lack of an obvious origin (compare Figures 12.2, 12.3, and 12.4).

In these three styles of stars, the plotting order needs to be defined by the user. In the above examples the order was chosen on the basis of prior geochemical knowledge and an appreciation of the structure of the data in order to obtain a particular graphical effect. An alternative procedure where no *a priori* knowledge is available, or where it is wished to avoid human bias, would be to carry out a cluster analysis (see Chapter 15) of the variables and plot the elements in their order of association. Thus closely-associated elements would be plotted adjacently.

12.3 Segments

The construction of segments is closely related to stars. Again the variables are arranged in regular arcs of a circle. Each variable is then represented by a segment and not by radii (see Figure 12.5). The area of each segment, proportional to its radius, is determined by the value of the sample.

Figure 12.5 again shows the same selection of elements as used for constructing stars. Once more a different visual impression is obtained. When compared to the other graphics (Figures 12.2, 12.3, 12.4, and 12.5) the segments provide one of the easier to appreciate

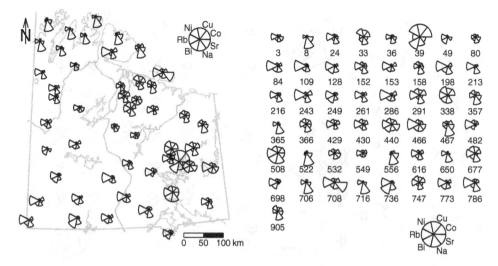

Figure 12.5 Segments constructed for 49 selected samples from the Kola O-horizon soil data set. The variables Bi, Co, Cu, Ni, Na, Rb and Sr were used to construct the segments. For sample locations see Figure 12.1 (left)

graphics. The inherent information content of the segment plot is, of course, the same as for all the other graphics (the various styles of stars) building on the same principle.

12.4 Boxes

Boxes are a powerful multivariate graphic for environmental applications. Here the variables first undergo a hierarchical cluster analysis (see Section 15.4). They are then assigned to one of

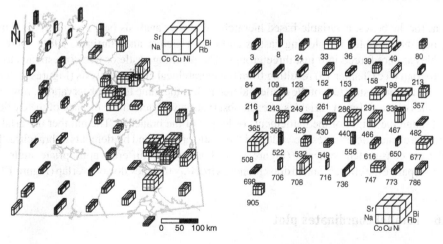

Figure 12.6 Boxes constructed for 49 selected samples from the Kola O-horizon soil data set. The variables Bi, Co, Cu, Ni, Na, Rb and Sr were used to construct the boxes. For sample locations see Figure 12.1 (left)

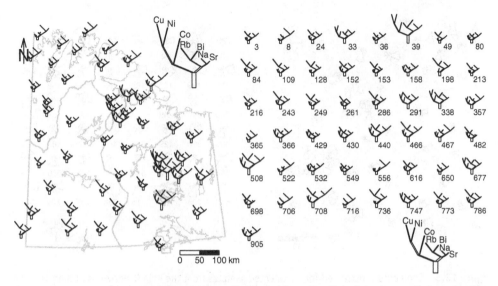

Figure 12.7 Trees constructed for 49 selected samples from the Kola O-horizon soil data set. The variables Bi, Co, Cu, Ni, Na, Rb and Sr were used to construct the trees. For sample locations see Figure 12.1 (left)

three clusters, corresponding to the three sides of a box (Figure 12.6). Therefore, this graphical approach is only applicable where the variables can be grouped into three clusters that reflect specific phenomena. The dimension of each axis of the box, representing one of the three clusters, is proportional to the sum of the values of the variables comprising the specific cluster.

12.5   Castles and trees

Just as for the boxes, a variable-based hierarchical cluster analysis (see Section 15.4) is used to group the variables for plotting castles and trees (Kleiner and Hartigan, 1981). Here the variables are not forced into just three groups. The resulting castle or tree corresponds to the dendrogram (see Chapter 15) resulting from the hierarchical cluster analysis (Figures 12.7 and 12.8). The general structure of the castle and tree is thus the same for all observations, however, the values of the different variables of each observation will determine the size of the castle or tree through the height of the battlements or length of the branches. The number of variables within the clusters determines the widths of the castle or trunk and battlements or branches. The castles are constructed vertically (Figure 12.8). For the trees different appropriate angles are used to separate the clusters and to draw branches in a way that they do not overlap (Figure 12.7).

12.6   Parallel coordinates plot

Instead of constructing one multivariate graphic for each observation, it is also possible to plot all observations onto one graphic. For constructing the parallel coordinates plot the variables are plotted along the x-axis just as for the profiles. The value of each observation for each

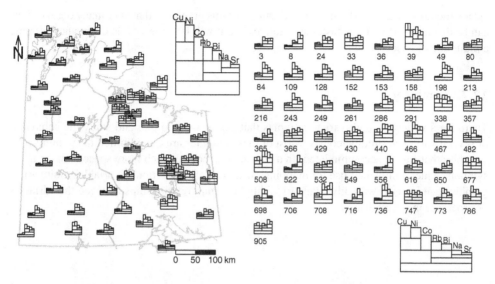

Figure 12.8 Castles constructed for 49 selected samples from the Kola O-horizon soil data set. The variables Bi, Co, Cu, Ni, Na, Rb and Sr were used to construct the castles. For sample locations see Figure 12.1 (left)

variable is plotted along the y-axis. A line is used to connect the values for each observation (Figure 12.9). The parallel coordinates plot is thus a graphical representation of all profiles of all observations in just one plot (Inselberg, 1985). The graphical impression of the parallel coordinates plot depends crucially on the sequence of the variables plotted along the x-axis – it can be advantageous to experiment with the sequence until an optimal visualisation, showing

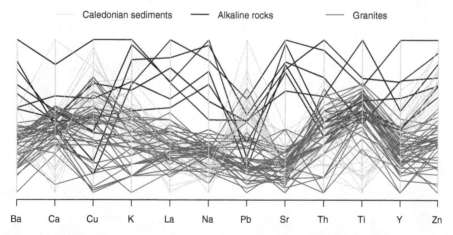

Figure 12.9 Parallel coordinates plot using three lithology-related data subsets (lithology 9, Caledonian sediments; lithology 82, alkaline rocks; and lithology 7, granites – see geological map Figure 1.2) from the Kola C-horizon soil data set. A colour reproduction of this figure can be seen in the colour section, positioned towards the centre of the book

clear structures, is achieved. The main advantage of this graphic is that very many observations can be displayed. If a clear cluster structure exists, it will become visible even when plotting a multitude of observations.

12.7 Summary

There are a variety of multivariate graphics that can be used to plot and study the behaviour of several variables simultaneously. Most of these graphics are quite limited in the number of variables that can be accommodated. In addition they require much more space than the usual symbols, which limits their use for mapping. With a careful selection of variables multivariate graphics can be used with advantage to highlight certain relationships between the variables or processes in a survey area.

13

Multivariate Outlier Detection

The detection of data outliers and unusual data structures is one of the main tasks in the statistical analysis of applied environmental data. Traditionally, despite the fact that environmental data sets are almost always multivariate, outliers are most frequently sought for each single variable in a given data set one at a time (Reimann *et al.*, 2005). The search for outliers is usually based on the location and spread of the data (see Chapter 7). The higher (lower) the analytical result of a sample, the greater is the distance of the observation from the central location of all observations; outliers thus, typically, have large distances. The definition of an outlier limit or threshold, dividing background data from outliers, has received much attention in the geochemical literature, and to date no universally applicable method of identifying outliers has been proposed (see discussion in Reimann and Garrett, 2005; Reimann *et al.*, 2005c). In this context, background is defined by the distributional properties, location and spread, of geochemical samples that represent the natural variation of the material being studied in a specific area, that are uninfluenced by extraneous and exotic processes such as those related to rare rock types, mineral deposit forming processes, or anthropogenic contamination. In geochemistry outliers are generally observations resulting from a secondary process and not extreme values of the background distribution(s). Samples where the analytical values are related to a secondary process – be it mineralisation or contamination – do not need to be especially high (or low) in relation to all values of a variable in a data set, and thus attempts to identify these samples with classical univariate methods commonly fail. However, this problem often may be overcome by utilising the multivariate nature of most geochemical data.

13.1 Univariate versus multivariate outlier detection

In the absence of a prior threshold (Rose *et al.*, 1979), a common practice of geochemists is to identify some fraction, often two per cent, of the data at the upper and lower extremes for further investigation (see Chapter 7 for discussions of methods to determine background and threshold). Dashed lines on Figure 13.1 indicate the two per cent limits. The points inside the rectangular area defined by the dashed lines identify the background population for the two variables, all data plotting outside the inner rectangle are "anomalies". However, this procedure ignores the bivariate structure of the data, in the example an elliptical shape, which is clearly visible in Figure 13.1.

Statistical Data Analysis Explained   Clemens Reimann, Peter Filzmoser, Robert G. Garrett, Rudolf Dutter
© 2008 John Wiley & Sons, Ltd.

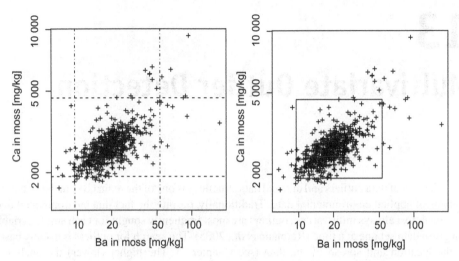

Figure 13.1 Scatterplot of Ba versus Ca in the Kola Project moss data. Lower and upper boundaries (2nd and 98th percentiles) for extreme values are indicated by stippled (Ba) and dashed (Ca) lines in the left plot. The right plot shows the resulting rectangle dividing bivariate anomalies from the bivariate background

To search for multivariate outliers, it would be sensible to use a method that takes into account the multivariate structure (relationships between all variables) of the data, and detects deviations from the overall shape. In Chapter 11 it was demonstrated that one method to describe the relationships between all variables is to calculate the correlation matrix. The Pearson correlation matrix is a standardised to unit variance form of the covariance matrix so that the coefficients are limited to the interval -1 to $+1$ and direct comparisons can be made between all possible pairs of variables.

The un-standardised underlying covariance between two variables is a measure of their linear relationship. Values obtained for different pairs of variables are no longer directly comparable, however, a positive value of the covariance indicates a positive linear relationship between the variables, a negative value a negative linear relationship, and the value of zero indicates no linear relationship. The covariance matrix has the advantage that information about the variance of each element is not lost via the standardisation. Based on a covariance matrix, it is possible to obtain the correlation matrix, however, it is not possible to back-calculate the covariance matrix from the correlation matrix without knowledge of the variances. Thus, when studying the multivariate structure of a data set it is advantageous to use the covariance matrix.

Figure 13.1 demonstrates that quantiles of the values of the single variables will not provide a good estimate of the shape of their distribution. Another possibility is to look at the distance of each data point from the data centre, resulting in a circle instead of the rectangle shown in Figure 13.1. Even when using a circle the shape of the bivariate data distribution is neglected: the data obviously have an elliptical structure. The reason for the elliptical structure is that the data approximate a bivariate normal distribution with a certain covariance structure resulting in an elliptical form. The covariance matrix, which includes this information, permits the estimation of distances from the centre of the data in the context of the data structure.

This results in an elliptical distance measure, called the Mahalanobis, or multivariate, distance (Mahalanobis, 1936). The advantage of this distance measure over the usual Euclidean distance is that all points lying on a certain ellipse have the same Mahalanobis distance from the centre. In terms of the bivariate normal distribution, points with the same distance have the same probability of occurrence. In the multivariate case the ellipse is replaced by a multi-dimensional ellipsoid. Points plotting on the surface of a certain ellipsoid define a certain distance from the centre, which corresponds to a certain probability of the multivariate normal distribution.

In Figure 13.1 quantiles of each single variable were used to identify the rectangular boundaries dividing outliers from non-outliers. When applying the covariance matrix to better approximate the data structure, again quantiles can be used to identify multivariate outliers. Now the ellipse is sought where, in the case of a multivariate normal distribution, 2 per cent of all data would fall outside of the ellipsoid and 98 per cent inside.

Figure 13.2 shows ellipses for defined quantiles (50, 75, 90 and 98 per cent) of the bivariate data distribution of the variables Ca and Ba of the Kola moss data (left plot). The ellipses illustrate the bivariate structure, in the case of a true bivariate normal distribution the ellipses include exactly 50, 75, 90 and 98 per cent of the data. The 98 per cent ellipse encircles all non-outliers, points plotting outside this ellipse are potential multivariate outliers. Several new points are now identified as multivariate outliers, while other points that are univariate outliers for one of the variables are no longer outliers because they fall within the dominant data structure. The new outliers are illustrated by a different symbol in Figure 13.2 (right), they do not show unusually high or low concentrations in terms of one or the other variable, or even both variables, they are, however, far removed from the centre of the data ellipsoid. Results of univariate and multivariate outlier detection methods can thus differ considerably; multivariate outliers are not necessarily univariate outliers, and vice versa.

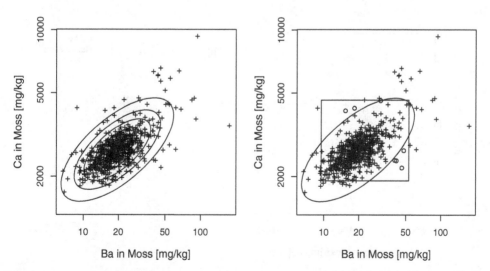

Figure 13.2 Scatterplot of Ba versus Ca for the Kola Project moss data with ellipses corresponding to the 50th, 75th, 90th, and 98th percentiles of the bivariate data distribution (left) and comparison of outliers as determined by the univariate 2nd and 98th percentiles (square) and the bivariate 98th percentile (ellipse – right plot). Newly identified outliers shown as circles (right)

13.2 Robust versus non-robust outlier detection

To compute the multivariate distance and the 98 per cent ellipsoid it is necessary to estimate the central location and covariance (spread) of the multivariate data set. It has been demonstrated (Chapter 4) that classical methods of estimating location and spread are very sensitive with respect to data outliers. Thus for the reliable detection of multivariate outliers, robust estimators for location and covariance need to be used. Several methods for the robust estimation of central location and covariance are available; most check for the majority of data points showing a common structure. These points, usually 50–75 per cent of all values, are used to estimate the data means and covariance. This permits the existence of 25–50 per cent of all data deviating from the main data structure. The examples in this and the following chapters use the Minimum Covariance Determinant (MCD) procedure of Rousseeuw and Van Driessen (1999) unless otherwise noted. Figure 13.3 (left) shows a scatterplot of Mg versus Sr in moss with two 98 per cent ellipses corresponding to robust and non-robust estimation of mean and covariance. It is clearly visible that the shape of the non-robust ellipse is influenced by data outliers and that quite a number of outliers would be masked when using non-robust estimators for the mean and covariance. The location of the bivariate data outliers can be directly plotted in a map (Figure 13.3, right). Most outliers mark the locations of alkaline intrusions, while some reflect the input of marine aerosols (high in both Mg and Sr) along the coast of the Barents Sea.

In the multivariate case it is impossible to plot the ellipsoids. However, the information as to which data points fall outside the 98 per cent ellipsoid can be extracted, i.e. those with a probability of membership of <2 per cent. Because each data point has associated spatial coordinates a map can be used to study the regional distribution of the resulting multivariate outliers.

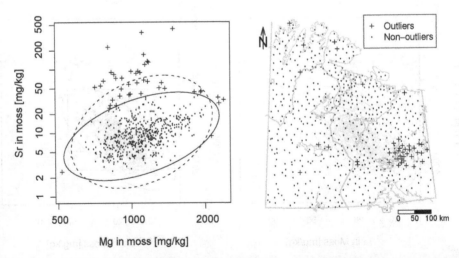

Figure 13.3 Scatterplot of Mg versus Sr in the Kola Project moss data set (left) with robust (solid line) and non-robust (dashed line) 98th percentile ellipse used for outlier detection. The location of the robust model outliers is shown in the accompanying map (right)

13.3 The chi-square plot

In Chapter 7 a variety of frequently used methods to define background and threshold were introduced. The 98th percentile was one of these methods. It has been shown that the 98th percentile has no inherent significance (Section 7.1.4), but simply defines an acceptable number of samples for further follow-up study (1 in 50). It has been demonstrated that visual inspection of the CP-plot for obvious breaks in the data structure is a more powerful though somewhat subjective method of finding "unusual data". For an improved identification of the threshold (better than the 98th percentile), it would thus be advantageous to have a multivariate equivalent of the CP-plot.

Garrett (1989) has used the so-called chi-square plot for this purpose. The underlying idea is that the non-outlying multivariate data follow approximately a multivariate normal distribution. In that case the robust distances (robust Mahalanobis distances) are approximately chi-square distributed (see Figure 4.1 middle right and Section 4.1.6) with degrees of freedom equal to the number of variables, two in the case of the bivariate example. It is thus possible to compare these distances with the chi-square distribution just like comparing data quantiles with the normal distribution in the CP-plot. Similarly as with the CP-plot, it is now possible to check for obvious breaks at the upper end of the plot. Points plotting above the percentile identifying such a break in the straight line are identified as multivariate outliers (Figure 13.4, upper left). Fewer outliers than in Figure 13.3 are identified and direct comparison of the map in Figure 13.4 (lower right) with the map from Figure 13.3 (right) shows that when using the chi-square plot only those samples that clearly deviate from the bivariate data structure (the samples identifying the alkaline intrusions near Apatity) are identified as outliers. However, in instances where there are many variables, it is necessary to ensure that there are a sufficient number of cases so that the estimation of the covariance matrix is stable. If it is unstable, the Mahalanobis distances will not be reliably estimated. Various empirical rules have been proposed for the minimum number of cases relative to the number of variables, e.g., $n > p^2$, $n > 8p$, $n > 5p$, or at a bare minimum $n > 3p$ (n = number of samples, p = number of variables); for a discussion see Le Maitre (1982) and Garrett (1993).

13.4 Automated multivariate outlier detection and visualisation

To avoid the subjectivity of the individual scientist looking at the data, the search for the break, the multivariate outlier boundary, can be automated (Filzmoser et al., 2005). To demonstrate this multivariate outlier detection method a selection of seven elements, Ag, As, Bi, Cd, Co, Cu and Ni, that are all emitted by the Russian nickel industry, are selected from the Kola moss data and used to search for multivariate outliers.

Figure 13.5 shows the resulting outlier maps. When the outliers are simply displayed (Figure 13.5, left), those caused by high average element concentrations cannot be distinguished from those identifying low element concentrations. The disadvantage of this map is that while it identifies many sample sites near the Russian nickel industry as outliers, it also identifies sites in north-west Norway and Finland as outliers, some of the most pristine parts of the survey area. Thus it is necessary to introduce a different symbol set for plotting these multivariate outlier maps so that the different types of outliers can be identified. One possibility would be to use different quantiles of the robust Mahalanobis distance (e.g., 25–50–75 per cent in addition

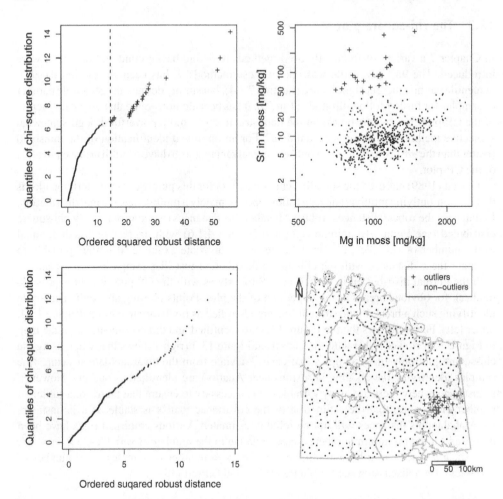

Figure 13.4 Multivariate outlier detection for the variables Mg and Sr for the Kola Project moss data. Chi-square plot with dashed line indicating selected break for removing outliers (upper left), and the resulting straight line in the chi-square plot when outliers are removed (lower left). Upper right: Mg-Sr scatterplot showing the identified outliers; lower right: map showing the geographical location of the outliers (alkaline intrusions near Apatity)

to the multivariate outlier break). In that case five symbols provide a better visualisation of the spatial structure of the multivariate distance from the data centre.

However, overall high element concentrations in the samples can still not be differentiated from overall low concentrations. This can be achieved by introducing colours (or grey scales), related to the average element concentration of each sample into the map (Filzmoser *et al.*, 2005). The resulting map (Figure 13.5, right) is now more informative. The outliers near the Russian nickel industry are identified as "high" (black), showing the extent of contamination around industry, while the outliers in north-west Norway and Finland mark unusually low average element concentrations (light grey), reflecting the extent of the pristine nature of these sample sites.

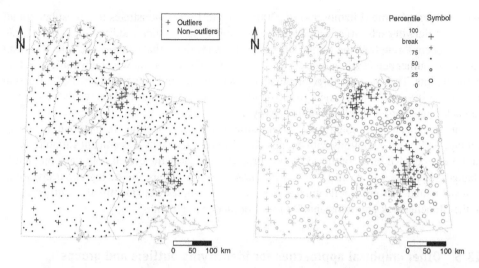

Figure 13.5 Maps identifying multivariate ($p = 7$) outliers in the Kola Project moss data. The left map shows the location of all samples identified as multivariate outliers, the right map includes information on the relative distance to the multivariate data centre (different symbols) and the average element concentration at the sample site (colour, here grey-scale, intensity). A colour reproduction of this figure can be seen in the colour section, positioned towards the centre of the book

As visible in Figure 13.2 (right) multivariate outliers are not necessarily univariate outliers. It may thus be interesting to investigate where the multivariate outliers occur in relation to each of the variables in the univariate space. This can be achieved by plotting a one-dimensional scatterplot for each variable with outliers encoded in a different symbol (Figure 13.6), using the same symbols as in Figure 13.5 (left). It would also be possible to prepare these plots with the symbol set used in Figure 13.5 (right). This, however, requires colour to be able to discern

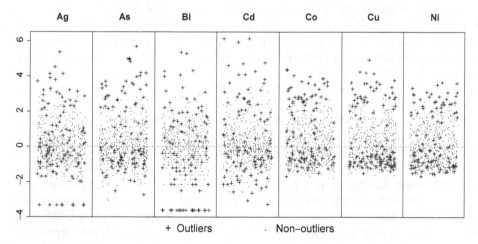

Figure 13.6 Univariate display of the identified multivariate outliers using univariate scatterplots (random horizontal position)

the resulting structure (Filzmoser *et al.*, 2005). Figure 13.6 demonstrates that most but not all extreme values of each variable are identified as multivariate outliers; additionally, multivariate outliers occur throughout the whole data range of each variable that could not be detected using univariate outlier detection methods. The elements Co, Cu and Ni show a comparable structure in these plots. These are the three elements that are emitted in large quantities by the Russian nickel industry. The other elements are co-emitted in minor amounts and many other processes can determine their distribution in the survey area.

Due to the general problems with compositional data (see Section 10.5) it could be an advantage to do the calculations with opened data. This is a necessity if dominant variables (e.g., SiO_2) are included in the outlier detection procedure because the correlations can completely change (see Figures 11.6 and 11.7). For the chosen example the isometric logratio transformation was tested. Throughout the map almost the same samples were identified as outliers, only in the Monchegorsk area substantially fewer outliers were found.

13.5 Other graphical approaches for identifying outliers and groups

While the computation of Mahalanobis distances reduces multivariate measures to a single dimension, the distance of an observation from the centroid of the data, it is also possible to project points in multivariate space onto a plane, i.e. into two dimensions for easy visualisation. A variety of procedures exist for this, examples are Principal Component Analysis (discussed in Chapter 14), Multidimensional Scaling (Kruskal, 1964), Minimum Spanning Trees (Friedman and Rafsky, 1981), and Sammon's Non-linear Mapping (Sammon Jr., 1969).

The algorithms for these procedures work in a generally similar way; they find a plane through the multidimensional data such that when the observations are projected onto it distances between the projected points are as similar as possible to the distances between the points in multivariate space. In general, Euclidean distances, i.e. the direct distances in the data space, are used as the distance measures. The differences between the visualisations generated by the procedures are a result of the use of different functions of the distances for minimisation.

Figure 13.7 (right) displays the non-linear map for the same Kola Project moss data: Ag, As, Bi, Cd, Co, Cu, and Ni (log-transformed and scaled), as used in Figure 13.5. The observations are coded similarly as in Figure 13.5 (left), where pluses are plotted for the detected outliers and points for the core background data. Figure 13.7 (left) is the chi-square plot corresponding to Figure 13.5 (left). It is immediately apparent that the outliers recognised by the automated multivariate outlier procedure fall as a halo around the background core as projected onto the two-dimensional plane (Figure 13.7, right).

Figure 13.8 (right) displays the minimum spanning tree (Friedman and Rafsky, 1981), which maps multivariate data in the two-dimensional space, for a subset of the Ca, Cu, Mg, Na, Sr and Zn data (log-transformed and scaled) for Kola Project C-horizon soils (this data set is later discussed in Chapter 15). The subset comprises the soil samples collected from sites where the underlying geology is either: alkaline rocks, unit 82; granites, unit 7; or sedimentary rocks, unit 9 (see geological map Figure 1.2). Figure 13.8 (left) displays, uniquely coded, the spatial distribution of the samples corresponding to each lithological unit. This same coding is used in Figure 13.8, right. Again, it is immediately apparent that most of the soils derived from alkaline rocks (82) plot as outliers in a cluster that is separated from the main mass of the data. A further alkaline rock derived soil plots as a discrete

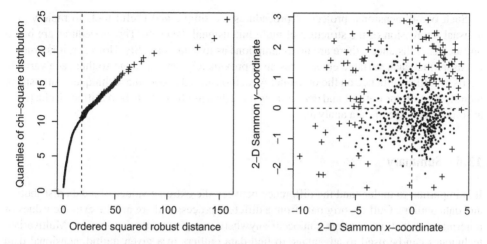

Figure 13.7 The Sammon non-linear map (right) for the Ag, As, Bi, Cd, Co, Cu and Ni (log-transformed and scaled) Kola Project moss data. Points are coded as outliers (pluses) detected by the automated outlier detection procedure, or core background observation (dots) (see Figure 13.5, left); the corresponding chi-square plot is displayed to the left

outlier. The remaining samples from areas underlain by granites (7) and Caledonian sediments (9) plot within the main mass of points, with a number of outliers, one being quite extreme (bottom centre of Figure 13.8, right), interestingly the samples for granitic (7) areas and sedimentary (9) areas plot in different parts of the main data mass with a "zone" of overlap.

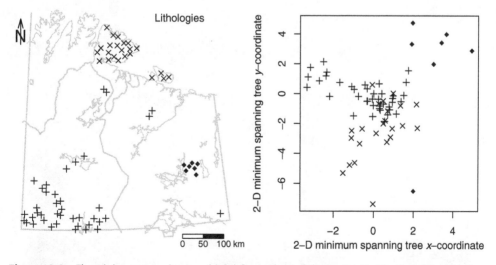

Figure 13.8 The minimum spanning tree (right) for a subset of the Ca, Cu, Mg, Na, Sr and Zn data (log-transformed and scaled) for Kola Project C-horizon soils representing three lithogical units: lithology 7, granites (+); lithology 9, sedimentary rocks (×); and lithology 82, alkaline rocks (◆). The spatial distribution of the members of the subset, coded by lithological unit, is displayed to the left

Such two-dimensional projection procedures are simple and useful tools to rapidly gain a visual impression of the structure of multidimensional data sets. The procedures are based on Euclidean distances, there are no assumptions as to data normality. However, it is prudent to ensure that each variable entered has an approximately similar range so that each variable exerts a similar influence on the outcome. As demonstrated above, the techniques can assist in identifying extreme outliers and determining if the data might usefully be partitioned, clustered, at some later stage of data analysis.

13.6   Summary

It is important to understand the difference between the extreme values of a data distribution and data outliers. Outliers originate from a different process and are not the extreme values of a normal distribution. Outliers can occur anywhere in a given data distribution. Multivariate techniques can be used to advantage to find data outliers in a given multidimensional data set. The interpretation of multivariate outliers can be a challenge. Maps and graphics help to understand the patterns of multivariate outliers.

14

Principal Component Analysis (PCA) and Factor Analysis (FA)

The principal aim of principal component analysis (PCA) is dimension reduction (Hotelling, 1933; Jolliffe, 2002; Johnson and Wichern, 2002; Jackson, 2003). In applied geochemistry and the environmental sciences most data sets consist of many variables (e.g., analysed chemical elements). For example, the Kola Project C-horizon soil data contains more than 100 variables. Graphical inspection of data is most easily accomplished with a scatterplot of two variables. To graphically inspect the main data structure of a multivariate data set it is thus desirable to find two components expressing as much of the inherent variability of the complete data set as possible. These are the first two principal components. PCA will generate as many components as there are variables; however, the majority of the inherent information in the data set is generally included in the first few components, resulting in a considerable reduction of the dimension of the data set.

PCA is often used as a first step for further multivariate data analysis procedures like cluster analysis, multiple regression, and discriminant analysis (see Chapters 15 to 17). By reducing the data dimensionality via PCA, it is often possible to remove "noise" in the multivariate data, thus allowing an improved prediction of another variable, or discrimination or classification of the multivariate data.

The principal aim of FA, which was developed by psychologists, is to explain the variation in a multivariate data set by as few "factors" as possible and to detect hidden multivariate data structures (Harman, 1976; Basilevsky, 1994; Johnson and Wichern, 2002). The term "factor" used by psychologists is equivalent to "controlling processes" in applied geochemistry. Thus, theoretically, FA should be ideally suited for the presentation of the "essential" information inherent in an applied geochemical data set with many analysed elements. One of the main differences of FA compared with PCA is that the number of axes no longer equals the number of variables, but is limited to a small number explaining a high proportion of the total data variability.

In geochemical mapping it would be advantageous if instead of presenting maps for 40 to 50 (or more) elements only maps of 4–6 factors containing a high percentage of the information of all the single element maps needed to be presented. It is even more informative if FA can be used to reveal unrecognised multivariate structures in the data that may be indicative of certain geochemical processes, or, in exploration geochemistry, of hidden mineral deposits. FA has been successfully used for this purpose (see, e.g., Garrett and Nichol, 1969; Chork and Govett,

Statistical Data Analysis Explained   Clemens Reimann, Peter Filzmoser, Robert G. Garrett, Rudolf Dutter
© 2008 John Wiley & Sons, Ltd.

1985; Chork, 1990; Chork and Salminen, 1993), but is still a controversial method. A focal point of critique is that too many different techniques are available, all giving slightly different results (Rock, 1988). It is argued that statistically-untrained users will thus always be tempted to experiment until finding a solution that fits their preconceptions. Another problem is that users are not always aware of some of the basic requirements for carrying out a successful FA. As a result, FA is often merely applied as an exploratory tool, and the results could frequently have been predicted using much simpler methods.

Both PCA and FA are based on the correlation or covariance matrix (see Chapters 11 and 13) and for compositional data appropriate transformations to open the data (Section 10.5) need to be considered. Both provide a summary of the inherent multivariate structure of the data sets. For FA two different approaches are the most frequently applied, the Maximum Likelihood (ML) method and Principal Factor Analysis (PFA). These two have different data requirements that are discussed below.

In geochemical textbooks and publications there is much confusion as to what comprises "factor analysis" (FA) and what is "principal component analysis" (PCA). Succinctly, PCA accounts for maximum variance of all the variables, while FA accounts for maximum intercorrelations. The model for FA allows that the common factors do not explain the total variability of the data. This implies that FA allows for the existence of some unique factors that have a completely different behaviour than the majority of all other factors. Thus unusual variables will not enter the common factors. In contrast PCA will always accommodate the total structure in the data (i.e. all variables are "forced" into the result). Thus, FA is better suited to detect common structures in the data. However, in practice, there are additional assumptions concerning the data that have to be made in using FA; these are discussed in Section 14.3. Thus, despite the advantages of FA, PCA is used more frequently than FA in environmental sciences.

14.1 Conditioning the data for PCA and FA

14.1.1 Different data ranges and variability, skewness

In the multi-element chemical analysis of geological and biological materials, the elements (variables) occur in very different concentrations. When dealing with biological materials, e.g., plants, these differences will be somewhat smaller, but usually essential nutrients occur in concentrations that are orders of magnitude larger than those of micronutrients and trace elements. This may become a problem in multivariate techniques simultaneously considering all variables because the variable with the greatest variance will have the greatest influence on the outcome. Variance is obviously related to absolute magnitude. As a consequence, data for variables quoted in different units should not be mixed in the same multivariate analysis (Rock, 1988). To directly enter raw geochemical data, including major, minor, and trace elements into PCA or FA is unproductive because it can be predicted that the minor and trace elements would have almost no influence on the result. The conclusion is that the data need to be standardised to comparable mean and variance (see Chapter 10). However, standardisation does not make much sense if the data distributions are very skewed – as is often the case with applied geochemical data (see Chapter 3). The effects of skew on covariance and correlation have been demonstrated in Chapter 11 (see, for example, Figure 11.1). Since correlation or covariance is the basis of FA and PCA, the results will be strongly influenced if highly skewed data are not conditioned (transformed) towards symmetry (see Chapter 10). Because in environmental sciences data frequently exhibit a strong right skew, log-transformation is often used to achieve symmetry (see Chapter 10). For an optimal result the distribution of each single variable should be checked for the optimal transformation. However, and very importantly, if compositional data are to be analysed, the use of a "data opening" transformation, e.g., additive, centred or isometric logratio transformation, must be considered and is strongly recommended (see Section 10.5).

Different trace and minor elements, i.e. those with concentrations of <1 per cent, exhibit widely different variabilities. Some cover only one order of magnitude while others may cover three or more. After a log-transformation, which has been suggested as advisable due to the large differences in magnitude observed for geochemical data, the elements with larger variation ranges will have greater weight in the PCA or FA. For example, in the Kola C-horizon soil data, Ag has a range (defined as maximum − minimum) of 0.119 in the raw data, while Mn has a range of more than 2000. After a log-transform Ag has a range of 7.1 while that for Mn is 1.8. If this effect of emphasising the influence of variables with a large variation is wanted, a log-transform of the data will suffice. If this effect is not wanted, the data should be standardised to zero mean and unit variance after the log-transformation.

14.1.2 Normal distribution

Before carrying out a classical FA, it should be determined whether or not all the variables follow a normal distribution, although this, even after log-transformation, is not a necessity (see Table 9.1). Just as for many other statistical techniques, FA is very sensitive to data non-normality (Pison et al., 2003). It is now well known amongst applied geochemists that regional geochemical data practically never show a normal distribution (Reimann and Filzmoser, 2000). When using PCA or principal factor analysis (PFA) data normality is not essential. However, these methods are based on the correlation or covariance matrix, which is strongly affected by non-normally distributed data and the presence of outliers. The maximum likelihood (ML) method used in FA requires not only a normal distribution for all the variables, but also a multivariate normal distribution (Lawley and Maxwell, 1963). This is completely neglected in many published examples of the use of FA. Continuing with FA when this assumption is not met will lead to biased results. The question of whether to enter FA with the original or somehow transformed data is easily answered. All the variables should come as close to a normal distribution as possible (Reimann and Filzmoser, 2000).

14.1.3 Data outliers

Regional geochemical data sets practically always contain extreme values and outliers. These should not simply be ignored, but they have to be analysed because they contain important information about unexpected behaviour in the study area. In fact, finding anomalies that may be indicative of mineralisation (in exploration geochemistry) or of contamination (in environmental geochemistry) is one of the major aims of any regional geochemical survey (Chapter 6). Extreme values as well as outliers can have a severe influence on PCA and FA, since they are based on the correlation or covariance matrix (Pison et al., 2003). Outliers should thus be removed prior to the statistical analysis, or statistical methods able to handle outliers should be employed, and the influence of extreme values needs to be reduced (e.g., via a suitable transformation). This has been rarely done. One reason may be lack of knowledge of the consequences, another that with the outliers removed and the extreme values down-weighted the anticipated "obvious" results will not be revealed in the factors, and thirdly, that software is not generally available to easily deal with outliers. Results of PCA and FA presented in environmental sciences are often mostly driven by outliers and do not focus on the main body of data. The real purpose of these methods, however, is to identify hidden data structures. As shown in Chapter 13, finding data outliers is not a trivial task, especially in high dimensions. A more elegant way to reduce the impact of outliers is to apply robust versions of PCA or FA (Filzmoser, 1999; Pison *et al.*, 2003). The aim is to fit the *majority* of data points, contrary to classical (least squares) estimation where *all* data values, including the outliers and extreme values, are fitted. When using a robust PCA or FA, the outliers can be identified and interpreted by studying the scores on the robust principal components or factors.

14.1.4 *Closed data*

The problem of data closure has already been discussed in Sections 10.5, 11.8, and above. This occurs when working with compositional, i.e. geochemical (environmental) data that sum up to a constant value (e.g., 100 wt%, or 1 000 000 mg/kg). Closure has a major influence on the covariance and correlation matrices (see Section 11.8), the very bases of PCA and FA. Three possible data transformations to solve this problem have been demonstrated in Section 10.5.

For example even the Kola moss data are from a closed number system, although the sums over all variables for all the individual cases (samples) range only from 0.7 to 3.2 per cent, with most falling between 1 and 1.5 per cent, see Figure 14.1 (left). The fact that the data are closed may thus not be immediately apparent. But if elements such as C and N had been analysed the sums would be close to 1 000 000 (100 per cent). It is thus advisable to open the data. The critical issue is that the sums of the measured variables are constrained. Applying a centred logratio transformation (see Section 10.5.2) opens the data, permitting the true variability to be expressed. Figure 14.1 (right) displays the anti-logs of the sums of each sample over all new centred logratio-transformed variables. The points in the plot are more scattered, as indicated by the span of the inner 50 per cent of the points (indicated by horizontal lines); compare with Figure 14.1 (left).

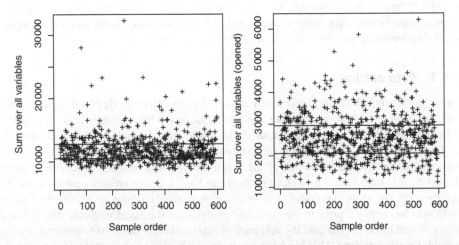

Figure 14.1 Plots of sums over all variables for each sample of the Kola Project moss data versus sample order. Closed data on the left and opened data on the right. Upper and lower quartiles are indicated by horizontal lines

14.1.5   Censored data

There is a further problem that often occurs when working with geochemical data, the detection limit issue (see Section 2.2). For some determinations a proportion of all results are below (or, rarely, above) the lower (upper) limit of detection of the analytical method. For statistical analysis these are often set to a value of half the detection limit. However, a sizeable proportion of all data with an identical value can seriously influence an estimate of correlation. It is very questionable whether or not such elements should be included at all in a PCA or FA. Unfortunately, it is often the elements of greatest interest that contain the highest number of censored data – the temptation to include these in PCA or FA is thus very high. Again the choice is to use robust methods that will be able to withstand a substantial amount of censored data.

14.1.6   Inhomogeneous data sets

Data sets in environmental sciences are often extremely inhomogeneous. Usually they consist of data drawn from a mixture of different populations. In different subsets antagonistic data structures may exist. In the case of the Kola Project data, there is some knowledge concerning the processes governing the data structure. For example, for the C-horizon soils the geological map could be used to construct more homogenous subsets of the samples that were collected. If this is done and the correlation matrix for each of these subsets is estimated, it becomes apparent that each of these lithologically-based data subsets has a quite different correlation structure (see Figures 8.14, 11.5). Entering PCA or FA with such inhomogeneous data will almost invariably lead to unstable results. Because FA aims at identifying some major processes in the multivariate data set, factors will be governed by the elements that show high (or low) values in one of the sub-populations. With PCA, often the first axis will simply help discriminate between the data subsets, and as a result the remaining axes, which are constrained to be orthogonal (perpendicular to each other), may not yield useful information concerning the internal structures of the subsets. In these instances it is predictable that FA and PCA will have as one of their main outcomes an elucidation of some of the more obvious geochemical differences related to the lithological units in the survey area. Such data behaviour can be expected in the vast majority of cases when working with regional geochemical or environmental data, and may be a useful as a first-order mapping tool. However, in such cases it is doubtful that FA or PCA are informative methods for discovering the more subtle internal relationships in the data.

   The correct data analysis procedure would be to first disaggregate the total data set into more structurally homogenous subsets, e.g., via a clustering procedure or by the use of prior knowledge (e.g., the geological map). However, in most cases, and definitely in the case of the Kola Project data, insufficient samples would remain in any of these subsets to carry out a meaningful FA.

14.1.7   Spatial dependence

PCA and FA assume that the data represent random, independent samples from a multivariate distribution. However, the measured variables in applied geochemical and environmental surveys usually have a spatial dependence, i.e. the observations are spatially correlated. PCA and FA of correlated observations will reflect such correlations, which cannot be removed by a random shuffling of the data points. PCA and FA carried out with spatially dependent samples will most often result in identifying spatial correlations. Users should be very aware of this fact. Strong spatial correlations may completely mask geochemical correlations. This clearly needs to be considered when interpreting the results of PCA or FA.

14.1.8 Dimensionality

One of the first requirements for stable results from a PCA or FA is that there are a sufficient number of samples for the number of variables, determining the dimensionality of the data. Different rules have been suggested (Le Maitre, 1982), e.g., $n > p^2 + 3p + 1$ (where n is the number of samples and p the number of variables). The Kola moss data contains 31 variables and the rule requires 1055 samples – but the data set consists of "only" 594 samples. If more tolerant rules are used, e.g., $n > p^2$, or $n > 9p$, or just $n > 8p$, the number of samples is still rather low in relation to the number of variables. Note that these rules are stricter than those proposed for multivariate outlier detection (see Section 13.3). The reason for this is that additional parameters are estimated in PCA and FA. Therefore, for many data sets, PCA or FA should preferably not be entered with the full set of elements.

The above cited rules suggest that for a data set consisting of 594 samples PCA or FA should not be entered with many more than about 23 variables. To overcome the situation of too many variables in relation to the number of samples some authors (see e.g., Böhm *et al.*, 1998) suggest the use of Monte Carlo techniques to reduce the number of variables entered into FA. Variable selection can of course also be based on geochemical reasoning (for example, why enter all the highly correlated Rare Earth Elements when one or two will do?), existing knowledge about the survey area, possible element sources, exploration models or collinearity studies. In these cases FA may be used to test various hypotheses.

When all variables are highly correlated (e.g., when analysing infra-red spectra) it is possible to use PCA even when the number of samples is much smaller than the number of variables.

14.2 Principal component analysis (PCA)

For illustrating the application of PCA, the most homogenous Kola Project data set with the smallest number of variables has been chosen, the moss data set. The complete data set has 31 variables and 594 samples (see above). Initially the data were log-transformed and standardised (centred and scaled, see Section 10.4) to yield a more symmetric data distribution and directly comparable data for all variables.

The goal of PCA is to explain as much information contained in the data as possible in as few components as possible. In statistics information content is expressed by variability. PCA searches for the direction in the multivariate space that contains the maximum variability. This is the direction of the first principal component (PC1). The second principal component (PC2) has to be orthogonal (perpendicular) to PC1 and will contain the maximum amount of the remaining data variability. Subsequent principal components are found by the same principle. They must be orthogonal to the previous principal components and contain the maximum of the remaining variability.

For the bivariate case this can be visualised in Figure 14.2, a scatterplot of Mg versus Na (log-transformed and standardised) for the Kola Project moss data. Figure 14.2 (left) shows the direction of the first principal component (PC1) following the direction of the main data

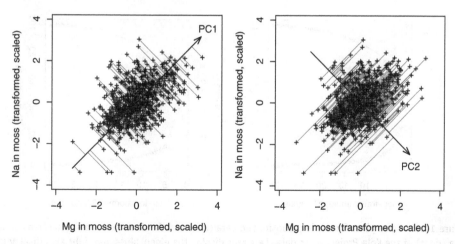

Figure 14.2   Scatterplot of Mg versus Na (both log-transformed and standardised) for the Kola Project moss data. The direction of the principal components and the construction of the scores are indicated

variability. All data points are orthogonally projected onto PC1, resulting in new data points, that are called the *scores* of the first principal component. The scores represent each data point by a single number along the new coordinate PC1.

PC2 is in the direction orthogonal to PC1, again all data points are orthogonally projected onto PC2, resulting in the scores of PC2. The direction of each principal component is expressed by its *loadings* which convey the relation to the original variables. In the two-dimensional example shown in Figure 14.2 (left) both variables show a positive relation to PC1. The direction of PC1 can now be characterised by two numbers, because the data are in a two-dimensional coordinate system determined by the variables Mg and Na. Because of the positive relation of PC1 with Mg and Na both loadings will be positive. For PC2 the loading for Mg is positive and the loading for Na is negative. In the multivariate case each principal component has as many loadings as there are variables.

Mathematically, PCA can be seen as a decomposition of the covariance matrix or correlation matrix (if the data are standardised) into its "eigenvectors" and "eigenvalues". The eigenvectors are the loadings of the principal components spanning the new PCA coordinate system. The amount of variability contained in each principal component is expressed by the eigenvalues which are simply the variances of the scores.

14.2.1   The scree plot

Since the principal aim of PCA is dimension reduction, i.e. to explain as much variability as possible with as few principal components as possible, it is interesting to study how much variability is explained by each single component. In the scree plot (Cattell, 1966) the component number is plotted against the explained variance.

If PCA is primarily used to reduce the dimensionality of the data for other multivariate techniques (multiple regression, discriminant analysis, cluster analysis) it is important to determine the optimum number of components for this purpose. In this case the cut-off is chosen at the point where the function displayed by the scree plot shows an elbow or the last clear break.

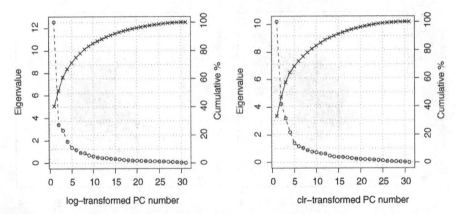

Figure 14.3 Scree plots for the log-transformed data (left) and the centred logratio (clr) transformed data (right) of the Kola Project moss data. Left axes display the eigenvalues and right axes display the cumulative percentages of explained variability

Scree plots were prepared for the log-transformed and the centred logratio transformed Kola moss data. In both cases the data were centred and scaled after the transformation, which corresponds to a PCA based on the correlation matrix. Figure 14.3 (left) shows the scree plot for the log-transformed data, and Figure 14.3 (right) the same plot for the centred logratio transformed data. The plots display both the eigenvalues from which the proportion of explained variability is derived (left y-axis) and the cumulative percentage of explained variability (right y-axis). For the log-transformed data the first component explains almost 40 per cent of the total data variability, and the second component explains a further 15 per cent. Both components together thus explain more than 50 per cent of the total variability. In contrast, for the centred logratio transformed data the first component only explains 33 per cent of the total variability, and the second component a further 14 per cent, for a total of 47 per cent. The reason that the first two components of the centred logratio transformed data explain a smaller percentage of the total variability is due to the fact that the variability in the data is no longer constrained and requires additional components for its explanation. After five components for the log-transformed data and six components for the centred logratio transformed data, the scree plots show an almost identical behaviour. Both scree plots flatten out after about ten components explaining about 85 per cent of the variability. One might then select the first ten components to use in other multivariate data analysis techniques if noise reduction is required. The plots both show clear breaks after four or five components, respectively, for the log-transformed and centred logratio transformed data, both explaining almost 70 per cent of the total variability. For visual inspection of the data set the first four or five components will probably be sufficient to reveal the main data structure.

Other criteria for the choice of the number of relevant principal components have been proposed, e.g.:

- The explained variability should be at least 80 per cent (see, e.g., Jackson, 2003);
- Only principal components with eigenvalues >1 should be considered (Kaiser, 1958);
- A formal statistical test to determine the number of principal components could be used (Barlett, 1950);

- Monte Carlo techniques like bootstrap or cross-validation can be used (Efron, 1979; Eastment and Krzanowski, 1982).

Unfortunately, the different criteria usually will not result in a unique solution.

14.2.2 The biplot

It is often desirable to simultaneously visualise scores and loadings of the principal components. This can be accomplished with a biplot (Gabriel, 1971). The plot is called a biplot because it contains information on loadings (arrows) and scores (data points or sample identifiers) and not because it plots two principal components against one another. Thus the relationships between variables (via the loadings) and observations (via the scores) become visible. The loadings (arrows) represent the elements (Figure 14.4). The lengths of the arrows in the plot are directly proportional to the variability included in the two components (PC1 and PC2) displayed, and the angle between any two arrows is a measure of the correlation between those variables. For example, a very short arrow in Figure 14.4 (left) represents Hg, indicating that the first two components contain almost no information about this element. Because a scatterplot is limited to two dimensions, only two components can be plotted against one another. In the biplot the scales for the scores are shown on top (PC1) and right (PC2) axes, and the scales for the loadings are shown at the bottom (PC1) and left (PC2) axes. Each arrow in the biplot represents the loadings of one of the elements on PC1 and PC2 and can be identified by the element (variable) name.

Figure 14.4 (left) displays the biplot for PC1 versus PC2 of the log-transformed data, while Figure 14.4 (right) displays the similar plot for the centred logratio transformed data, both containing roughly 50 per cent of their respective total data variabilities. The effect of opening the data is immediately clear, the variables in the centred logratio transformed analysis are more evenly spread across the component space, facilitating the recognition of element associations characteristic of different biogeochemical processes.

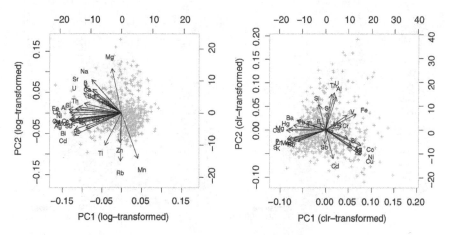

Figure 14.4   Biplot of PC1 versus PC2 for the log-transformed (left) and centred logratio (clr) transformed (right) Kola Project moss data set

Figure 14.4 (right) shows an association of smelter-related elements, Co, Ni, Cu, Bi, As, Ag (lower right quadrant); an association of Al, Th, U, Si, Fe and V (upper right quadrant) which reflects the deposition of terrigenous dust on the mosses; and a K, S, P, Zn, Mn, Ca and Mg association characteristic of biological processes in the mosses. The first principal component is mainly influenced by biological processes (negative loadings) and to a lesser extent by anthropogenic smelter processes (positive loadings), while the second principal component is dominated by terrigenous dust (positive loadings). Thus, major spatial and biological processes in the survey area determine the first two principal components, for instance the smelter-related elements contribute to both positive PC1 loadings and negative PC2 loadings.

The representation of the scores by a biplot is incomplete because the geographical information about the location of each sample is missing. A combination of the biplot with geographical mapping is an exceptionally powerful data analysis tool, especially if Spin and Brush (as provided by programs like GGOBI: – see http://www.ggobi.org/) techniques are applied (see also discussion in Section 8.5).

14.2.3 Mapping the principal components

Figure 14.5 shows maps of the scores of the two first principal components of the centred logratio transformed data. The first component clearly indicates contamination from the Kola smelters (high scores) as already indicated in the biplot by the Cu, Ni, Co association (Figure 14.4, right). The lowest scores on PC1 are present north of the tree-line in Norway and Finland where the deposition of terrigenous dust is greater than in forested areas to the south where PC1 scores are not as low. Near-zero PC1 scores occur near the coast, in forested areas, and peripherally to the smelters. The marine influence, characterised by a Na, B, Sr association is not revealed in the first two components, but is the dominant driver of the third principal component and its scores (see below). The highest scores on PC2 occur north of the tree-line and close to the alkaline intrusions mined near Apatity where there is extensive deposition of

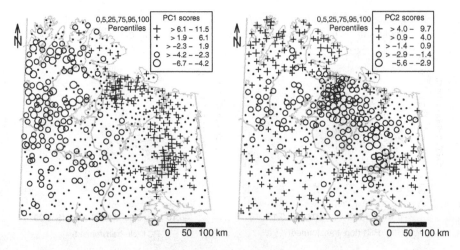

Figure 14.5 Maps of the scores of the first two principal components for the centred logratio transformed Kola Project moss data

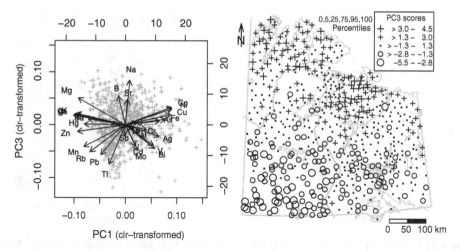

Figure 14.6 Biplot of PC1 versus PC3 of the centred logratio (clr) transformed Kola moss data (left), and a map of PC3 (right)

terrigenous dust relative to the rest of the project area. The lowest scores are present in the vicinity of the smelters, indicating the dominance of particulate deposition from them relative to naturally sourced dust.

Figure 14.6 (left) presents the biplot of PC1 versus PC3 of the centred logratio transformed data set. High scores on PC3 are driven by Na, Mg, Sr and B (and to some extent by Ni and Co). The former elements are characteristic of marine sea spray. Figure 14.6 (right) displays the scores on PC3 as a map. It is immediately apparent that the highest scores occur in a belt across the northern part of the project area adjacent to the Barents Sea, and the lowest scores deep in the forested south and western parts of the project area.

14.2.4   Robust versus classical PCA

As has been discussed previously there are often advantages to employing robust procedures. The computation of robust estimates for centred logratio transformed data will usually encounter numerical problems, leading to a failure to complete the calculations. Thus, a useful alternative is the isometric logratio transformation (see Section 10.5.3) which avoids these numerical problems. Figure 14.7 displays the biplots for classical (left) and robust (right) estimation of the principal components for the isometric logratio transformed data. The robust analysis is based on covariance estimation using the MCD estimator (see Section 13.2). When performing an isometric logratio transformation, "artificial variables" are constructed and thus the direct relation to the original variables is lost. Plotting the resulting loadings and scores in a biplot results in a plot that cannot be interpreted. Thus the results of PCA need to be back-transformed to obtain the original variables and then plotted. The back-transformation is made to the space of the centred logratio transformed data (for details see Egozcue *et al.*, 2003).

The resulting Figure 14.7 (left) can now be directly compared to Figure 14.4 (right). It appears that the isometric logratio transformation results in an even better "opening" of the

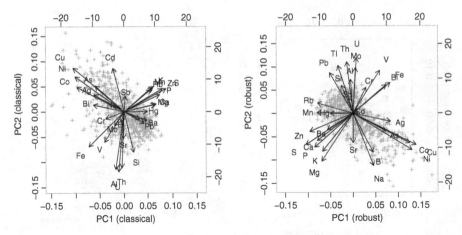

Figure 14.7 Biplots of the classical (left) and robust (right) PCAs based on isometric logratio transformed Kola Project moss data

data than the centred logratio transformation used for the results displayed in Figure 14.4 (right).
Figure 14.7 (right) shows the results for robust PCA of the isometric logratio transformed data.
The robust PCA is less influenced by the outliers visible in Figure 14.7 (left) and the loadings
are almost symmetrically spread around the plot. Several groups emerge, e.g. a "contamination
group" (As, Co, Cu, Ni), a "sea spray group" (Sr, B, Na) which plots close to the "contamination
group" (because Nikel is close to the coast of the Barents Sea and the alkaline rocks mined and
processed near Monchegorsk have very high concentrations of Na and Sr), a "plant nutrient
group" (Mg, K, S, P, Ca, Zn), and two different "dust groups" (Fe, Bi, Cr and V – probably
"industrial dust" and a "geogenic dust group" (U, Mo, Al, Th Si, Tl, etc.)). This is clearly the
best result obtained with PCA so far and demonstrates the need for opening the data and for
using robust techniques with geochemical data.

14.3 Factor analysis

Factor analysis (FA) and PCA both have the same objective: reduction of dimensionality.
Factor analysis works in principle like PCA; the multivariate data are decomposed into scores
and loadings. However, the more strict definition of factor analysis implies that the number
of factors to be extracted is defined at the beginning of the procedure. In addition an estimate
of the proportion of variability for each element that is not to be included in the factors but is
considered unique to that element has to be provided (see, e.g., Harman, 1976). This unique
variance is also called *uniqueness*.

The interpretability of the components of PCA is often difficult because the components
are determined by a maximum variance criterion. One solution to improve interpretability is
to rotate the principal components (Figure 14.8) such that the variables load predominantly
onto one component. Many practitioners refer to the rotated PCA solution as a factor analysis
as they have specified the number of components to be rotated. In effect they have chosen a
model by specifying the number of components.

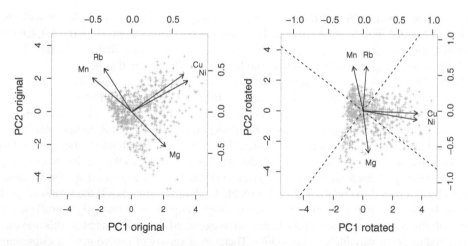

Figure 14.8 Biplots of results of a robust PCA of log-transformed and standardised Mn, Rb, Ni, Cu and Mg from the Kola Project moss data. The left diagram shows the "original" biplot, the right diagram shows the biplot for the Varimax rotated principal components. To better visualise the effect of rotation the old axes (dashed lines) before rotation are displayed in the right plot

For an example of the rotation procedure to improve interpretability see Figure 14.8, where a robust PCA was undertaken with five log-transformed and standardised variables from the moss data. In the unrotated example both components are influenced by all elements (Figure 14.8 left). Using the Varimax criterion of Kaiser (1958), the first rotated component is, however, exclusively determined by Cu and Ni, and the second rotated component by Mn, Rb and Mg (Figure 14.8, right). Rotation thus results in the separation of the two main processes.

In contrast to PCA, FA permits excluding parts of the variability of some variables because their variability is unique and may disturb the interpretability of the results. FA uses a different terminology from PCA: components are now named factors. Figure 14.9 shows biplots for

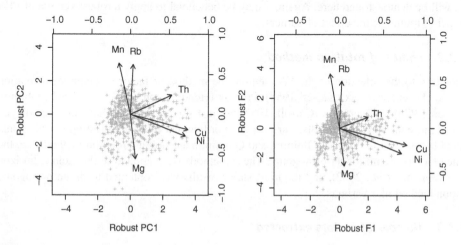

Figure 14.9 Biplots for robust PCA and robust FA for log-transformed and standardised Mn, Rb, Th, Cu, Ni and Mg from the Kola Project moss data

the results of robust PCA and robust FA for the example used in Figure 14.8, but with the element Th added. For the PCA solution Th influences both components (Figure 14.9, left) as its variability, indicated by the length and position of the arrow, contributes to both PC1 and PC2. Thus the two groups of variables are no longer as well separated as before.

In the FA solution the arrow representing the variability of Th is much shorter, therefore far less of its variability is projected onto and contributes to F1 and F2. The main part of the variability of Th was excluded by the method before the factors were calculated. The two factors have the same interpretation as in the first example (Figure 14.8, right) without Th; the disturbing effect of Th is almost excluded. Thus, in theory FA should be better suited to determine the main processes in a complex data set. FA accomplishes this by first determining if the variability of each single variable (element) can be well expressed by the resulting factors by computing the relation of each variable to all other variables. If the relationship of a variable with all others is weak, it is assigned a high uniqueness so that only a small part of its variability contributes to any of the factors to be extracted. In the FA literature this measure is referred to as the variable's *communality*. There are a variety of procedures for estimating this communality. Frequently the squared multiple correlation coefficient, also called multiple R-squared measure (see Section 16.3.2), is employed (see, e.g., Johnson and Wichern, 2002).

14.3.1 Choice of factor analysis method

A variety of methods exist for carrying out a FA. The two most frequently used methods are principal factor analysis (PFA) and the maximum likelihood (ML) method. A reasonable question is whether the choice of method will have any fundamental influence on the results of the procedure. The mathematical background is well documented in a number of specialised textbooks (see, e.g., Harman, 1976; Seber, 1984; Basilevsky, 1994; Mardia *et al.*, 1997; Johnson and Wichern, 2002). In contrast to PCA, FA uses a statistical model with quite restrictive assumptions. In addition, the ML method requires multivariate normally distributed data, a state that is rarely achieved with applied geochemical (environmental) data. Thus PFA may be the "safer" method for many applications. Both methods have been applied to the Kola C-horizon data, and PFA resulted in clearly superior results (Reimann *et al.*, 2002). Thus only PFA will be demonstrated here. Again, it may be beneficial to apply a robust version of PFA in order to minimise the effect of outliers.

14.3.2 Choice of rotation method

In addition to the selection of the FA method, the method of factor or component rotation must be chosen: Varimax (Kaiser, 1958), Promax (Hendrickson and White, 1964), Oblimin (Harman, 1976) or Quartimin (Carroll, 1953) are just some examples that can be employed to yield more interpretable results. Varimax is an orthogonal rotation, resulting in orthogonal rotated factors, while Promax, Oblimin, and Quartimin are oblique rotation methods, i.e. the rotated factors are no longer orthogonal. These methods were applied to the Kola C-horizon data (Reimann *et al.*, 2002), and the most stable results were obtained using an orthogonal rotation method like Varimax.

14.3.3 Number of factors extracted

In contrast to PCA, FA requires that the number of factors to be extracted be determined in advance. There are different procedures to determine the "optimum" number of factors to be extracted, including a number of statistical tests (see Basilevsky, 1994). However the procedures most often used are more of the "rule of thumb" type. One such possibility is to select the number of factors such that they explain a certain amount of variance (e.g., >70 per cent). The scree plot, based on a PCA, is probably the most widely used tool to determine an optimum number of factors (see Section 14.2.1 and Figure 14.3). The scree plot (Figure 14.3) suggests that at least four factors should be extracted.

The results of FAs can drastically change with the number of factors extracted. This effect is especially pronounced when FA is undertaken with only a few variables. Experience from studying many factor maps suggests that it might in general be better to extract a low number of factors (see, e.g., Reimann *et al.*, 2002).

14.3.4 Selection of elements for factor analysis

One of the main critiques of FA is that the selection of variables will exert considerable influence on the results. Because of the many statistical parameters required for the FA model, the results are extremely data sensitive, and an FA may result in completely different factors when only a few variables are added or deleted (see examples in Reimann *et al.*, 2002).

Because PCA and FA are dimension reduction techniques it is logical to undertake them with all available variables, after taking censoring into consideration, as long as the number of samples is considerably larger than the number of variables (see Section 14.1.8). When some factors clearly explain specific processes in the data, the explanatory variables for those factors could be removed. A further FA could then be carried out with the remaining variables to see if additional structure could be identified. Selection of variables may thus be an iterative process (van Helvoort *et al.*, 2005).

As mentioned above, some authors (e.g., Bohm *et al.*, 1998) suggest the use of Monte Carlo methods to extract those elements that have a major influence on the total variance of the data set. However, the result of this approach can be that the regional distribution patterns of some of the most, *a priori*, interesting elements are never studied. Another approach could be to remove all elements that have loadings within the range -0.5 to $+0.5$ on all factors, or that have a uniqueness of >0.5 (or <0.5 for the communality). Again this approach may result in removing some interesting elements. A third approach could be to prepare a combination of some of the elements that have high absolute loadings on the same factor and carry out a further FA. A sub-selection of variables could also be motivated by geochemical reasons/models. The possibilities are limitless.

14.3.5 Graphical representation of the results of factor analysis

Results of FAs are usually presented in biplots and factor maps exactly as demonstrated above for PCA (Sections 14.2.2 and 14.2.3). The loadings, on which the interpretation of the factors is based, are often presented in tabulations. It is, however, possible to plot the factor loadings as a simple graph, where the x-axis is scaled according to the relative amount of variability explained by each single factor in the FA model, excluding the unexplained part of the variability (uniqueness) of each variable. In addition, the percentages at the top (see Figure 14.10) display the cumulative explained variance for the total data variability. It is thus possible to easily appreciate how much of the total variance is explained and how important the single factors are for this explanation. The y-axis is scaled from $+1$ through 0 to -1 and shows the factor loadings of the different variables entering each factor. Dashed lines at values of $+0.5$ and -0.5

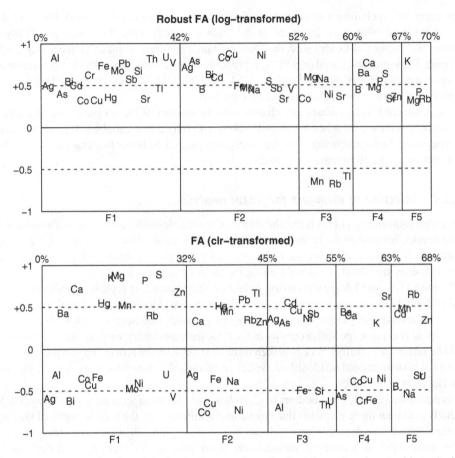

Figure 14.10 Factor loading plots for the log-transformed robust (upper) and centred logratio (clr) transformed (lower) five-factor models (PFA and Varimax rotation) for the Kola Project moss data

help distinguish the important ($> +0.5$, < -0.5) from the less important variables. Names of variables with absolute loadings of < 0.3 are not plotted because their contributions to the factors are negligible (Reimann *et al.*, 2002). The plot facilitates an appreciation of the differences between FAs with different selections of elements, different data transformations (see Figure 14.10), different methods of FA (e.g., ML versus PFA), different numbers of factors extracted, or different rotation methods.

Figure 14.10 presents the factor loading plots from PFAs with Varimax rotation for log-transformed robust and centred logratio transformed five-factor models of the Kola Project moss data. Both five-factor models explain some 70 per cent of the total data variability, with five factors being chosen on the basis of the scree plot (see Figure 14.3). Comparing the two solutions, and that for a similar PCA (see Section 14.2), the centred logratio transformation leads to factors that are both positively and negatively influenced by the variables. Factors 1, 2, 4, and 5 of the log-transformed data are good examples where all significant loadings are only positive, whilst with centred logratio transformed data the loadings are broadly distributed between positive and negative. This should improve interpretability. The two different factor solutions are generally similar in the processes they identify, but are different in detail and can be explained in terms of several important processes determining the chemistry of the moss samples in the survey area.

For the robust FA of the log-transformed data, Factor 1 is a dust factor, as indicated by the presence of elements like Al, Fe, Th and U. Particulate input will be the main process determining the concentration of all these elements in the moss. Factor 2 is dominated by Cu, Co and Ni, the three main elements emitted by the Russian nickel industry, and is thus a contamination factor. Elements like Na and Sr have entered this factor because they are emitted by the mining and transport of alkaline rocks at Apatity, which is close to Monchegorsk. Factor 3 shows positive loadings of Mg, Na and Sr and negative loadings of Mn, Rb, and Tl. This factor shows a strong north-east to south-west gradient in the survey area due to the input of marine aerosols near the coast of the Barents Sea. There is an opposite trend of increasing element concentrations from the coast inland due to the three vegetation zones that are crossed (Mn, Rb, Tl – but also Ag, Bi, Pb and Sb, which did not enter the factor – compare Section 6.4). Ni and Co enter this factor because, by chance, the Russian nickel industry is predominantly located in the northern part of the survey area. They do not enter the factor for geochemical reasons but because of the spatial dependencies in the survey area. Factors 4 and 5 both collect together a number of important plant nutrients (Ca, P, S, Mg and K, P, Mg) and are probably an indication of the nutrient status of the mosses from the survey area. It could also be interesting to compare their distribution against the distribution of the two moss species (*Hylocomium spendens* and *Pleurozium schreberi*) that were collected. However, Ca, S, and Mg are also major ingredients of marine aerosols and thus this factor may represent a mixture of two different processes.

For the FA of the centred logratio transformed data (Figure 14.10, lower plot) Factor 1 is dominated by a biological association (Ca, P, S, Mg, K and Zn). In contrast the smelter releases from Monchegorsk are characterised by negative loadings (Cu, Ni, Co, Ag, Bi, Mo and V). Factor 2 is dominated by smelter releases from both Monchegorsk and Nikel/Zapoljarnij due to negative loadings of only Cu, Ni, and Co (apparently Nikel/Zapoljarnij are not associated with major Ag, Bi, Mo and V releases). The elements Mg, Rb, Th and Pb indicative of vegetation changes in robust Factor 2 are present here as positive loadings. Factor 3 appears to reflect the source of particulates to the mosses. Positive loadings for Cd, Cu, Sb, Ag and As reflect releases from sulphide processing facilities, while negative loadings of Al, Th, U, Fe, and Sr reflect releases of oxide and silicate minerals that could be derived naturally or from mining activities such as those at Apatity. Factor 4 further reflects the presence of anthropogenic smelter sources. Finally, Factor 5 reflecting marine influence is dominated by negative loadings of B, Na, Sr and U, and positive loadings of Rb, Mn, Cd and Zn associated with increasing terrestrial biological influence.

As can be seen the dominant anthropological and biogeochemical processes in the region are reflected in both factor models. Interestingly, the model for the centred logratio transformed data appears to be better at differentiating between the different anthropogenic sources, but possibly less effective at reflecting the biological changes across the region.

Figure 14.11 illustrates the regional distribution of the factor loadings of the first four factors the robust PFA with Varimax rotation for the log-transformed Kola Project moss data. The maps support the above interpretation based on the factor loadings. They even show some interesting additional details. For example, the plot for F1 (Figure 14.11, upper left) shows that it is even possible to detect the roads/railroads from Apatity/Monchegorsk to Murmansk and Kovdor as important sources of dust input. In northern Norway along the coast of the Barents Sea dust plays

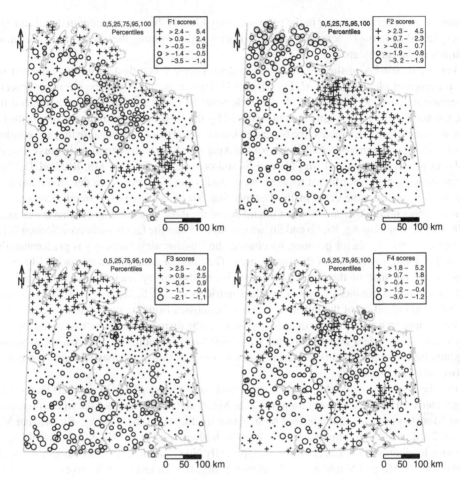

Figure 14.11 Results of robust PFA with Varimax rotation for the log-transformed Kola Project moss data presented as factor score maps for the first four factors

an important role due to harsh climate and the resulting poor vegetation cover. Surprisingly Nikel/Zapoljarnij does not show up in this map. Factor 2 (Figure 14.11, upper right) provides an informative differentiation between contaminated, including Nikel/Zapoljarnij, and non-contaminated parts of the survey area. Factor 3 (Figure 14.11, lower left) shows, as expected, the strong north-east to south-west gradient in the survey area, that is only locally disturbed (Apatity and Kovdor). The map for Factor 4 (Figure 14.11, lower right) is much noisier (high local variability) than any of the other maps and probably presents a mixture of different processes.

Figure 14.12 displays maps for Factors 1, 2, 3 and 5 for the centred logratio transformed data. Again, the maps support the interpretation based on the factor loadings. In the map of Factor 1 (Figure 14.12, upper left), Monchegorsk and the pattern of dispersed particulates, dominantly to the north, is clearly visible and characterised by low scores, while to the north-west high scores indicate areas of relatively pristine tundra. In contrast, in the map for Factor 2

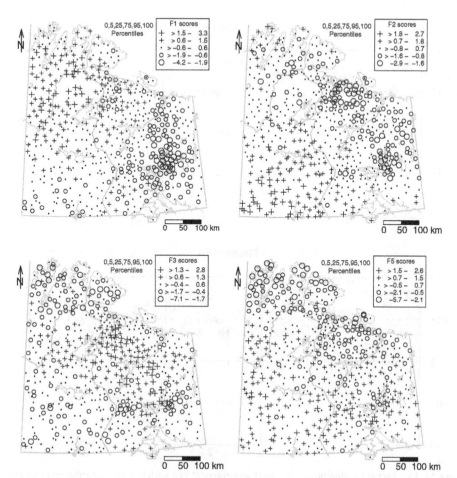

Figure 14.12 Results of PFA with Varimax rotation for the centred logratio transformed Kola Project moss data presented as factor score maps for Factors 1, 2, 3 and 5

(Figure 14.12, upper right) both smelter complexes are clearly identified by negative scores, and the pristine forested area in the south-west of the project area is emphasised by positive scores. In the map for Factor 3 (Figure 14.12, lower left) areas influenced by the smelter complexes are characterised by high scores, while areas in the tundra and close to Apatity and Kovdor appear to be dominated by the deposition of silicate and other non-sulphide dust. Finally, the influence of the Barents Sea is clearly visible in the map for Factor 5 (Figure 14.12, lower right) where low scores occur proximal to the coastline.

14.3.6 Robust versus classical factor analysis

In the above comparison of factor models for log-transformed and centred logratio transformed data the robust model for log-transformed data was selected in preference over the non-robust classical approach.

Figure 14.13 shows the five-factor model loading plots for the robust and a classical PFA (Varimax rotation) for the log-transformed Kola Project moss data set. The main differences

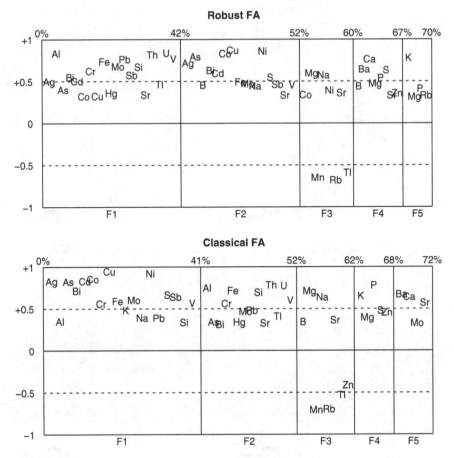

Figure 14.13 Factor loading plots for robust and classical PFA with Varimax rotation and five factors for the log-transformed and scaled Kola Project moss data. Note that the x-axis is scaled according to the relative amount of variability (excluding the uniqueness) explained by each factor, while the percentages at the top dispay the cumulative variance for the total data variability

include:

- The position, i.e. order and therefore importance, of several factors has been reversed (F1 and F2, and F4 and F5);
- F1 has a larger explained variability in the classical FA;
- For the five-factor model the classical FA results in a higher total explained variability;
- Cu and Ni are more dominant in F1 of the classical FA than in F2 of the robust FA; and
- F2 of the classical FA contains fewer elements than the corresponding F1 of the robust FA.

All in all the results are not that different, and the decision as to which result is the best or more reliable is difficult on the basis of the plots. However, there are some indications. The exchange of position of F1 and F2 and the increased loadings of Cu and Ni indicate that the outliers caused by the emissions from the Russian nickel industry have attracted the classical

F1. Hidden processes in the main body of the data may stay invisible if outliers are allowed to overly influence (lever) the result. There are considerably easier techniques to identify outliers than FA, and in that sense the robust FA is strongly preferable.

This discussion may appear academic at first glance. However, in the scientific literature FA is often misused as a "proof" for certain element associations and to draw far-reaching conclusions from the factor loadings. FA cannot be used as a "proof" for the existence of processes – it can help identify relationships in the data and stimulate ideas; they have to be proven in a different way. Many justifiable critiques of FA as applied to geochemical data are due to the misuse of the technique. Factor analysis can be used to explore the data for hidden multivariate structures, which then have to be explained by different means. If used correctly it may not result in a reduction of geochemical maps but rather in additional maps, but those maps should be more informative and relatable to known physical and geochemical processes. To objectively judge the quality of these maps is very difficult. In regional geochemical mapping good results are probably best indicated by stable geochemical maps displaying broad and homogeneous patterns. Real success is achieved if new patterns are revealed that do not necessarily fit the established geological maps and concepts, and that may be difficult to see in single element maps. However, even such wide-scale patterns may often be difficult to interpret based on current knowledge. They can be used to develop new ideas about processes influencing element distribution on a regional scale or to re-interpret the geological map. To interpret the results beyond pure speculation may require considerable new fieldwork; the original geological mapping at a regional scale may have missed features picked up by more detailed geochemical mapping. Thus FA may be useful in guiding field activities into especially interesting areas.

The example presented here is based on the moss data set that has the lowest number of variables of all the data sets and the fewest processes influencing the samples. Results become far more varied and difficult to interpret when applying FA to any of the other layers (see results for the C-horizon presented in Reimann *et al.* (2002).

14.4 Summary

When carrying out PCA and FA with applied geochemical (environmental) data, which may, for many reasons, not be ideally suited for this method, the following points should be considered:

1. Before entering PCA or FA the distribution of each of the variables must be carefully studied. A transformation may be required to remove skewness (improve symmetry) or to approach normality.
2. When working with compositional data, a transformation should be undertaken to "open" the closed array (Section 10.5). Even when the measured variables only sum to a few per cent, it will be advantageous to open the data. For an example see Section 14.2.
3. If minor and trace elements are mixed in one and the same factor, log-transformation may not reduce the data to comparable scales and ensure that each variable has equivalent weight in the analysis. In this case standardisation to zero mean and unit variance guarantees an equal influence of all variables. Note that many software packages automatically carry out this standardisation, which does not allow carrying out a factor analysis based on the covariance matrix but only based on the correlation matrix.

4. Furthermore, it may be advisable to subdivide a large regional geochemical data set into the most obvious data subsets showing different geochemical behaviour, either based on pre-existing knowledge of the regional geology or following a cluster analysis, prior to undertaking a PCA or FA. If this is not done, the analysis may simply highlight differences between the subsets rather than their internal structures, though this can be a valid objective.

5. The safest method for undertaking FA with applied geochemical data is principal factor analysis (PFA). Advanced techniques like the maximum likelihood (ML) method are more dependent on the data being normally distributed, and given the nature of regional geochemical data this is rarely achieved. Failing to meet this assumption can produce misleading results. Robust methods of FA will very likely perform better, but still they cannot overcome problems related to multimodality (i.e. clear evidence of multiple processes influencing the data).

6. With a centred logratio transformation, in some instances the computation of FA (classical or robust) can fail due to a lack of full rank of the covariance matrix. In contrast, this usually does not occur with classical PCA.

7. Note that data outliers are a characteristic of many regional geochemical data sets – when using non-robust methods of factor analysis outliers should be identified and removed prior to undertaking the PCA or FA.

8. For factor rotation an orthogonal method should be considered first, e.g., Varimax.

9. The number of factors to extract is difficult to determine. The most practical approach is to use a scree plot for guidance, and then experiment with different numbers of factors and study the results (loadings and maps) in detail, even though this introduces a lot of subjectivity. In most cases a low number of extracted factors will give the best results in terms of interpretability.

10. The selection of the elements entered will influence the results of a PCA or FA. Firstly, it should always be remembered that there must be sufficient samples for the number of elements being entered. If the data set is large enough, it may be informative to first enter all elements, and afterwards test different combinations of elements. Obviously redundant or poorly conditioned (too many censored values) measurements should be considered for immediate or early exclusion. Just one element more or less may substantially alter the results of an analysis. It is a fact that in applied geochemistry the elements available for study will very often arbitrarily depend on what data were available from the analytical laboratory. This is often governed by price, methods available, and/or detection limits, and not by science.

Finally, PCA and FA are widely applied data analysis procedures, they can reduce dimensionality, and in doing so this may generate new variables more closely related to the underlying processes controlling the data distribution. However, both procedures need to be undertaken with care ensuring that their assumptions are met as closely as possible, that the data are appropriately conditioned, that outliers are handled judiciously, and that the problems posed by data closure were considered.

15
Cluster Analysis

The principal aim of cluster analysis is to split a number of observations or measured variables into groups that are similar in their characteristics or behaviour (see, e.g., Hartigan, 1975; Everitt and Dunn, 2001; Kaufman and Rousseeuw, 1990, 2005). A favourable cluster analysis outcome will result in a number of clusters where the samples or variables within a cluster are as similar as possible to each other while the differences between the clusters are as large as possible. Cluster analysis must thus determine the number of groups as well as the memberships of the observations or variables in those groups. To determine the group memberships, most clustering methods use a measure of similarity between the measurements. The similarity is usually expressed by distances between the observations in the multivariate data space.

Cluster analysis was developed in taxonomy. The aim was originally to get away from the high degree of subjectivity when different taxonomists performed a grouping of individuals. Since the introduction of cluster analysis techniques, there has been controversy over its merits (see Davis, 1973; Rock, 1988, and references therein). It was soon discovered that diverse techniques lead to different groupings even when using exactly the same data. Furthermore, the addition (or deletion) of just one variable in a cluster analysis can lead to completely different results. Workers may thus be tempted to just experiment with different techniques of cluster analysis and the selection of variables entered until the results fit their preconceived ideas. Practitioners employing cluster analysis should be aware of these problems. Cluster analysis should be applied as an "exploratory data analysis tool" to elucidate the multivariate behaviour of a data set. Comparing results of different techniques can be useful for obtaining a deeper insight into the multivariate data structure. However, like PCA and FA (see Chapter 14) it cannot itself provide a "statistical proof" of a certain relationship between the variables or samples and their groupings.

While PCA and FA (see Chapter 14) use the correlation matrix for reducing dimensionality or extracting common "factors" from a given data set, most cluster analysis techniques use distance measures to assign observations to a number of groups. Correlation coefficients lie between -1 and $+1$, with 0 indicating linear independence. Distance coefficients lie between 0 and ∞, with 0 indicating complete identicality (Rock, 1988). Some distance measures do not make, *a priori*, any statistical assumptions about the data, an ideal practical situation when working with geochemical data. Distance measures will also be essential for cluster validation, i.e. measuring the quality of a clustering. In theory, it should be advantageous to first use cluster analysis on a large geochemical data set to extract more homogeneous data subsets (groups) and

Statistical Data Analysis Explained   Clemens Reimann, Peter Filzmoser, Robert G. Garrett, Rudolf Dutter
© 2008 John Wiley & Sons, Ltd.

to then perform other data analysis procedures, e.g., univariate or multivariate investigations, on these homogeneous data subsets to study their data structure. In particular, for data sets with many variables it has been suggested (see, e.g., Everitt, 1974) to first use principal component analysis to reduce the dimensionality of the data and to then perform cluster analysis on the first few principal components. This approach has been criticised because clusters embedded in a high-dimensional variable space will not be properly represented by a smaller number of orthogonal components (see, e.g., Yeung and Ruzzo, 2001).

There are clustering methods that are not based on distance measures, like model-based clustering (Fraley and Raftery, 1998). These techniques usually find the clusters by optimising a maximum likelihood function. The implicit assumption is that the data points forming the single clusters are multivariate normally distributed, and the algorithm tries to estimate the parameters of the normal distributions as well as the membership of each observation to each cluster.

With applied geochemical and environmental data, cluster analysis can be used in two different ways: it can be used to cluster the variables (e.g., to detect relationships between them), and it can be used to cluster the samples into more homogeneous data subsets for further data analysis. Historically these have been referred to as R-mode and Q-mode, respectively.

15.1 Possible data problems in the context of cluster analysis

15.1.1 Mixing major, minor and trace elements

In multi-element analysis of geological materials one usually deals with elements occurring in very different concentrations. In rock geochemistry, the chemical elements are divided into "major", "minor" and "trace" elements. Major elements are measured in per cent or tens of per cent, minor elements are measured in about one per cent amounts, and trace elements are measured in ppm (mg/kg) or even ppb (μg/kg). This may become a problem in multivariate techniques considering all variables simultaneously because the variable with the greatest variance will have the greatest influence on the outcome. Furthermore, when variables are being clustered, special attention needs to be taken when working with closed compositional data (see Section 10.5). Variance is obviously related to absolute magnitude. As a consequence, one should not mix variables quoted in different units in one and the same multivariate analysis (Rock, 1988). Transferring all elements to just one unit (e.g., mg/kg) is not a complete solution to this problem, as the major elements occur in so much larger amounts than the trace elements. The data matrix will thus need to be "prepared" for cluster analysis using appropriate data transformation and standardisation techniques (see Chapter 10).

15.1.2 Data outliers

Regional geochemical data sets practically always contain outliers. The outliers should not simply be ignored, they have to be accommodated because they contain important information about the data and unexpected behaviour in the region of interest. In fact, identifying data outliers that may be indicative of mineralisation (in exploration geochemistry) or of contamination (in environmental geochemistry) is a major objective of any geochemical survey and many environmental surveys. Outliers can have a severe influence on cluster analysis, because they can affect distance measures and distort the true underlying data structure. Outliers should thus be removed prior to undertaking a cluster analysis, or statistical clustering methods able to handle outliers should be used (Kaufman and Rousseeuw, 1990, 2005). This is rarely done. Removing outliers that occur as single points or groups of several points does not infer that the reason for their being outliers does not have to be explained. In effect a manual clustering process is being undertaken to create an "external" cluster whose reason for existence has to be explained. However, their removal facilitates the subsequent formation of clusters by the chosen criteria. Finding data outliers is not a trivial task, especially in high dimensions (see Chapter 13).

15.1.3 Censored data

A further problem that often occurs when working with applied geochemical and environmental data is the presence of censored values, i.e. values below the detection limit. For some determinations a proportion of all results are below the lower limit of detection of the analytical method. For statistical analysis these are often set to a value of half the detection limit, resulting in (left) censored data (see Section 2.2). However, a sizeable proportion of all data with an identical value can seriously influence any cluster analysis procedure. For the Kola study data sets several variables have more than 25 per cent of the data below detection. It is very questionable whether such elements should be included at all in a cluster analysis. Unfortunately it is often the elements of greatest interest that contain the highest number of censored values (e.g., Au) – the temptation to include these in a cluster analysis is thus high. In the following examples all elements with more than 3 per cent censored values have been omitted from the cluster analyses.

15.1.4 Data transformation and standardisation

Some cluster analysis procedures do not require the data to be normally distributed. However, it is advisable that heavily-skewed data are first transformed to a more symmetric distribution. If a good cluster structure exists for a variable, a distribution which has two or more modes can be expected. A transformation to a more symmetrical form will preserve the modes but remove excessive skewness.

Most geochemical textbooks, including this one, claim that for applied geochemical and in general environmental data, a log-transformation is generally appropriate. Each single variable needs to be considered for transformation (see Section 10.2.4), with the Box–Cox transformation being the most flexible for minimising skewness.

An additional standardisation is needed if the variables show a striking difference in the amount of variability, i.e. when mixing data for major, minor, and trace elements. Thus centring and scaling the data (see Section 10.4) is usually a necessity for cluster analysis. When working with applied geochemical or environmental data, a robust version, using MEDIAN and MAD, should be preferred over the traditional MEAN and SD.

15.1.5 Closed data

Closed data (see Sections 10.5, 11.8, 14.1.4) can also be a problem when carrying out cluster analysis. The distance between the observations changes if a transformation to open the data is applied. Consequently results of cluster analyses will change. It may thus be a good idea to try cluster analysis with opened data in addition to cluster analysis on the simply log-transformed and standardised data. A general problem of the additive logratio transformation is the choice of the ratio-variable; results will change substantially when different variables are used.

15.2   Distance measures

A key issue in most cluster analysis techniques is how to measure distance between the samples (or variables). Note that "distance" in cluster analysis has nothing to do with geographical distance between two sample sites but is rather a measure of similarity between samples in the multivariate space defined by the variables, or a measure of similarity between variables. Many different distance measures exist (Bandemer and Näther, 1992). Modern software implementations of cluster algorithms can accommodate a variety of different distance measures because the distances rather than the data matrix are taken as input, and the algorithm is applied to the given input.

For clustering samples the Euclidean or Manhattan distances between the samples are often used. The latter measures the distance parallel to the variable axes, whilst the former measures the direct distance; usually both measures lead to comparable results. Other distance measures like the Canberra distance (Lance and Williams, 1966) or a distance based on the random forest proximity measure (Breiman, 2001) can give completely different cluster results or produce highly unstable clusters. For studying the clustering of variables, additionally, correlation measures may be used.

15.3   Clustering samples

An important and confusing task for the practitioner in using cluster analysis is choosing between the various different clustering methods. The samples need to be grouped into classes (clusters). If each sample is allocated into only one (of several possible) cluster(s), this is called "partitioning". Partitioning will result in a pre-defined (user-defined) number of clusters. It is also possible to construct a hierarchy of partitions, i.e. group the samples into 1 to n clusters (n = number of samples). This is called hierarchical clustering (see, e.g., Kaufman and Rousseeuw, 1990, 2005). Hierarchical clustering always yields n cluster solutions and based on these solutions the user has to decide which result is most appropriate.

Principally, there are two different procedures. A sample can be allocated to just one cluster (hard clustering) or be distributed among several clusters (fuzzy clustering). Fuzzy clustering allows that one sample belongs, to a certain degree, to several groups (see, e.g., Dumitrescu et al., 2000). In terms of applied geochemistry and environmental studies this procedure will often yield the more interesting results because it reveals if several processes, as characterised by the clusters, have influenced one sample. The cluster solution will then indicate to what degree the different processes influence the samples.

15.3.1   Hierarchical methods

The input to most of the hierarchical clustering algorithms is a distance matrix (distances between the observations). The widely used agglomerative techniques start with single sample clusters (each sample forms its own cluster) and then they enlarge the clusters stepwise. The computationally more intensive reverse procedure starts with one cluster containing all samples and then successively splits this into groups step by step. This procedure is called divisive clustering.

In the agglomerative algorithm, where initially each sample forms its own group, the number of clusters is reduced one by one by combining (linking) the most similar groups at each step of the algorithm. At the end of the process there is one single cluster left containing all the samples. A number of different methods exist to link two clusters, the best known are average linkage, complete linkage, and single linkage (see, e.g., Kaufman and Rousseeuw, 1990, 2005). The average linkage method considers the averages of all pairs of distances between the samples of two clusters. The two clusters with the minimum average distance are combined into one new cluster. Complete linkage looks for the maximum distance between the samples of two clusters. The clusters with the smallest maximum distance are combined. Single linkage considers the minimum distance between all samples of two clusters. The clusters with the smallest minimum distance are linked. Single linkage will result in cluster chains because for linkage it is sufficient that only two samples from different clusters are close together. Complete linkage will result in very homogeneous clusters in the early stages of agglomeration, however, the resulting clusters will be small. Average linkage is a compromise between the two other methods and usually performs best in typical applied geoscience and environmental applications.

Because the cluster solutions grow tree-like (starting with the branches and ending with the trunk) results are often displayed in a graphic called the *dendrogram* (Figure 15.1) (see,

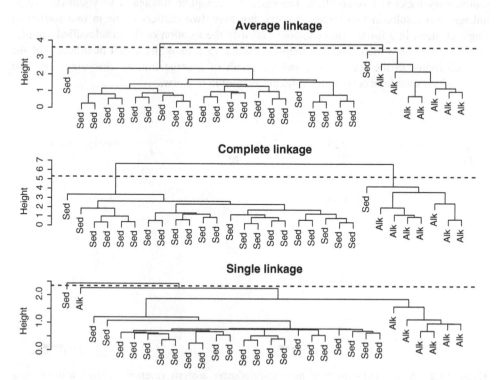

Figure 15.1 Dendrograms obtained from hierarchical cluster analyses of the log-transformed and standardised Al, Ca, K and Na data for Kola C-horizon soils overlying two discrete lithological units (lithology 82, alkaline rocks; and lithology 9, sedimentary rocks)

e.g., Kaufman and Rousseeuw, 1990, 2005). Horizontal lines indicate linking of two samples or clusters, and thus the vertical axis presents the associated height or similarity as a measure of distance. The samples are arranged in such a way that the branches of the tree do not overlap. Linking of two groups at a large height indicates strong dissimilarity (and vice versa). Therefore, a clear cluster structure would be indicated if observations are linked at a very low height, and the distinct clusters are linked at a greater height (long branches of the tree). Cutting the dendrogram at such a greater height permits assigning the samples to the resulting distinct clusters. Visual inspection of a dendrogram is often helpful in obtaining an initial estimate of the number of clusters for partitioning methods.

To demonstrate the procedure, a subset of samples of the Kola C-horizon soils from two lithological groups, one developed on alkaline rocks (lithology 82 – see Figure 1.2, the geological map) and the other on late Proterozoic sedimentary rocks which occur on the Norwegian and parts of the Russian coast (lithology 10) was created. Dendrograms based on the Al, Ca, K, and Na analyses of these samples are shown in Figure 15.1. There are clearly two groups that correspond well, but not perfectly, to the soils from each lithology. Hierarchical clustering based on Euclidean distances for the log-transformed and standardised data was used in these analyses. The resulting dendrograms for average, complete and single linkage are shown (Figure 15.1). A dashed line indicates the cut-off for the solutions resulting in two clusters.

The performance of the three procedures can be directly judged and compared. Average linkage results in two clear clusters, and only one of the samples collected from over the sedimentary rocks is misclassified. The result for complete linkage is very similar. Single linkage also results in two strong clusters, however, two outliers occur in two additional single clusters. In a further step one could identify the location of the misclassified samples or the outliers on a map. The two maps in Figure 15.2 show the actual distribution of the two lithologies in the survey area and the result of average linkage clustering. The one misclassified sample occurs on the Varanger Peninsula in Norway.

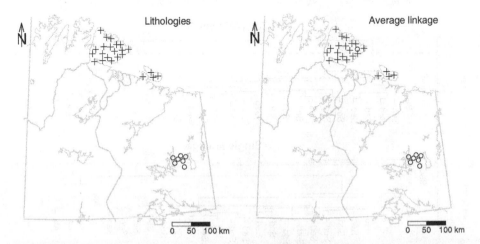

Figure 15.2 Results obtained from hierarchical cluster analysis (average linkage) with the log-transformed and standardised Al, Ca, K and Na data for a subset of the Kola C-horizon soils (right map). The location of the samples comprising the data from two lithological units is shown in the left map: lithology 82, alkaline rocks (circles); and lithology 9, Caledonian sediments (crosses)

The problem with dendrograms is that the number of samples needs to be small (e.g., $n < 100$) to obtain "readable" dendrograms. With hundreds of samples, as is the case for the complete Kola data set, dendrograms can be hard to read and appreciated as they become overly cluttered. Maps showing the cluster memberships can profitably be used to display the results of the analyses, and when the multivariate clusters form coherent spatial patterns, confidence in the procedure and the validity of the results is increased.

15.3.2 Partitioning methods

In contrast to hierarchical clustering methods, partitioning methods require that the number of clusters be pre-determined (see, e.g., Kaufman and Rousseeuw, 1990, 2005). In the case of the above example, it is known *a priori* that two groups should result. In the case where nothing is known about the samples, it is good practice to investigate the multivariate structure of the data using projection methods (see Section 13.5) and/or undertake a hierarchical clustering to obtain an idea of the likely number of groups. The other possibility is to partition the data into different numbers of clusters and evaluate the results (see Section 15.5). For regional-scale survey data a more subjective evaluation, but still useful approach, is to visually inspect the location of the resulting clusters in a map. This exploratory approach can often reveal interesting data structures.

A very popular partitioning algorithm is the k-means algorithm (MacQueen, 1967). It attempts to minimise the average of the squared distances between the observations and their cluster centres or centroids. This can usually only be approximated via an iterative algorithm leading to a local rather than the global optimum. Starting from k given initial cluster centroids (e.g., random initialisation by k observations), the algorithm assigns the observations to their closest centroids (using, e.g., Euclidean distances), recomputes the cluster centroids, and iteratively reallocates the data points to the closest centroid. Several algorithms exist for this purpose, those of Hartigan (1975), Hartigan and Wong (1979) and MacQueen (1967) are the most popular. There are also some modifications of the k-means algorithm. Manhattan distances are used for k-medians where the centroids are defined by the medians of each cluster rather than their means. Variations on the k-means procedure are discussed by Templ *et al.* (2006). A new highly extensible toolbox for centroid clustering was recently implemented in R (Leisch, 2006) where the user can easily experiment with almost any arbitrary distance measures and centroid computations for data partitioning.

Kaufman and Rousseeuw (1990, 2005) proposed several clustering methods which became popular because they are implemented in a number of software packages. The partitioning method PAM (Partitioning Around Medoids) minimises the average distances to the cluster medians. It is thus similar to the k-medians method but allows the use of different distance measures. A similar method called CLARA (Clustering Large Applications) is based on random sampling. It saves computation time and can therefore be applied to larger data sets.

Figure 15.3 (left) shows a map of the results of k-means clustering for two clusters based on Euclidean distances for the same log-transformed and standardised data used previously (Section 15.3.1). Note that a dendrogram cannot be constructed for partitioning-based cluster methods because there is no hierarchy. The right map shows the results of PAM clustering with two clusters (Figure 15.3, right) based on Manhattan distances. The k-means method results in the same sample being misclassified as by hierarchical clustering (compare Figure 15.2, right, and 15.3, left). The PAM method is successful in classifying the samples into the known lithological units.

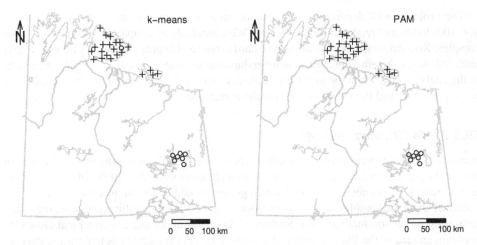

Figure 15.3 Results obtained from k-means (left) and PAM (right) cluster analyses of the log-transformed and standardised Al, Ca, K and Na data for a subset of the Kola C-horizon soil samples. Compare to Figure 15.2

The results of all these algorithms depend on randomly selected initial cluster centres. If bad initial cluster centres are selected, the iterative partitioning algorithms can lead to a local optimum that can be far away from the global optimum. This can be avoided by applying the algorithms with different random initialisations and by taking the best (according to a validity measure, see below) or most stable result. Additional procedures for finding the global optimum have been described by Breiman (1996) and Leisch (1998, 1999).

15.3.3   Model-based methods

Model-based (Mclust) clustering (Fraley and Raftery, 2002) is not based on distances between the samples but on models describing the shape of possible clusters. The Mclust algorithm selects the cluster models (e.g., spherical or elliptical cluster shape) and determines the cluster memberships of all samples for solutions over a range of different numbers of clusters. The estimation is done using the Expectation Maximisation (EM) algorithm (Dempster *et al.*, 1977). The EM algorithm is performed on the specified numbers of clusters and with several sets of constraints on the covariance matrices (shapes) of the clusters. Finally, the combination of model and number of groups that leads to the highest BIC (Bayesian Information Criterion) value can be chosen as the optimum model (Fraley and Raftery, 1998).

Figure 15.4 shows the results of model-based clustering for a subset of the Kola C-horizon soil data overlying "alkaline" rocks (lithology 82), "sediments" (lithology 9), and "granites" (lithology 7) using the elements Ca, Cu, Mg, Na, Sr and Zn (log-transformed and standardised). To display the results, a PCA (Section 14.2) was used to reduce the six-dimensional space (six variables) to two dimensions. These two first principal components reflect the main variability of the data set (80 per cent explained variance). The main structure in the six dimensions can thus be projected into two dimensions. Figure 15.4, left, shows the location of the three different lithologies (three different symbols) in the reduced two-dimensional PCA space (for comparison, see Section 13.5 for a display of these data as a minimum spanning tree). The cluster memberships of the samples are displayed by different symbols in Figure 15.4, right.

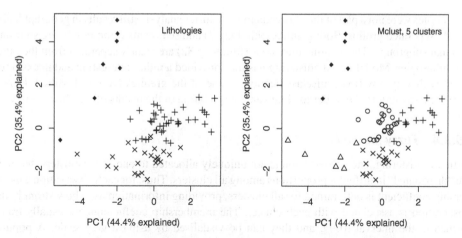

Figure 15.4 Results obtained from model-based cluster analysis with the log-transformed and standardised variables Ca, Cu, Mg, Na, Sr and Zn for a subset of the Kola C-horizon soil samples derived from three lithologies: lithology 7, granites (+); lithology 9, sediments (×); and lithology 82, alkaline rocks (•). Left plot: projection of the data on the first two principal components. Right plot: the same projection as used for the left plot, but symbols according to the cluster result

The model-based method suggests two additional clusters beyond the expected three, and that five clusters are the optimal solution (resulting in five symbols in the plot). Two of the lithologies are split into two groups each. A few samples are obviously misclassified. As the next step of data analysis, it is necessary to investigate where in the map these five clusters plot.

Figure 15.5, left, shows the location of the three selected lithologies in the survey area, Figure 15.5, right, shows the spatial location of the five clusters. Note that the coordinates of

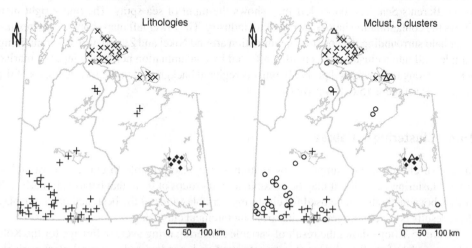

Figure 15.5 Results obtained from model-based cluster analysis with the log-transformed and standardised variables Ca, Cu, Mg, Na, Sr and Zn for a subset of the Kola C-horizon soils derived from three lithologies: lithology 7, granites (+); lithology 9, sediments (×); and lithology 82, alkaline rocks (•). Left plot: location of the three lithologies. Right plot: distribution of the five clusters using the same symbols as in Figure 15.4, right

the samples were not a part of the cluster analysis. Cluster analysis still results in geographically clearly separated groups. Both granites (lithology 7) and sediments (lithology 9) are split into regional subgroups. The alkaline intrusions (lithology 82) are clearly separated from the other two lithologies. Model-based clustering has thus provided a rather interesting and unexpected result with very few (two) misclassifications; none of the granites are misclassified. Here a map is clearly required in order to obtain an interpretation of the results of the cluster analysis.

15.3.4 Fuzzy methods

In fuzzy clustering, each observation is not uniquely allocated to one of the clusters, but they are "distributed" in different proportions among all clusters. Thus, for each observation a membership coefficient is determined for all clusters, providing information as to how strongly the observation is associated with each cluster. The membership coefficients are usually transformed to the interval [0, 1], and they can be visualised by using a grey scale. A popular fuzzy clustering algorithm is the fuzzy c-means (FCM) algorithm, developed by Dunn (1973) and improved by Bezdek (1981). The number of clusters has to be determined by the user. The algorithm calculates the cluster centroids and the membership coefficients for each observation. Alternative fuzzy clustering algorithms have been developed, e.g., Gustafson–Kessel (Gustafson and Kessel, 1979) and Gath–Geva (Gath and Geva, 1989), which are more flexible in detecting different cluster sizes and shapes.

Figure 15.6 displays the results of fuzzy clustering with the FCM algorithm into four clusters using the Kola moss data set and the elements Cu, Mg, Ni and Sr (log-transformed and standardised). Each cluster is represented on its own map. A dark symbol indicates a high likelihood of group membership, a light symbol a low likelihood of membership of the observation to the cluster displayed on the map. The maps show the distribution of two main processes determining the regional distribution of elements in this data set: contamination from the Russian nickel industry (Cu, Ni) and the steady input of sea spray (Mg, Sr) along the coast of the Barents Sea. The upper left map shows the input of sea spray. The upper right map shows the contamination haloes surrounding industry. The lower left map shows an interesting outer halo surrounding the core of contamination around Nikel and Zapoljarnij. The remaining samples fall into a cluster that is neither affected by contamination nor by the input of marine aerosols along the coast, and probably reflects regional background in the forested parts of the project area (Figure 15.6, lower right).

15.4 Clustering variables

Instead of clustering the samples (observations), it is also possible to cluster the variables. When clustering variables, it may be possible to find groups of them that behave similarly. All of the above methods can be used for clustering variables. One of the best methods to display these results is the dendrogram arising from hierarchical clustering.

Figure 15.7 (top) shows the result of variable clustering using average linkage for the Kola O-horizon data for all variables (log-transformed and standardised). Several clear variable associations emerge that can be explained from their geochemical behaviour or on the basis of knowledge concerning important processes in the survey area. For example, Cu, Ni and Co are the three elements with very high emissions from the Kola smelters. Arsenic (As), Mo and Cd are inferred to be comparatively minor components of the emissions by their greater

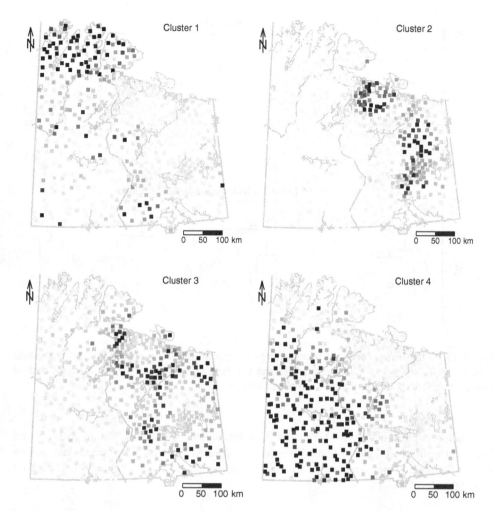

Figure 15.6 Results obtained from FCM clustering into four clusters based on the log-transformed and standardised variables Cu, Mg, Ni and Sr for the Kola C-horizon moss data. A dark symbol colour indicates a strong group membership of the sample in the respective cluster

distances from Cu, Ni and Co in the dendrogram. The branch linking LOI, S, Sb and Hg indicates the major role of organic material in sequestering these elements. Na, Mg, Ca and Sr are all important ingredients of marine aerosols, playing a determining role for the chemical composition of the soils collected near the Arctic coast.

Figure 15.7 (bottom) displays the similarly computed dendrogram but using centred logratio transformed and standardised data in order to remove the effects of closure. The two dendrograms are almost identical, however, there are a few interesting differences. The elements associated with emissions from the Kola smelters, Cu, Ni, Co, Mo and As still form a unique group. However, Cd has now moved to be associated with Sb, Bi and Pb, which is a more geochemically explainable association. The cluster including Ca, Mg, K, P, Hg, LOI and S reflects the abundance of organically derived material. The marine aerosol Na–Sr

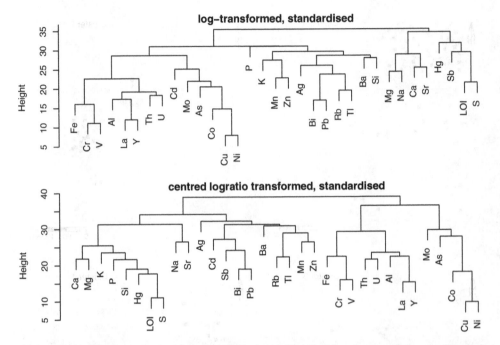

Figure 15.7 Dendrogram obtained from hierarchical variable clustering (average linkage) of the Kola O-horizon data set, using all log-transformed and standardised variables (top) and using all centred logratio transformed and standardised variables (bottom)

association is more clearly differentiated from the organic matter association with centred logratio transformed data. In several respects the centred logratio transformed data yield a clearer set of elemental associations with larger geochemically coherent clusters being linked at a higher level.

15.5 Evaluation of cluster validity

Because there is no universal definition of clustering, there is no universal measure with which to compare clustering results. However, evaluating cluster quality is essential since any clustering algorithm will produce some clustering for every data set. Validity measures should support the decision made on the number of clusters, and they should also be helpful for evaluating the quality of the single clusters.

As mentioned in Section 15.3.2, a rather simple method of evaluating quality of clustering for regionalised data is to check the distribution of the resulting clusters on a map. The distribution of the clusters can then be evaluated against knowledge concerning the survey area. It is also likely that clusters resulting in spatially well-defined subgroups are more likely to have a meaning than clusters resulting in "spatial noise".

There are many different statistical cluster validity measures. Two different concepts of validity criteria – external and internal criteria – need to be considered.

External criteria compare the partition found with clustering, with a partition that is known *a priori*. The most popular external cluster validity indices are Rand, Jaccard, Folkes and

Mallows, and the Hubert Indices (see, e.g., Hubert and Arabie, 1985; Gordon, 1999; Haldiki *et al.*, 2002; Milligan and Cooper, 1985).

Internal criteria are cluster validity measures that evaluate the clustering result of an algorithm by using only quantities and features inherent in the data set. Most of the internal validity criteria are based on within and between cluster sums of squares, i.e. Multivariate Analysis of Variance (MANOVA) procedures. Well-known indices are the Calinski–Harabasz index (Calinski and Harabasz, 1974) and Hartigan's indices (Hartigan, 1975), or the Average Silhouette Width of Kaufman and Rousseeuw (1990, 2005).

Figure 15.8 shows the "average silhouette width" plot for the results of the PAM algorithm (Figure 15.3, right), one plot (Figure 15.8, left) for two PAM clusters, and the other (Figure 15.3, right) for three PAM clusters. The sample numbers are shown along the y-axis. A grey stripe, indicating how well the sample is represented in the cluster, is displayed for each sample. An optimal classification of the sample will be represented by the stripe reaching 1. If a sample does not really belong to the cluster, the stripe will be rather short, and if the sample should in fact belong to another cluster, a stripe in the negative direction may result. Figure 15.8, left, shows that two good clusters were obtained. Sample 406 shows a poor group membership in the first cluster, and sample 262 does not really fit into either of the two clusters but is placed into cluster 2. The quoted average silhouette width of 0.59 indicates the overall quality of the clustering. It can aid the decision as to how many clusters should be selected. If PAM is run for 3 clusters the diagram shown in Figure 15.8, right, is generated. It is obvious that the overall cluster quality is much worse than in the two-cluster case; the average silhouette width has declined to 0.32.

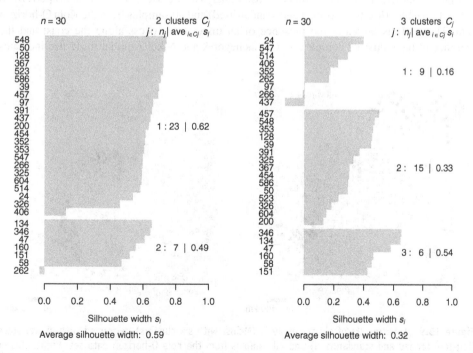

Figure 15.8 Average silhouette width plot (Kaufman and Rousseeuw, 1990, 2005) of the results of PAM clustering. The left plot relates to Figure 15.3, right, for two clusters, the right plot presents the result if three clusters are selected

From a practical point of view, an optimal value of the validity measure does not imply that the resulting clusters are meaningful in geochemical or environmental contexts. Some of these criteria evaluate only the allocation of the data to the clusters. Other criteria evaluate the form of the clusters or how well the clusters are separated. The resulting clusters are the best for the partitioning algorithm and parameters employed and are evaluated according to the validity measure selected. The measures provide good results when a very clear cluster structure exists in the data. Unfortunately, when working with applied earth science and environmental data, such good clusters are rare. Thus cluster quality measures often fail to be usefully informative when working with such data, and the best approach to evaluating cluster quality is often the most simple one: look at the results on a map.

15.6 Selection of variables for cluster analysis

Multivariate techniques may be used in order to reduce the dimensionality of a data set or to learn something about the internal structure within the variables and/or samples. It may thus appear desirable to perform cluster analyses with all available samples and variables. However, as already observed when performing principal component or factor analysis (Chapter 14), the addition of only one or two irrelevant variables can have drastic consequences on defining the clusters. The inclusion of only one irrelevant variable may be enough to hide the real clustering in the data (Gordon, 1999). The selection of the variables to enter into a cluster analysis is thus of considerable importance when working with applied earth science and environmental data sets containing a multitude of variables.

Figure 15.9 left shows the results of cluster analysis (Mclust with six clusters) performed with all variables (log-transformed and standardised) and all samples from the Kola O-horizon data. The analysis indicates the presence of an important process along the coast and the presence of the industrial complexes at Monchegorsk and Nikel/Zapoljarnij. It also indicates

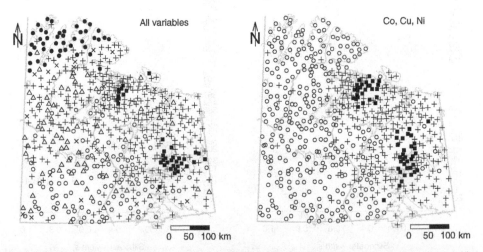

Figure 15.9 Left: results of cluster analysis (Mclust with six clusters) performed with all variables (log-transformed and standardised) and all samples from the Kola O-horizon data set. Right: cluster analysis (Mclust with three clusters) performed with three key variables (Co, Cu and Ni – log-transformed and standardised) indicative of industrial contamination only

the existence of an important north–south and east–west change in cluster membership related to biological processes.

Another reason for variable selection may be a desire to focus the analysis on a specific geochemical process. Such a process is usually expressed by a combination of variables, and using these variables for clustering facilitates the identification of those samples or areas where the process is present or non-existent. The variables could simply be chosen based on expert knowledge. It is also possible to apply variable clustering (see above) and select variables which are in close association (i.e. on one branch of the cluster tree) to highlight a certain process or to study in more detail. Figure 15.9 right shows the results of cluster analysis when only the key elements for contamination from the nickel industry, Co, Cu and Ni (see Figure 15.7), are included. Here the clusters identify a contaminated core area in the near vicinity of the smelters at Nikel/ Zapoljarnij and Monchegorsk, and a halo of lesser contamination covering most of the Russian survey area. The non-contaminated areas in Finland and northern Norway emerge as the third cluster.

Variable clustering can of course also be used to select single key variables from each important cluster to simply reduce dimensionality for clustering samples. This approach may be particularly useful with data sets that have an insufficient numbers of samples, n, to support the number of variables, p (see discussions in Sections 13.3 and 14.1.8).

15.7 Summary

Cluster analysis can be used to structure the variables or to group the observations. Variable selection is crucial to the outcome of cluster analysis. Just one variable more or less entered into cluster analysis can drastically change the results. As such cluster analyses are an exploratory method to better understand the inner structure of the data; they should not be misused as a "statistical proof" for any identified associations. As a general rule, ensuring symmetry of the variable distributions (e.g., via using a log-transformation for each variable) and standardisation is a necessary part of data preparation before applying cluster analysis. Depending on the clustering method, outliers can heavily affect the results. It may thus be necessary to remove outliers, to form "special" clusters requiring explanation, prior to carrying out a cluster analysis. Closure is usually not as serious a problem for cluster analysis as for correlation-based methods. It is still advisable to also try cluster analysis with opened data.

It is difficult to provide a general recommendation concerning the "best" cluster method. With the Kola Project data the most interesting results were obtained with model-based clustering (algorithm Mclust), but also other simpler algorithms led to useful interpretations and maps. If observations are clustered (and not variables) the visualisation of the clusters in maps is useful. Such maps are often far more helpful than plots of validity measures or dendrograms, the latter being particularly difficult to read for applications with many observations.

16

Regression Analysis (RA)

Regression analysis is used to predict values of a dependent or response variable from the values of one or more independent or explanatory variables (see, e.g., Fox, 1997; Draper and Smith, 1998). The dependent variable y is the variable that will be predicted using the information of the independent, explanatory, variable(s) $x_1, \ldots, x_p$ with $p \geq 1$. When dealing with only one x and one y variable for correlation analysis, which is somewhat related to regression analysis, the choice of the x and y variable has no influence on the resulting correlation coefficient. For regression analysis, however, this choice is of paramount importance. Figure 16.1 shows this for the example of Ca and Sr used in Figure 6.3, right. In general, the higher the Pearson correlation coefficient between the x and y variable, the more similar regression lines will result. When plotting the scatterplot matrix with additional regression lines, it is thus important to look at the mirrored matrix as well or to plot the regression lines for x versus y and for y versus x (for an example see Figure 8.14).

Regression analysis could, for example, be interesting if one variable is very difficult or very expensive to analyse or the determinations were of poor quality (see Chapter 18). It could thus be tempting to predict values of this variable using analytical results of (an)other variable(s). In practice a number of measurements for the x and y variable are needed to build a model for prediction. Such a relation could take the simple form of a linear model, or more generally of a non-linear model (see, e.g., Seber and Wild, 2003).

The close relation between Cu and Ni was highlighted in Figure 6.3 by adding a least squares regression line (Gauss, 1809; Draper and Smith, 1998). Fitting the line also accentuated that the relation is not really linear but rather curvilinear. Figure 16.2 shows both cases, however, for selected data points only, to be able to better demonstrate the parameter estimation for the model. A linear model was fitted in the left hand plot and a quadratic model was used in the right hand plot. The regression equations take the following form:

Linear: $\log \text{Ni} = b_0 + b_1 \log \text{Cu} + \varepsilon$,
Quadratic: $\log \text{Ni} = b_0 + b_1 \log \text{Cu} + b_2 (\log \text{Cu})^2 + \varepsilon$,

where ε represents the error, which is mathematically minimised using some criterion, frequently least squares.

These equations are then used to estimate the unknown regression parameters b_0 and b_1 in the linear model and b_0, b_1, and b_2 in the quadratic model, from the measured values of

Statistical Data Analysis Explained   Clemens Reimann, Peter Filzmoser, Robert G. Garrett, Rudolf Dutter
© 2008 John Wiley & Sons, Ltd.

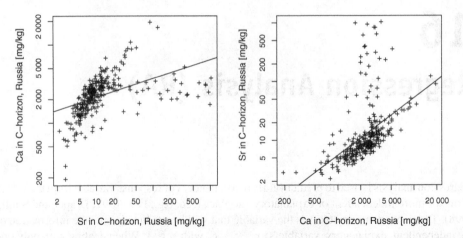

Figure 16.1 Sr versus Ca and Ca versus Sr from the Kola C-horizon soil data (only samples from Russia used) with fitted linear (least squares) regression lines

Cu and Ni. To get a good fit of the model, the error for each single point has to be as small as possible. There are many different methods to estimate the regression parameters such that the error is minimised, the most frequently chosen approach is the least squares method. It minimises the sum of the squared errors. The resulting least square regression parameters lead to the line and curve shown in Figure 16.2. The vertical lines shown in the figure indicate the magnitude of the errors. The errors are the differences between the observed y-values and the predicted values and they are usually called the residuals. Another minimisation procedure is, for example, the trimmed sum of squared residuals (Rousseeuw, 1984), where the largest squared residuals are trimmed (down-weighted). This leads to the well-known least trimmed

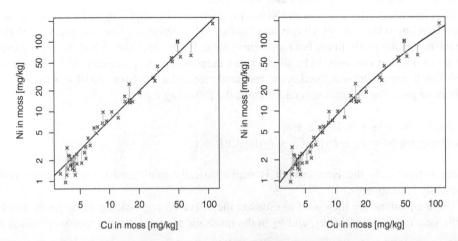

Figure 16.2 Cu versus Ni, Kola moss data set, with linear (left) and quadratic (right) regression models. Although the model was built using all measurements, only 50 points are shown in the diagrams to be able to plot the residuals (errors that have been minimised) with vertical lines from data points to regression lines

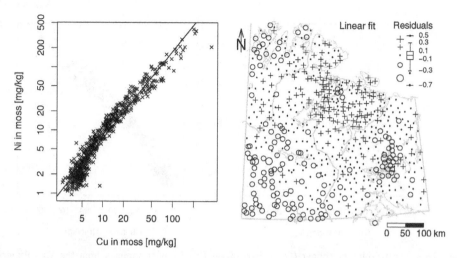

Figure 16.3 Cu used to predict Ni, Kola moss data. Left plot: linear regression model with all data points; right plot: map showing the regional distribution of the residuals

sum of squares (LTS) regression method, which is highly robust (see Figure 6.4). Other alternatives are not discussed here; the interested reader is referred to, for example, Rousseeuw and Leroy (1987), Maronna *et al.* (2006), and Section 16.6.

Once the model is established the regression parameters are known and new measured x-values can then be used to predict corresponding y-values. For the above example we obtain the following formula:

Linear: predicted $\log \mathrm{Ni} = -0.52 + 1.39 \log \mathrm{Cu}$,
Quadratic: predicted $\log \mathrm{Ni} = -0.90 + 2.12 \log \mathrm{Cu} - 0.30(\log \mathrm{Cu})^2$.

Another application of regression analysis would be to compare the measured y-values with the predicted y-values, i.e. to study the residuals (and their location in space).

In the above examples one x variable was used to predict the values for the y variable. For building a better model, several variables could be used to predict the y variable. Thus multiple regression uses two or more explanatory variables. Furthermore, there are different computational procedures that can be applied to multiple regression. The term "regression model" is frequently used to describe the result of an RA, and the computational process is referred to as fitting a model to the data. In this sense a data-based model has been constructed to predict the unknown value of a variable from measurements of other known variables. This chapter focuses on multiple regression and describes several of the available procedures.

16.1 Data requirements for regression analysis

16.1.1 Homogeneity of variance and normality

The most important requirement for undertaking practical regression analyses is that the variances of the response variable across the range of the data are similar. This property of homogeneity of variance is also referred to as homoscedasticity, and lack of homogeneity

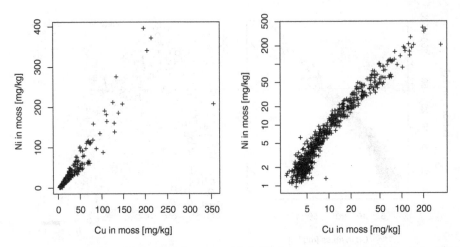

Figure 16.4 Scatterplot of copper (Cu) versus nickel (Ni) in moss samples from the Kola Project. Left side: original data scale; right side: log-scale. Ni is the dependent and Cu is the independent (or explanatory) variable

as heteroscedasticity. Figure 16.4 demonstrates the visual impact of heteroscedasticity for Cu and Ni determined in Kola Project mosses. In Figure 16.4, left, the scatterplot is for the un-transformed data; the visual expression of heteroscedasticity is the "fanning out" out of the points as they become more distant from the origin. Thus for higher values of Cu the variance of Ni is higher than for low values, i.e. in a linear regression model the residuals will increase with increasing values of Cu.

Figure 16.4, right, displays the same data plotted with logarithmic scaling, the same as plotting the logarithms of the values. Applying a logarithmic transformation has removed the "fanning" so that the data largely fall within a parallel band, which in the case of the moss data is slightly curved.

In many statistical texts a normal distribution of the residuals is stated as the most important consideration. However, as the main objective of RA is prediction, homogeneity of variance is the more important for these analyses to ensure that predictions across the range of the explanatory variables are made with similar precision. Normality of the residuals becomes important if inference tests or confidence intervals based on normal theory are to be computed.

Transformations are thus an important consideration in RA; the paper by Bartlett (1947) is still an important contribution, more modern treatments are available in text books such as Cook and Weissberg (1982) or Johnson and Wichern (2002). However, in contrast to other multivariate techniques, many practitioners do not routinely apply transformations to the variables prior to commencing an RA, and then only consider transformations depending on the results of regression diagnostics (see below). It is also possible to visually inspect scatterplots of the dependent variable against all response variables for homogeneity of variance and decide on required transformations before commencing the RA. Note that the x-variable(s) as such do(es) not need to follow any statistical distribution. Transformations are used to reach a linear relation to the y-variable and ensure homogeneity of variance rather than to approach normal distributions.

16.1.2 Data outliers, extreme values

Data outliers and extreme values take on a special importance in regression analysis. Their presence at large distances and away from the apparent trend in the data can "lever" the regression line (simple regression) or (hyper)plane (multiple regression) away from its "true" position (see, e.g., Rousseeuw and Leroy, 1987; Maronna *et al.*, 2006). In simple regression outliers and extreme values can easily be detected by inspecting a scatterplot (xy-plot, see Figure 16.4, right) of the two variables. Outliers and extreme values acting as possible *leverage points* will immediately be visible. In the example (Figure 16.4) it is questionable whether the two observations deviating from the linear trend are in fact influential on the estimation of the regression line as they lie relatively close to the apparent trend in the data. In multiple regression it is more difficult to graphically identify outliers due to the number of variables that need to be inspected simultaneously; the scatterplot matrix (see Section 6.7, Figure 6.14) can help in this respect. Additionally, the statistical distributions of the response and explanatory variables can be visually inspected to detect any obvious outliers or extreme values. Alternatively, outliers and extreme values can be left in the data and regression diagnostics computed to identify them. These values may be removed at a later stage, or robust procedures employed that down-weight outliers and extreme values, so minimising their influence (see, e.g., Rousseeuw and Leroy, 1987; Maronna *et al.*, 2006).

Simple regression and its visualisation were introduced in Section 6.2 where adding linear regression lines to scatterplots was discussed. Figure 6.4 (right) clearly illustrates the impact of using a robust fitting procedure; the position of the robust regression line is dominantly controlled by the main mass of the data, and the impact of unusual values at higher levels of Sr but lower levels of Ca has been significantly reduced.

16.1.3 Other considerations

RA is based on an underlying correlation between the response and explanatory variables. If no such correlation exists, an effective regression model for prediction cannot be derived. Conversely, if the data are spatially correlated (see Section 14.1.7), certain relationships may be over-represented as the attributes of proximal samples are more similar to each other than to the attributes of samples located more distantly. This can lead to biased estimates. This problem is fundamental to spatial data as used in applied geochemical and environmental studies, and is dependent on sample density (see discussion in Section 5.7.2 concerning spatial dependency).

Censored data (Section 2.2) should be avoided, as in PCA, FA, and Discriminant Analysis (DA, see Chapter 17); a large number of observations with similar low values set to half (or some similar proportion) of the detection limit can become highly influential and distort the prediction model being estimated.

RA should not be applied to inhomogeneous data. Examples are data sets that are separated in the data space and have different internal (covariance) structures. The resulting regression model will be attracted to the line joining the two centroids if there are two groups or clusters of data. If there are more groups, the position of the regression line in the multivariate space is probably meaningless in the context of the subject matter.

Dimensionality is an important consideration as in other multivariate data analysis techniques. The number of observations must be larger than the number of explanatory variables. If they are not, computational problems arise and a regression model cannot be computed using standard procedures. In most instances these problems can be avoided by having the number of observations exceed the number of explanatory variables by at least a factor or three, and five is better. In addition, when some of the explanatory variables are highly correlated, a property named collinearity, computational problems may also be encountered and subsequent tests on the regression coefficients will be unreliable (see, e.g., Draper and Smith, 1998). Approaches to this problem will be discussed in the final section of this chapter.

Closure can be an issue in regression analysis. It can be advisable to try an additional regression analysis with the opened data (see Section 10.5). The final decision on whether the closed array or the opened data should be used can then be based on the quality of the results of the RA (see below).

16.2 Multiple regression

As an example, using the Kola Project C-horizon soil data, Co determined by ICP-AES after an aqua-regia digestion is predicted from the Cr data determined by both ICP-AES after the same aqua-regia digestion and by INAA. The INAA determination is total and measures the Cr in all mineralogical forms, whereas the ICP-AES data only measures the Co and Cr liberated from minerals that can be decomposed by aqua-regia. When dealing with three variables it is no longer possible to fit a single line or curve to the data, but rather a plane or a surface.

Figure 16.5 (left) displays the plane through the data based on the logarithms of the two Cr-determinations in an attempt to predict the logarithm of the Co values of the soil samples. There are a number of observed Co values that are distant from the plane, i.e. they have large *residuals*. The regression procedure has found the plane for which the sum of the squares of the distances from the points to the plane parallel to the Co axis is a minimum. This is also known as a Least Squares (LS) fit. The distances between the observations and the plane, the residuals, are visualised by lines in Figure 16.5. An inspection of the plane reveals that it is

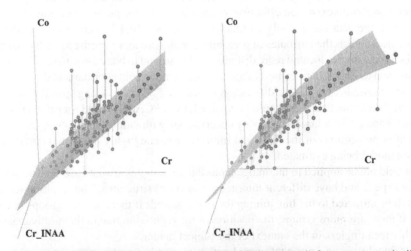

Figure 16.5 Cobalt (Co) (response, or dependent, variable) determined after an aqua-regia digestion versus Cr determined both after an aqua-regia digestion and by INAA (all variables log-transformed), Kola C-horizon, shown in 3-D plots. Left diagram, linear model; right diagram, second-order polynomial surface

intersecting a curved surface of points. At low levels of Co there appear to be more points below the plane than above, and at higher (medium) levels of Co perhaps more points are above the plane than below. To accommodate this apparent curvature a more complex model can be estimated that includes a quadratic term. This is achieved by using a second order polynomial, which not only includes squared terms for the two Cr determinations but also their cross-product. The result of fitting this polynomial surface is illustrated in Figure 16.5 (right). The main curvature introduced accommodates the variability of Cr determined by ICP-AES on the same solution in which Co was determined. The other curvature related to the total (INAA) Cr and the cross-product is minimal. A question arises as to whether this increase in fit is meaningful, and the polynomial terms that lead to a significant improvement, reduction in the residual sums of squares, of the fit can be identified by an ANOVA (see, e.g., Scheffé, 1999). If some polynomial terms do not contribute significantly to the model they can be dropped and the model recomputed. Any "overfitting" just for the sake of improving the fit should be avoided. The question in the context of the problem is always whether any additional terms lead to a better understanding? Can a meaningful interpretation be provided for model coefficients including powers and cross-products? If the answer is "no", then it is better to stay with a simple model that can be explained in terms of simple physical and chemical processes, and try to understand any other processes that lead to the large residuals, situations where the model does not fit the data well.

In regression modelling when there are more than two explanatory variables, the plane becomes a hyperplane that cannot be graphically visualised. The next step in regression modelling is to determine the quality of the fit. Can it be improved by some simple means, are some of the residuals significantly large in the context of the study being undertaken to warrant further attention, and are any of them influential on the orientation of the hyperplane in the data space; do they "lever" it?

16.3 Classical least squares (LS) regression

In previous chapters (e.g., Chapter 7) it has been stressed that (1) geochemical background is a range; and (2) different geochemical or environmental entities are characterised by different ranges and central tendencies (MEANS or MEDIANS, etc.). In some instances such a simple discrete, "pigeon hole", approach is not the most appropriate. In reality changes may be gradational, and some sort of sliding scale might be advantageous. The use of RA is one approach to accommodating this reality.

The fundamental rules of crystal chemistry control the levels of trace elements that occur in minerals and rocks. In general, an "expected amount" can be determined based on the major element geochemistry that controls the presence and absence of the minerals that might compose a rock, soil, or sediment. Observations that contain significantly more or less of a trace element than might be expected are particularly interesting. These must have come under the influence of processes other than the main one(s) explained by the model.

16.3.1 Fitting a regression model

As an example, the geochemistry of Be can be studied in the context of the major elements that reflect the mineral composition of the C-horizon soil. A regression model was developed for logarithmically transformed Be, and the major and minor elements in the soil as determined

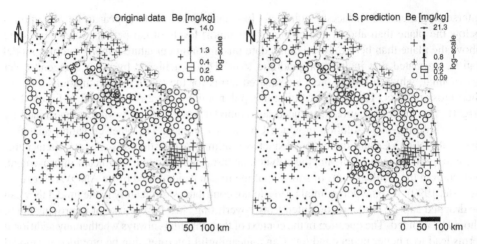

Figure 16.6 Map of Be in Kola Project C-horizon soil with Tukey boxplot-based divisions for log-transformed data (left). Similar Tukey boxplot-based plot for back-transformed estimates of log10(Be) using a least squares (LS) regression model employing the major elements (right)

by XRF, i.e. Si, Al, Fe, Mg, Ca, Na, K, Ti, Mn and P. The major and minor elements were not transformed as most of the major elements were not positively skewed, in fact many were negatively skewed, and the central parts of the distributions for the minor elements were unskewed. The original Be data and the predicted values from the least squares (LS) regression model are presented as maps in Figure 16.6 (left and right panels, respectively). A visual inspection indicates the similarity between the observed and predicted data.

The quality of the fit of the model for Be may be quantified by determining the proportion of the variance of the original data that is explained by the variance of the predicted values. For the model in Figure 16.2 (right) this is 69.5 per cent (see Table 16.1); the higher the value of the fit the closer the predicted values come to the original observed values.

It can be argued that the above model should have been developed using "opened" major and minor element data. The application of that approach is demonstrated in the context of robust regression modelling (see Section 16.4).

16.3.2 Inferences from the regression model

A typical output of RA is shown in Table 16.1. The estimated LS regression coefficients of the explanatory variables in the model and the inference statistics of the model, indicating the influence of each variable on the model, are displayed (see, e.g., Fox, 1997; Draper and Smith, 1998). The intercept, which is an estimated additive parameter of the regression model, ensures that the hyperplane passes through the average of both the response and explanatory variables. The remaining regression coefficients determine the tilt of the hyperplane. Due to the fact that the explanatory data have not been scaled, the absolute values of the regression coefficients cannot be compared. However, the signs indicate whether increasing or decreasing levels of the explanatory variables lead to increasing or decreasing levels of the response variable.

Table 16.1 Table of Least Squares (LS) regression coefficients and inference statistics for the log10(Be) versus major element model

	Estimate	Std. Error	t value	Pr(>|t|)	
(Intercept)	4.435839	0.504842	8.787	< 2e-16	***
Al_XRF	-0.099572	0.015680	-6.350	4.27e-10	***
Ca_XRF	-0.149608	0.020062	-7.457	3.14e-13	***
Fe_XRF	-0.080976	0.015005	-5.397	9.81e-08	***
K_XRF	0.139508	0.022843	6.107	1.83e-09	***
Mg_XRF	-0.052704	0.024974	-2.110	0.03524	*
Mn_XRF	1.934213	0.401166	4.821	1.81e-06	***
Na_XRF	-0.005727	0.021113	-0.271	0.78627	
P_XRF	-0.731348	0.325637	-2.246	0.02508	*
Si_XRF	-0.127679	0.010734	-11.895	< 2e-16	***
Ti_XRF	0.345275	0.111709	3.091	0.00209	**

```
Signif.codes:0'***'0.001 '**' 0.01 '*' 0.05 '.' 0.1 ' ' 1
```

```
Residual standard error: 0.19 on 595 degrees of freedom
Multiple R-Squared: 0.695, Adjusted R-squared: 0.6899
F-statistic: 135.6 on 10 and 595 DF, p-value: < 2.2e-16
```

The positive coefficient for K reflects the elevation of Be in potassic rocks and their derivatives, whereas the negative coefficients for Fe, Mg and Ca reflect the tendency of Be levels to be lower in mafic rocks. The negative coefficients for Al and Si, and positive coefficients for Ti and Mn may be related to soil-forming processes rather than the relationship of Be to these elements in source rocks. The standard errors and t-values are intermediate steps to determining the significance, p-values, of the regression coefficients (see penultimate column of Table 16.1). For these significance determinations to be correct, the regression residuals should be normally distributed.

This will be checked at a later stage of graphical inspection of the model fit (see Figure 16.7, lower left). The p-values indicate the probability that the value of the regression coefficient could have arisen purely by chance. Commonly, a cutoff of 0.05 (5 per cent) is used to determine if the variable concerned contributes significantly to the regression model. To aid appreciation of the table, the significances are coded with asterisks; the more asterisks the greater the significance of the coefficient. Table 16.1 indicates that the intercept and all the major and minor elements except Na contribute significantly to the regression model. The conclusion is that Na could be dropped from the model as its variability does not contribute significantly to the estimation of Be.

Additional information in Table 16.1 concerns the quality of the fit and the overall validity of the regression model. The quality is expressed by the measures, multiple R-squared and adjusted R-squared. Values close to 1 indicate excellent fit. The fit expressed by the multiple R-squared measure of 69.5 per cent is dependent on the number of explanatory variables. The adjusted R-squared measure penalises for the number of explanatory variables. The adjusted R-squared measure is displayed so that it is possible to be able to compare

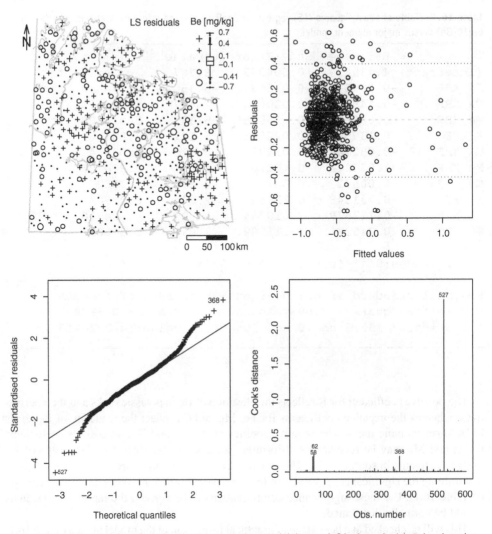

Figure 16.7 Map of LS residuals for the Be regression model (upper left) plotted with Tukey boxplot-based divisions. Plot of residuals versus fitted (predicted) values (upper right); QQ-plot of standardised residuals (lower left); and plot of Cook's distances versus observation number (lower right)

between models based on different numbers of explanatory variables, should that be required. Finally, an F-statistic is computed to indicate whether the model has satisfactorily explained the data variability. If the associated *p*-value exceeds 0.05, it is an indication that the model has failed to adequately explain the data variability. However, in the example the *p*-value is extremely small, as could be expected based on the high multiple R-squared measure.

16.3.3   Regression diagnostics

As has been pointed out, the magnitude of the residuals is of importance in determining the validity of the model, and their location is of interest in the subject matter context. Figure 16.7 (upper left) displays a map of the LS residuals using a Tukey-based EDA symbology. The adjacent plot (upper right) is of the residuals (y-axis) versus the predicted (fitted) values (x-axis); the dotted lines correspond to the upper and lower inner fences of the Tukey boxplot. As such, points falling outside these limits may be considered outliers, and be of particular interest. The negative outlying residuals display little spatial structure, though there tends to be some clustering north-west of Monchegorsk. In contrast, the positive outlying residuals are clustered around Apatity, south-west of Lake Inari in northern Finland and in northern Norway. These indicate areas with abnormally high Be levels (compare Figure 16.7, left). It is known that the alkaline intrusive rocks at Apatity and the soils derived from them are elevated in Be. The areas of high Be residuals in northern Finland and Norway may represent the presence of similar rocks or sedimentary units containing detritus derived from the weathering of such rocks. The absence of observations with larger absolute residuals at higher predicted levels may indicate that the LS regression has been levered too close to those points, with the result that the residuals are small.

Figure 16.7 (lower right) displays a QQ-plot of the standardised residuals, i.e. residuals divided by their standard error (see Table 16.1). The central part of the distribution of the standardised residuals follows a normal distribution quite closely. However, there are clearly outliers occurring at both ends of the distribution. The distribution lacks observations with low residuals, and there appears to be an excess of residuals with high values, indicating that the residuals are positively skewed. This finding indicates that the statistical inferences drawn from Table 16.1 should be treated with caution. Finally, Figure 16.7 (lower right) presents the Cook's distances (Cook and Weissberg, 1982) for the observations. The value of the Cook's distance is high (e.g., >1) when the individual observation for which the Cook's distance is computed exerts a high influence on the estimation of the regression coefficients. The conclusion from this is that individuals with Cook's distances >1 have high leverage and are capable of inducing large changes in the regression coefficients. Therefore, an improved model may be derived by deleting observations with large Cook's distances from subsequent regression analyses. In practice, when outliers are clustered in the data space, measures such as Cook's distance may not always be effective in the detection of the most influential observations in the data. Figure 16.7 (lower right) indicates that the 527th observation in the data set is the most influential; and the QQ-plot (Figure 16.7, lower left) indicates that the particular Be value is abnormally low.

16.3.4   Regression with opened data

The above example is a typical case of using closed data in multivariate data analysis. The major elements from total analysis (XRF results, recalculated to element percentage) were used in the regression model. Recalculated to the oxides (Al_2O_3, CaO, Fe_2O_3, K_2O, ...) these variables sum up to 100 per cent (with LOI which was not included here). According to Aitchison (1986) this is a classical case where the data have to be opened. In this example only the additive logratio transformation can be used. The centred logratio transformation cannot be used in regression analysis due to singularity problems (Section 10.5.2). When using the isometric logratio transformation the relation to the original variables is lost and the inference statistics cannot be interpreted. Ti is chosen as ratioing variable for the additive logratio transformation.

Table 16.2 Table of least squares (LS) regression coefficients and inference statistics for the log10(Be) versus additive logratio transformed major element data model

	Estimate	Std. Error	t value	Pr(>|t|)	
(Intercept)	3.40045	0.18098	18.789	< 2e-16	***
Al_XRF	-0.04559	0.16833	-0.271	0.786586	
Ca_XRF	-0.32076	0.06853	-4.680	3.55e-06	***
Fe_XRF	-0.68390	0.16367	-4.179	3.37e-05	***
K_XRF	1.14638	0.08416	13.622	< 2e-16	***
Mg_XRF	0.30751	0.08036	3.827	0.000144	***
Mn_XRF	0.63046	0.07903	7.977	7.68e-15	***
Na_XRF	0.27470	0.08527	3.222	0.001344	**
P_XRF	-0.06331	0.05100	-1.241	0.214987	
Si_XRF	-1.85900	0.10342	-17.974	< 2e-16	***

```
Signif.codes:0 '***' 0.001 '**' 0.01 '*' 0.05 '.' 0.1 ' ' 1

Residual standard error: 0.2 on 596 degrees of freedom
Multiple R-Squared: 0.6618, Adjusted R-squared: 0.6567
F-statistic: 129.6 on 9 and 596 DF, p-value: < 2.2e-16
```

Table 16.2 shows the regression coefficients and the inference statistics for regression of log-transformed Be versus the additive logratio transformed major element data. Figure 16.8 shows the corresponding plots with the regression diagnostics. Although some changes are obvious (Al_XRF and P_XRF leave the model, Na_XRF enters the model) the adjusted R-squared is almost the same (0.69 vs. 0.66) and the prediction map (Figure 16.8 upper left) is fairly comparable. A problem with the interpretation is of course that it is no longer possible to discuss the variables as such but only their ratios to Ti. It can be discussed whether the minor changes are worth losing one variable; the direct interpretability and the risk that results will change with the choice of another ratioing variable.

16.4 Robust regression

The analysis above indicates the presence of one or more outliers that could have levered the regression hyperplane. To limit the effect of these outliers, a robust regression procedure was applied to the data (compare Section 6.2). Of the various procedures available, least trimmed squares (LTS) was selected because it is highly robust, and a fast algorithm for its computation is available (Rousseeuw and Van Driessen, 2002). In contrast to LS regression, where the sum of all squared residuals is minimised, in LTS regression the sum of some proportion of the smallest squared residuals is minimised. This proportion may be varied from 50 to 100 per cent, with a default of 50 per cent that provides the maximum protection against outliers.

In the following, LTS is used for the opened data based on the additive logratio transformation with Ti as divisor.

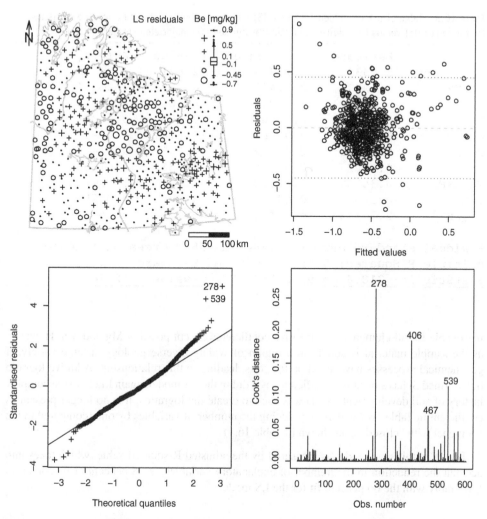

Figure 16.8 Map of LS residuals for the Be regression model with opened major element data (upper left) plotted with Tukey boxplot-based divisions. Plot of residuals versus fitted (predicted) values (upper right); QQ-plot of standardised residuals (lower left); and plot of Cook's distances versus observation number (lower right). Compare with Figure 16.7

16.4.1 Fitting a robust regression model

The estimated LTS regression coefficients of the additive logratio transformed elements in the model are displayed in Table 16.3. Al and P now enter the model, the only coefficient that is not statistically significant (five per cent) is Fe. In comparison with the LS model for the opened data (see Table 16.2) one of the coefficients has changed sign (Al, from negative to positive). Positive coefficients indicate that as the element forming the numerator of the additive logratio transformation increases relative to Ti (indicating increasing amounts of minerals associated with the weathering of felsic rocks), the predicted Be levels increase. This is mineralogically

Table 16.3 Table of least trimmed squares (LTS) regression coefficients and inference statistics for the log10(Be) versus the additive logratio transformed (Ti) major element data model

	Estimate	Std. Error	t value	Pr(>|t|)
Intercept	2.13331	0.16903	12.621	< 2e-16
Al_XRF	1.52971	0.14478	10.566	< 2e-16
Ca_XRF	-0.99380	0.06826	-14.560	< 2e-16
Fe_XRF	-0.25991	0.13805	-1.883	0.06028
K_XRF	0.73345	0.06869	10.678	< 2e-16
Mg_XRF	0.39079	0.06702	5.831	9.51e-09
Mn_XRF	0.17227	0.06806	2.531	0.01165
Na_XRF	0.35567	0.08380	4.244	2.58e-05
P_XRF	-0.12674	0.04183	-3.030	0.00256
Si_XRF	-2.37134	0.09423	-25.165	< 2e-16

Residual standard error: 0.1484 on 538 degrees of freedom
Multiple R-Squared: 0.7851, Adjusted R-squared: 0.7815
F-statistic: 218.4 on 9 and 538 DF, p-value: < 2.2e-16

reasonable for all elements with positive coefficients except possibly Mg and Mn. However, as the sample material is soil from a region of widely diverse geology, a large variety of geochemical processes have acted on the soils, leading to this relationship. Additive logratio transformed Si has a negative coefficient, as it did in the LS model, again likely to the complex history of soil development. As Ti was used to create the logratios, it is no longer present as coefficient in Table 16.2 and 16.3, reducing the number of variables by one (compared to the LS model with "closed" data shown in Table 16.1).

The fit of this new model as measured by the adjusted R-squared value, which takes into account the reduction in the number of explanatory variables, is 78 per cent. This compares favourably with the 66 per cent fit for the LS model.

16.4.2 Robust regression diagnostics

Regression diagnostics for the new LTS model for the variable Be are presented in Figure 16.9. The QQ-plot includes the residuals for both the observations included in the final LTS model (circles) and those excluded (pluses). In comparison with the LS model (see Figure 16.8, lower left), the included observations come much closer to a normal distribution. The adjacent plot (Figure 16.9, upper right) uses the same symbology and plots the standardised LTS residuals versus the predicted (fitted) values. The distribution of the observations included in the final LTS model (i.e. those plotted as circles) shows no systematic relationship or structure, indicating the appropriateness of the model. However, negative outlying residuals are clustered at higher predicted values for Be, while positive outlying residuals are spread across the range of the predicted values. The fact that the negative residuals are clustered at high fitted values may indicate the final model does not completely accommodate all the variation due to natural variability. The robust LTS regression model provides a fit that accommodates the dominant geochemical processes, and this may lead to exclusion of variability due to rarely occurring rock types.

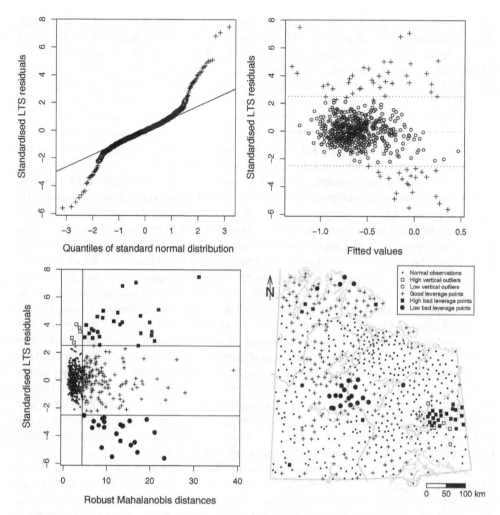

Figure 16.9   Robust regression diagnostics for the Be regression model (compare Figure 16.8). QQ-plot of the standardised LTS residuals (upper left); plot of the LTS residuals versus fitted (predicted) values (upper right); plot of standardised LTS residuals versus robust Mahalanobis distances (lower left); and map (lower right) indicating the location of residuals with different characteristics using the symbology of the residuals versus distances plot (lower left)

The diagnostic plot of standardised LTS residuals versus robust Mahalanobis distances (Figure 16.9, lower left) provides insight into the data structure (see also Sections 13.3 and 13.4). The leverage any observation can exert on a regression model is related to both its position in the data structure of the explanatory variables, as measured by the Mahalanobis distance, and the magnitude of the residual (Rousseeuw and Van Driessen, 2002; Maronna *et al.*, 2006). Thus, observations may be divided into six groups. The majority of observations have both low absolute residuals, and low Mahalanobis distance (small black dots). These observations have low leverage. Similarly, observations with larger absolute residuals but low Mahalanobis distances (*vertical outliers*) exert little leverage on the model (open squares for positive residuals, open circles for negative residuals (in the example no observations occur in this class)). The remaining observations are all capable of exerting leverage on the model. Those with small absolute residuals (pluses) are *good leverage points* and anchor the regression hyperplane in the data space, improving the precision of the estimates of the regression coefficients. In contrast, observations that have a large Mahalanobis distance and large absolute residuals are *bad leverage points* (black squares for positive residuals, black bullets for negative residuals). The LTS procedure excludes these observations from the estimation of the regression coefficients, making the procedure robust to their presence. Without such a procedure the inclusion of these bad leverage points in a classical LS model will lead to the model failing to represent the dominant underlying data relationships (compare Figure 6.4, right).

Figure 16.9 (lower right) presents a map where the observations have been divided into the six groups according to the magnitude of their residuals and their ability to lever the classical LS model. Of particular interest to applied geochemists and environmental scientists are observations associated with large absolute residuals. Those with low Mahalanobis distances are outliers for the response variable, while those with high Mahalanobis distances are outliers in both the response and explanatory variable space. Presentation of this information in a map permits their spatial distribution to be displayed and the presence of any spatial clustering determined.

The map draws attention to the two alkaline intrusions near Apatity, an almost circular feature along the Finnish–Russian border within the granulite belt (compare Figure 1.2, the geological map) and to the sediments along the Norwegian coast. These are the areas where the original data show strong deviations from the regression model. It is thus informative to compare the map in Figure 16.9 with the map of the raw Be data (see Figure 16.6, left). Many of the observations identified as high bad leverage points are characterised by background or high background Be levels. Only those samples from around Apatity and one site in Finland were classed as high outliers in the original data. The low bad leverage points occurring in northern Norway are all classed as background or high background samples together with one positive outlier. The leverage map provides a completely different story indicating that relative to their major element composition, these soil samples are depleted in Be. The second group of low bad leverage points that occur along the Russian–Finnish border is also interesting. These form a coherent pattern of 17 observations with a wide range of Be levels from low to high background. Clearly, the soils of this area have been identified as being Be depleted. Whether this is due to the properties of the soil parent material, or due to weathering and soil-formation processes is not known. Possibly, the most significant finding from this regression analysis are the soil samples from northern Norway exhibiting large positive residuals that may indicate the presence of either unmapped higher Be rocks (e.g., hidden alkaline intrusions?) or identify an area of unusual Be accumulation in the soils.

16.5 Model selection in regression analysis

If one variable is for some reason of special interest the arising question is which combination of the other variables can best explain its behaviour? The task then is to develop the best model of the available explanatory variables for the variable under study. For example it could be interesting to study which combination of other parameters (e.g., temperature, solar radiation, wind speed, precipitation, SO_2 concentration, CO_2 concentration) have the highest influence on the measured ozone concentration in the atmosphere.

The variable Bi in the Kola O-horizon data displays an unusual spatial distribution with a strong north–south gradient (see Figure 8.5). Are there variables in the data set that explain this phenomenon? In a first step the variable Bi is regressed against all other available variables (see Section 16.3.1). The adjusted R-squared measure is 88.6 per cent. This indicates a very good explanation of the variability of the log-transformed values of Bi. Of the 38 entered variables (log(Ag), log(Al), log(As), log(B), log(Ba), log(Be), log(Ca), log(Cd), log(Co), log(Cr), log(Cu), log(Fe), log(Hg), log(K), log(La), log(Mg), log(Mn), log(Mo), log(Na), log(Ni), log(P), log(Pb), log(Rb), log(S), log(Sb), log(Sc), log(Si), log(Sr), log(Th), log(Tl), log(U), log(V), log(Y), log(Zn), C, H, N, and pH), 19 explanatory variables (log(Ag), log(Al), log(As), log(B), log(Ba), log(Ca), log(Cd), log(Fe), log(Hg), log(La), log(Mn), log(P), log(Pb), log(Sb), log(Tl), log(U), log(V), log(Y), and N) and the intercept were significant. This is not a very useful model, because it still contains 19 variables and it is still impossible to obtain a better interpretation of the north–south gradient.

Stepwise regression (Hocking, 1976; Draper and Smith, 1998) can provide a solution to this problem. The basic idea of stepwise regression is to first include only the most important variable in the model and to add step by step further important variables until a certain statistical criterion is met (see, e.g., Draper and Smith, 1998). The criterion indicates that no further essential improvement of the model fit can be reached by adding additional variables. This procedure is called forward selection. Alternatively it is possible to start with all variables in the model and to exclude variables with the least contribution step by step. This procedure is called backward selection. The two procedures can result in completely different models, which is an unfortunate situation. The reason is that different combinations of variables are tested. Forward and backward selection can be combined to reach an optimal model. In the above example 25 explanatory variables remain in the regression model, 20 of them are significant. Thus the situation has not improved.

The problem is that not all possible combinations of variables are tested by these procedures. With 38 variables there exist 2^{38} (= 274877906944) possible combinations of variables. It is impossible to test all these combinations. However, there exists an algorithm, called the "leaps-and-bounds algorithm" (Furnival and Wilson, 1974), which performs an intelligent search over all possible models. The quality of the resulting models can be displayed in a simple graphic (Figure 16.10). The variables are plotted on the x-axis and the adjusted R-squared, as a measure of the quality of the fit, is plotted on the y-axis. The graphic is read from bottom

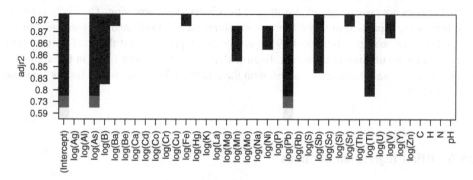

Figure 16.10 Results of regression using the leaps-and-bounds algorithm. Grey scale indicates the R-squared measure of quality of fit. adjr2 = adjusted R-squared measure

Table 16.4 Table of least trimmed squares (LTS) regression coefficients and inference statistics for the best model explaining log10(Bi)

	Estimate	Std. Error	t value	Pr(>|t|)
Intercept	-3.71107	0.10817	-34.309	< 2e-16
log(Pb)	0.77031	0.02470	31.192	< 2e-16
log(As)	0.26071	0.01419	18.370	< 2e-16
log(Tl)	0.14793	0.01720	8.599	< 2e-16
log(B)	-0.11755	0.01615	-7.277	1.10e-12

```
Residual standard error: 0.1504 on 585 degrees of freedom
Multiple R-Squared: 0.9089, Adjusted R-squared: 0.9083
F-statistic: 1460 on 4 and 585 DF, p-value: < 2.2e-16
```

to top with indications of the variables entering the regression model. The grey scale provides an indication of the adjusted R-squared measure. Dark values show a high adjusted R-square. The best model with only one variable is a model with log(Pb) plus intercept and results in an adjusted R-square of 0.59. In the example the first satisfying model is reached with 4 variables (log(Pb), log(As), log(Tl), log(B)) plus intercept resulting in an adjusted R-square of 0.83 (see Figure 16.10).

At present the leaps-and-bounds algorithm is based on performing least squares regressions. Data outliers could thus influence the fit. Therefore it is a good idea to check the model by performing a robust regression. Table 16.4 shows the usual output for the resulting robust regression model (compare Table 16.3). Surprisingly the adjusted R-square increases to 0.91. The reason is that four outliers exist in the variable Pb, strongly influencing the least squares fit. In this special case the outliers resulted in a deterioration of the LS fit.

With just four explanatory variables left it is much easier to consider the process causing the regional distribution of Bi. When studying maps of the explanatory variables, Pb, As, and Tl, all three show a north–south gradient, while B has the opposite behaviour with the highest concentrations along the coast in the north. This combination of elements and observed spatial pattern appears likely to be related to a climatic and bio-productivity gradient (Reimann *et al.*, 2001a). The model thus identifies four variables that follow the same gradient as Bi, but does not provide any explanation of the observed phenomenon. The likely explanatory processes, expressed in some form as variables, are "lurking" and not contained in the data set (see discussion in Section 11.9). It is a common problem that the "real" explanatory variables may be missing in a data set. Even with a very good model fit, as in the above example, it is necessary to be quite careful with the interpretation of statistical results (see also Section 11.9).

16.6 Other regression methods

In the investigation of the Kola data, LTS regression was selected as the method of choice due to its ability to attain maximum resistance to outliers. Other methods include Least Median

Squares (LMS) (Rousseeuw, 1984), regression methods using M- and S-estimators and combinations thereof (Maronna *et al.*, 2006). However, as LTS is relatively widely available, it was selected over other methods.

The problems of collinearity arise when the explanatory variables are highly correlated. Several different approaches exist to undertake regression analysis in these circumstances. Ridge Regression (Hoerl and Kennard, 1970) minimises the sum of squared residuals, similarly to LS regression, but an additional constraint is introduced which limits the sum of the squared regression coefficients. This shrinkage method prevents the inflation of the regression coefficients during the computational procedure. Lasso Regression (Hastie *et al.*, 2001) works in a similar way, but it is the sum of the absolute regression coefficients that is limited. As a consequence of this, some of the coefficients may be reduced to zero, indicating that the explanatory variable in question contributes no useful information to model. As such, lasso regression can also be used as part of a model selection procedure.

An alternative approach is to reduce the dimensionality of the explanatory variables by undertaking a PCA (see Chapter 14). The result of this provides a set of new explanatory variables, the scores of the principal components that are uncorrelated. A subset of these is then selected for inclusion in the regression model, leading to a Principal Components Regression (PCR), see for example Garrett (1972) or Jolliffe (2002). In addition to removing the collinearity, this procedure has the advantage that it reduces the dimensionality of the explanatory variables. Also, when "noise", random variability, partitions into the last few principal components, their removal could lead to an improved model with reduced prediction errors.

Partial Least Squares (PLS) Regression (Tenenhaus, 1998), a very prominent tool in chemometrics (Wold, 1975), achieves a similar objective to PCR by deriving a set of components that capture the variability in the explanatory variables that is most related to the response variable. This is in contrast to PCR where the components reflect only the variability in the explanatory variables. A subset of these new variables is then used in the PLS regression with the response variable. Cross-validation or bootstrapping is then used to determine the optimum number of components leading to a minimum prediction error.

All of the above regression methods are considered linear. Some physical or chemical relationships are inherently non-linear. Fitting models (often exponential) to these relationships requires special numerical procedures. Examples are the exponential decay in the deposition of atmospheric particulates from a contamination source (de Caritat *et al.*, 1997b; Bonham-Carter *et al.*, 2006), and the modelling of the uptake of Cd by wheat (Garrett *et al.*, 1998). Details of methods for undertaking non-linear regressions are discussed in Hastie *et al.* (2001).

Other methods have been developed for a wide range of regression problems. A few examples are Regression Trees, Generalised Additive Models and Neural Networks. These are not discussed further here, interested readers are referred to Hastie *et al.* (2001).

For the special case where the response variable is a binary variable indicating membership (1) or non-membership (0) in a group, Logistic Regression (LR) is the recommended procedure (see, e.g., Kleinbaum and Klein, 2002). As a regression method, LR provides inference statistics that can be used to evaluate the model and to determine the relative importance of the explanatory variables for estimating group membership. Having developed a reliable model, the probability that a new observation is associated with membership or non-membership can be directly estimated.

16.7 Summary

In regression analysis the relationships between a variable of interest and other measured variables are modelled. Once a mathematical model has been established, it can be used to predict values for the variable of interest, using the information of the other variables. For the sake of simplicity and interpretability of the model, usually only linear relationships are considered.

A good and reliable model leading to a good prediction quality requires certain statistical assumptions. The most important requirement is that the residuals (deviations between the measured and the predicted values of the variable of interest) have about the same spread across the whole range of the predictor variables. Classical least squares based methods also require that the residuals be normally distributed. Outlying observations typically lead to deviations from these assumptions, and robust regression methods should then be preferred.

Even if many predictor variables are available, it is usually still desired to find a prediction model that is as simple and as efficient as possible. Fast search procedures, such as the leaps-and-bounds algorithm, have been developed to identify the best-suited variable combinations. When working with compositional data the opened data can be added to the predicting variables. The algorithm for model selection will then automatically select the explanatory variables that are best suited for prediction.

If it is not necessary to keep the original variables as predictor variables one can use regression methods like PCR (principal component regression) and PLS (partial least squares) regression. The "predictors" are linear combinations of the original variables, and sometimes it is impossible to find a clear interpretation of the resulting regression model if there has not been a successful interpretation at the earlier PCA stage of the analysis. However, for the purpose of prediction such models can still be very useful.

17
Discriminant Analysis (DA) and Other Knowledge-Based Classification Methods

Cluster analysis may be used to identify groups in a multivariate data set (Chapter 15). The groups describe certain processes or phenomena in the data set. They can also be used to create data subsets for further statistical investigation.

In contrast, discriminant analysis (DA) (Fisher, 1936) uses *a priori* knowledge about existing group memberships. This knowledge is used to develop a function that will result in an optimal discrimination of the groups. Once the function is estimated, the group membership of new observations can be predicted. For example, samples from different vegetation zones, boreal-forests, forest-tundra, and different gradations of shrub tundra to tundra (shrub tundra, dwarf shrub tundra, and tundra) were collected for the Kola Project. These can be used to define three main "vegetation zone" groups for DA: boreal-forest, forest-tundra, and tundra. Development of a discriminant function will then make it possible to classify any new observation collected as belonging to one of the vegetation zones based on its chemical composition alone. However, misclassifications are also possible, but it is the general objective of DA to keep the number of misclassifications as low as possible.

17.1 Methods for discriminant analysis

There are different methods for undertaking DA. The most often applied methods are linear (LDA) and quadratic (QDA) discriminant analysis (see, e.g., Huberty, 1994; Johnson and Wichern, 2002). LDA separates the groups using linear functions, while QDA uses quadratic (curvilinear) functions to achieve separation. For construction of both the linear or quadratic discriminant functions, it is necessary to estimate the central location, and multivariate spread – covariance – for each group. LDA uses a simplification by assuming that the group covariances are equal. Classically, the group centres are estimated using the mean of each variable for each group. The group covariances are estimated using the empirical (classical) covariance matrix of the data for each group (the difference between correlation and covariance is discussed in

Statistical Data Analysis Explained   Clemens Reimann, Peter Filzmoser, Robert G. Garrett, Rudolf Dutter
© 2008 John Wiley & Sons, Ltd.

Section 13.1). It is also possible to estimate centre and covariance for each group using robust methods (see Section 13.2) resulting in robust LDA and robust QDA. Due to the presence of outliers in applied geochemistry and environmental data, the use of robust procedures beneficially reduces their impact on the results of the DA. Finally, the probabilities of group membership for each observation, or new observations, can be estimated from the discriminant functions.

17.2   Data requirements for discriminant analysis

Discriminant analysis, like all statistical procedures, is underlain by certain assumptions concerning the distribution of the data in the different groups. For optimal discrimination, the data of the different groups should be distributed as multivariate normal. To approximate multivariate normality the distribution of all variables has to be checked (graphically, see Section 3.4, or using statistical tests, Section 9.1) and transformed if necessary (Section 10.2). Note that a univariate normal distribution for each variable does not guarantee a multivariate normal distribution (see, e.g., Gnanadesikan, 1977), which was also demonstrated by Reimann and Filzmoser (2000). Furthermore, DA is built on the further assumption of the independence of the observations. However, geochemical data are usually spatially dependent, and thus when carrying out DA, it is necessary to be aware that spatial correlations may drive the results of the analysis.

It is not necessary to standardise the data because DA can account for different data ranges. However, DA is a covariance-based method, and thus most of the points discussed in Section 14.1 for PCA and FA need to be considered.

Data outliers influence the estimation of the group centres and covariances. Because geochemical and environmental data frequently contain outliers, robust methods of DA are preferable (He and Fung, 2000; Croux and Dehon, 2001; Hubert and Van Driessen, 2004). For all covariance-based methods, compositional data (closed data) will cause artificial correlations that do not reflect the underlying real correlations due to geochemical processes (Section 10.5). Although discrimination of different groups can be undertaken with closed data, an improved result will generally be obtained when the data array is opened using an appropriate transformation (Section 10.5).

Censored data are a general problem for covariance-based methods. Again the best choice is to use robust methods that will be able to withstand a substantial amount of censored data (Section 15.1.5). Whenever possible, variables with censored data should be excluded from DA.

DA is based on the presence of dissimilarities between the groups. However, within each group the data should be homogeneous, otherwise the estimation of the group covariance matrix may not be representative. Again robust methods, concentrating on the (homogeneous) majority of the data, are recommended.

Dimensionality is an even greater problem than described for PCA and FA (see Section 14.1.8) because the rules discussed there should be fulfilled for each of the data groups in order to derive reliable discriminant functions. If the number of observations for one group is fewer than the number of variables in the entered data set, most software implementations of DA will break down due to numerical problems. A method to overcome this problem is to first carry out a PCA for dimension reduction and establish the discriminant functions on the first few principal components (see Chapter 15).

17.3   Visualisation of the discriminant function

To visualise the discriminant function for classical and robust LDA and QDA, a two-dimensional example from the Kola C-horizon data set is used. Figure 17.1 shows two scatter-plots of the log-transformed elements Cr and K. The task is to discriminate the C-horizon soil samples collected on top of Caledonian sediments, occurring along the Norwegian coast (see geological map – Figure 1.2) and identified as subset 10 in the variable "LITHO" of the data set, against all other observations. The samples collected on top of the Caledonian sediments are identified with "×" in the plot, all others are shown as points (Figure 17.1).

The plot of the results of classical LDA (Figure 17.1, left) shows that a considerable over-lap exists between the two groups. A complete discrimination based on a linear or quadratic function will therefore be impossible. The task then is to minimise the number of misclas-sifications rather than to completely separate the two groups. The basis of DA is the group centres and covariances. The latter can be visualised as ellipses (see Section 13.2). In the plots the 90 per cent ellipses of each group are displayed. In the left plot the classical means and covariances are used for constructing the ellipse, while in the right plot their robust counter-parts are employed. It is noticeable that outliers determine the form of the ellipses in the left plot. The robust counterparts provide the better basis for building the discriminant function. The resulting LDA functions are shown as solid lines in both plots. In both cases LDA is not an optimal solution; the covariances of the two groups as displayed by the ellipses are quite different. LDA cannot account for this difference, and the number of misclassified samples is much higher for non-robust LDA.

Figure 17.2 shows the same comparison for classical and robust QDA. The separating line now follows a quadratic function. The ellipses remain the same as displayed in Figure 17.1, the difference between LDA and QDA is that now the different covariances are accounted for. Again the robust procedure performs better than the non-robust procedure because in the non-robust case the outliers have inflated the covariance estimates. Overall, in terms of the percentage of misclassified samples, robust QDA performs best (LDA, 7.8 per cent; robust LDA, 6.8 per cent; QDA, 7.4 per cent; robust QDA, 6.1 per cent).

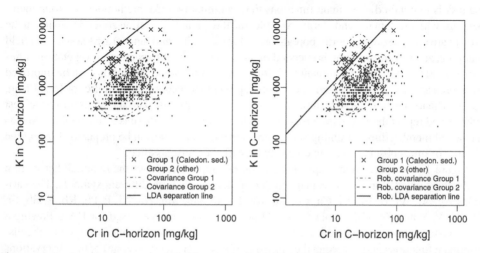

Figure 17.1   Results of non-robust (left) and robust (right) LDA with Cr and K from the Kola Project C-horizon data set

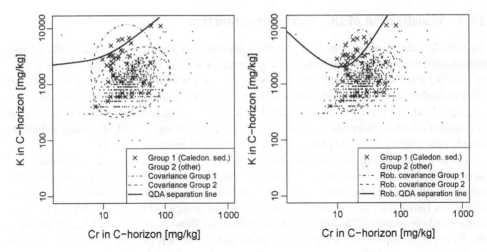

Figure 17.2 Results of non-robust (left) and robust (right) QDA with Cr and K from the Kola Project C-horizon data

For the location of the separating lines, DA takes the size of the groups into account. This is the reason that the lines are moved towards the smaller group, i.e. in Figures 17.1 and 17.2 towards the upper left of the diagrams.

17.4   Prediction with discriminant analysis

In the above example DA was used for classification. When new observations are available, their data can be entered into the resulting discriminant functions. DA will then be used to determine to which of the two groups the new observations most likely belong.

In some instances no new observations are available, only the existing data. In these cases the task is to obtain discriminant functions that can reliably make predictions of group memberships with the data in hand. As an example, the three main vegetation zones occurring in the project area (from south to north: boreal-forest, forest-tundra, and tundra) are known from field mapping, and this information is encoded in the data set. The information on vegetation zones can be used to estimate discriminant functions. The quality of the predictions can be improved by approaches like cross-validation or bootstrapping that increase the stability of DA (Efron, 1979). The idea behind these techniques is to repeatedly split the data set into training and test subsets and repeat the DA many times. The outcome of each of these is a "truth table", and as the actual membership of training set members is known a table can be prepared showing the number of times observations are correctly classified or misclassified.

Figure 17.3 displays four maps; the map on the upper left shows the regional distribution of the three vegetation zones, as mapped during field work, with different symbols. The variables Ag, Al, As, B, Ba, Bi, Ca, Cd, Co, Cu, Fe, K, Mg, Mn, Na, Ni, P, Pb, Rb, S, Sb, Sr, Th, Tl, V, Y and Zn of the Kola Project O-horizon data were then used for DA following a log-transformation of all variables. LDA with cross-validation was applied to estimate the discriminant functions and to predict the group membership (vegetation zone) of the observations based on their chemical composition. The remaining three maps (upper right and lower row)

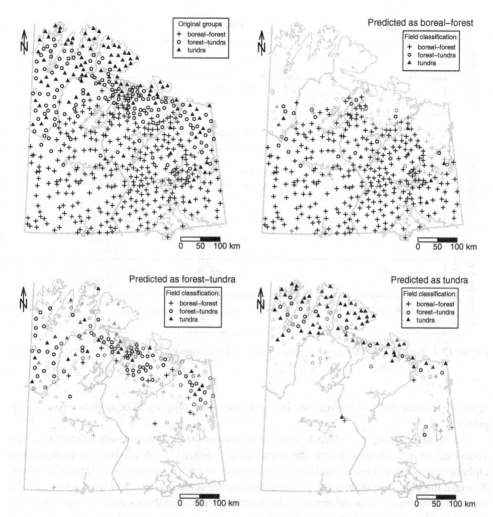

Figure 17.3 Map showing the distribution of the vegetation zones in the Kola Project survey area (upper left) and maps of the predicted distribution of the three vegetation zones in the survey area using LDA with cross-validation. The grey scale corresponds to the probability of assigning a sample to a specific group (boreal-forest, upper right; forest-tundra, lower left; tundra, lower right). A colour reproduction of this figure can be seen in the colour section, positioned towards the centre of the book

show the resulting predictions for each of the groups. The symbols are the same as in the map of the vegetation zones. The additional grey scale is used to display the probability of the sample falling into this group (boreal-forest: upper right, forest-tundra: lower left, and tundra: lower right). Black indicates high probabilities and decreasing shades of grey lower probabilities of group membership. Black circles and triangles in the upper right map are thus misclassified samples assigned to the group "boreal-forest". The grey scale permits visualisation of the quality of discrimination in the border zones between the groups where the majority of misclassifications occur. The overall misclassification rate is 20 per cent. Thus the chemical composition of the O-horizon soils at each sample site can be used to predict membership

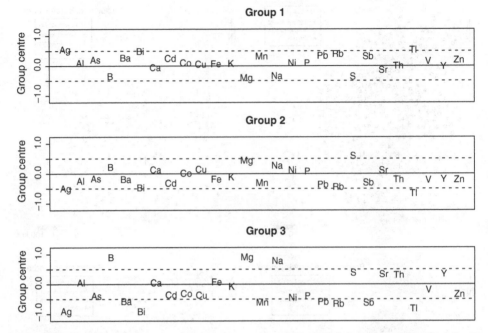

Figure 17.4 Element plot showing the group centres of each variable following completion of the DA

in the vegetation zones; however, one in every five observations will be assigned to a wrong group.

The relative importance of each variable in the computed discriminant functions can be visualised in a scatterplot where the variables are plotted in an arbitrary sequence, here alphabetical, along the x-axis, and the group average for each respective variable on the y-axis. To make the plot comparable, the variables have to be centred and scaled (see Section 10.4); this standardisation has no influence on the discrimination. Variables with a high or low group mean are relevant for characterising the groups.

Figure 17.4 shows this plot for the above example of discriminating between the three vegetation zones using the O-horizon data. Group 1 (boreal-forest) is characterised by high means of Ag, Bi, Mn, Pb, Rb and Tl and relatively low means of B, Mg, Na and S. Group 2, forest-tundra, is characterised by high means of Mg and S and low means of Ag, Bi and Tl. Group 3 (tundra) is characterised by high values of B, Mg and Na and low values of Ag, Bi, Mn, Pb, Rb, Sb and Tl. Group 1 is very different from the other two groups and shows almost the exact opposite behaviour of Group 2 and 3. This behaviour was observed in Section 6.4 when studying the spatial trends in a north–south direction along the western project boundary. The elements B, Mg, Na and S are characteristic of the input of marine aerosols near the coast. The elements Ag, Bi, Mn, Pb, Rb, Sb and Tl in contrast show a characteristic increase from north to south, following the vegetation zones and the increasing bio-productivity. One could assume that the coastal position rather than the vegetation zones is the driving factor behind the discrimination of the vegetation zones. However, when all sea spray related elements are omitted from the DA, the resulting discrimination rates do not deteriorate, an indication that the vegetation zones are the real driving force behind the discrimination.

A further application of the element plots would be to determine if any of the elements could be dropped from the DA. Elements with means close to zero for all groups are likely not contributing useful information to the discrimination, in fact they may be contributing multivariate noise degrading its quality. Inspecting Figure 17.4 leads to the identification of Ca and K as candidates. Interestingly, these are two elements that are bio-essential to plants, which control the uptake of these and other bio-essential elements. Thus, irrespective of vegetation zone, the plants contain similar levels of these elements over the whole project area, making them poor contributors to discrimination. Other variables like Al, Fe and V may be worthy of further investigation for their utility, or rather lack of it, in the DA.

Prior to undertaking DA, Tukey boxplots (Section 3.5.1) for each variable can be prepared for the data divided by the *a priori* groupings. In instances where the MEDIANS and the spread of the central 50 per cent of the data are very similar, these variables may not be useful discriminators and could be candidates for elimination. In instances where variables display contrary behaviour between different groups, the calculation of ratios and their inclusion in the DA may be advantageous. For example, the Mg/Mn ratio will be uniquely low in Group 1 data (see Figure 17.4).

17.5   Exploring for similar data structures

The Kola Project area is characterised by the occurrence of several world-class mineral deposits (e.g., Cu-Ni, Fe, apatite). Samples collected in the surroundings of these deposits could be used to construct discriminant functions that allow exploring the remaining project area for similar features. As an example, the alkaline intrusions near Apatity that host the largest known apatite deposit on Earth (lithologies 82 and 83 in the geological map – Figure 2.1) have been used to search for similar geochemical signatures in the whole data set. Only 11 samples in the data set were collected from the C-horizon soils developed on top of these two lithologies. Thus only very few variables could be employed for DA because the number of observations should be considerably greater than the number of variables (see also discussions in Sections 13.3 and 14.1.8). For demonstrating the procedure, the ten major elements determined by X-ray Fluorescence (XRF) were used for the DA. When using the major elements, closure is a serious problem, thus the additive logratio transformation using Ti as the divisor (see Section 10.5.1) was applied to the data prior to carrying out the DA. This also has the beneficial effect of reducing the number of variables to 9. Following an LDA with cross-validation for all available data, a prediction of the occurrence of the group "alkaline intrusions" in the survey area can be made. It should be noted that while $n = 11$ and $p = 9$ for the observations overlying lithological units 82 and 83, the use of cross-validation provides a more reliable estimation of the discriminant function. Additionally, the use of LDA rather than QDA requires that fewer parameters be estimated, and in general additional parameters require additional data to ensure reliable estimation. Figure 17.5 shows the result.

The surroundings of Apatity are clearly indicated, and 10 out of the 11 observations are correctly identified. Some additional sites in the surroundings of the intrusions, as well as three points with high probabilities of an "alkaline intrusions" affinity and four points with lower probabilities in northern Norway, are also identified. Reimann and Melezhik (2001) have already reported geochemical similarities between the alkaline intrusions and areas within the Caledonian sediments occurring along the Norwegian coast. These authors had speculated whether a deep seated alkaline intrusion or alkaline clasts in the sediments could be the reason

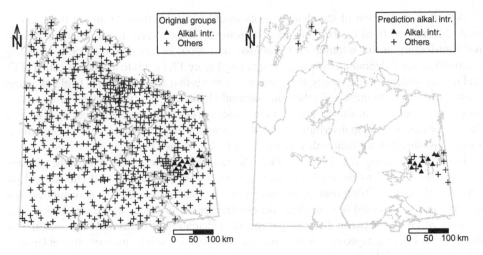

Figure 17.5 Location of the alkaline intrusions near Apatity (lithologies 82 and 83 – left map) and the locations for the samples predicted as members of the group "alkaline intrusions" (right map) based on LDA with cross-validation

for the unusual geochemical signature. The three high-probability points identified in northern Norway may thus be worthy of further investigation.

Due to the low sample density used in the Kola Project, the problem of dimensionality is serious for such interesting applications of DA. Mineral deposits are usually characterised by an insufficient number of samples to carry out DA.

17.6   Other knowledge-based classification methods

17.6.1   Allocation

The task of classifying individuals into one or more groups can be undertaken directly without the use of discriminant functions through the use of allocation procedures, see Campbell (1984) and Garrett (1990) for a regional geochemical application. As with DA, the key to this approach is quantifying the central location and multivariate spread of the *a priori* data groups. Thus everything discussed concerning outliers and transformations earlier in this chapter also applies here, and it is desirable to use robust estimates of the group means and covariance matrices. Allocation can be carried out with a single group or with multiple groups.

As has been demonstrated, the probability of the membership of any individual in a group, assuming a consistent suite of measured variables, can be estimated from the Mahalanobis distance of the individual from the group centroid by taking into account the shape of the group through its covariance matrix. In DA the problems with inhomogeneity of covariance, i.e. the representative group ellipses being of different sizes, shape, and orientation, are addressed through the application of QDA. In allocation this is handled through a modification to the Mahalanobis distances used in Generalised Discriminant Analysis (GDA) (Gnanadesikan, 1977), and the probabilities of group membership are then estimated (Kshirsagar, 1972). The observations are allocated to the group for which they have the smallest modified Mahalanobis distance. A further step, if some critical level of group membership is chosen, is to determine if any observation has a lower probability than this critical level in all of the groups. If that is the case, the observation can be allocated as "undefined", a member of no group. These individuals are multivariate outliers with respect to the defined groups. The multivariate outlier detection methods described in Sections 13.3 and 13.4 are the graphical and automated, respectively, one group-approaches to this task.

For the maps in Figure 17.6, the same Kola Project O-horizon data as used for the LDA with cross-validation (see Figure 17.3) were employed. They are: Ag, Al, As, B, Ba, Bi, Ca, Cd,

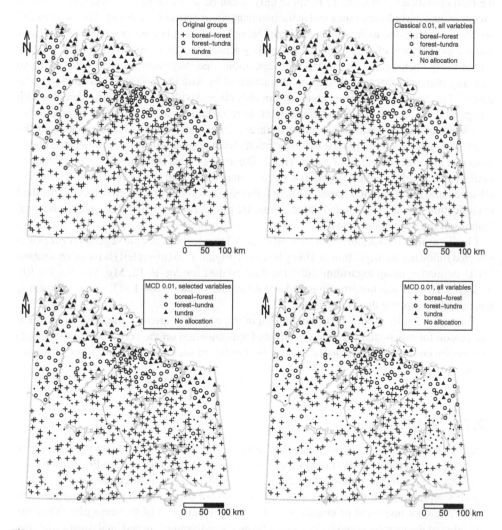

Figure 17.6 Map showing the distribution of the vegetation zones in the Kola Project survey area (upper left) and maps of the predicted distribution of the three vegetation zones in the survey area using allocation. Classical non-robust estimates were used in the upper right map, robust estimates (MCD) were used for the map below (lower right), and in the lower left map the allocation was based on a subset of 11 elements using robust estimates. A colour reproduction of this figure can be seen in the colour section, positioned towards the centre of the book

Co, Cu, Fe, K, Mg, Mn, Na, Ni, P, Pb, Rb, S, Sb, Sr, Th, Tl, V, Y and Zn. Three data subsets corresponding to the field classification of the three main vegetation zones in the project area ($n = 568$) were created. For each of these, robust MCD estimates of the locations and covariances (see Section 13.2) were made following a logarithmic transform. These parameter estimates were then used to allocate the survey samples into the three groups, with those observations that had a probability of less than one per cent in any of the groups classified as "undefined".

The results of this operation are displayed in Figure 17.6, lower right. Comparing this with the field classification (Figure 17.6, upper left), it can be seen that there is general agreement except around the industrial sites and in the boundary areas between the zones. The "undefined" observations arise due to the robust estimates being based on only core groups of observations that are most typical of the vegetation zones they represent. The surroundings of the smelters do not really "belong" to their respective vegetation zone. Vegetation near the smelters is so seriously damaged that these areas are characterised by Russian colleagues as an "industrial desert". Furthermore, vegetation zones do not only change with distance to coast but also with topography; this can result in undefined points in mountainous areas within the large groups.

When classical estimation procedures are used, the geochemical characterisation of the vegetation zones is contaminated by observations from around industrial facilities and due to other rare processes (Figure 17.6, upper right). The results of this are that only the most extreme outliers associated with the vegetation zones are "undefined", but less extreme outliers are now allocated into the vegetation zones. Although this allocation is correct with respect to the field classification, the newly included observations are not representative of the core geochemical features of the vegetation zones.

Finally, a subset of variables that showed clear differences between the vegetation zones was selected following an inspection of Tukey boxplots. Figure 17.6 (lower left) is based on a robust MCD estimation using logarithmically transformed data for Ag, B, Bi, Mg, Mn, Na, Pb, Rb, S, Sb and Tl. This was undertaken in order to eliminate variables that had little discriminating power and to improve the n/p ratio.

This approach leads to a clearer definition in the boundary areas of the vegetation zones. The reason for this is that the Tukey boxplots focus attention on the core features of the data distribution and not the tails where data from contaminated sites and due to other rare processes occur.

17.6.2   Weighted sums

Weighted sums are not a true multivariate approach, they treat measurements (elements) one at a time, not simultaneously (Garrett and Grunsky, 2001). Techniques like PCA and FA are data based, if a rare but important process, e.g., a mineral deposit or contamination source, is poorly represented in the data, it will only be reflected in one of the minor components or factors often not inspected or visualised as it falls at the tail end of the scree plot. Thus any indication of the process could be lost in the "noise" due to sampling and analytical variability in one of the minor components or factors.

However, knowledge may exist from other investigations that a particular process is reflected by abnormally high or low values of some measured variables, i.e. it has a characteristic "fingerprint". For example, it may be known that a mineral deposit and its weathering products dispersed into the environment are characterised by a particular suite of "pathfinder" elements.

Similarly, it may be known that emissions from an industrial source are characterised by a particular suite of contaminants. The relative importance of the variables comprising the suite are defined simply as numbers like one, two, or three, where three would indicate that a particular variable is three times as important as the least important in the suite. The assignment of the relative importances is based on expert knowledge. In instances where a low value of a variable is considered favourable, the relative importance is defined as a negative number. These relative importances are then normalised into coefficients so that the final values to be computed, weighted sums, are approximately centred on zero.

The coefficients for the variables are applied to the robustly centred and scaled data, where this standardisation is in terms of "average background" for the variables in the survey area. In this context, "average background" refers to an amalgamation of the several background populations that may exist and reflects the sum of natural processes independent of the process being sought.

Figure 17.7 (right) displays the weighted sums computed for the elements Si, Al, K, Ti, Mn and P. These elements were selected following the inspection of Tukey boxplots of the same ten major and minor element determinations of C-horizon soils that were used in the LDA prediction example (Figure 17.5, right). The selection was based on how well the elements could be visually differentiated between the alkaline intrusions (lithological units 82 and 83) and the remaining eight rock types. Soils developed over the alkaline intrusions were recognised in this process as having elevated levels of Al, K, Ti, Mn and P, while levels of Si were lower. This led to a set of relative importances of 1.5, 2, 2, 3, 2 and −2, respectively. The resulting coefficients were applied to the robustly centred and scaled, but not log-transformed, data. A logarithmic transformation was not employed as the Tukey boxplot inspection did not indicate any major skew in the central parts of the soil data distributions associated with the different rock types. Figure 17.7 (left) displays as triangles the location of sample sites underlain by alkaline intrusions. The symbols for the weighted sums (Figure 17.7, left) are coded both according to underlying bedrock and value. Darker symbols indicate higher values of the weighted sums.

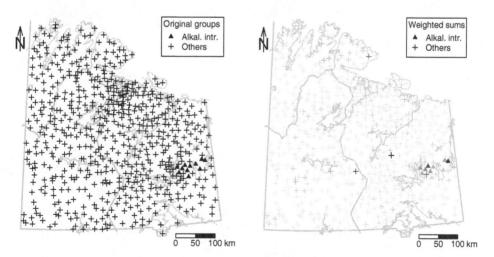

Figure 17.7 Location of the alkaline intrusions near Apatity (lithologies 82 and 83 – left map) and the weighted sums represented as a grey scale (right)

While most of the soils overlying the alkaline intrusions have high weighted sums and are dark, a considerable number of soils from non-alkaline intrusive areas are recognisably dark. Several sample sites in northern Norway within the Caledonian sediments have been reported with geochemical similarities to the alkaline intrusions (see discussion in Section 17.5), and observations with high weighted sums in that area may be similarly indicative.

Although this application of weighted sums does not yield an as-conclusive outcome as LDA described in Section 17.5, this procedure is simple, fast, is relatively assumption free, and is not accompanied by the computational difficulties encountered with the LDA.

17.7 Summary

The methods described in this chapter assume that subsets of "training data", whose true memberships are known, are available. This information is then used to estimate discriminating functions that permit the classification of "test data".

The most commonly used methods, linear and quadratic discriminant analysis (LDA, QDA) are based on estimates of the group centres and covariances. Thus, like other multivariate statistical methods, they are sensitive with respect to severe deviations from multivariate normality. Such deviations can be caused by skewed data distributions, by outliers, or by the closure effect. Appropriate data transformations and/or robust methods will be helpful in such cases for improving the analyses. This also holds for more advanced methods like allocation. Weighted sums have less strict data requirements and can be used to advantage where a process is being sought that is only represented by a few observations in the full data set, or the elemental patterns being sought are known though "external" *a priori* knowledge.

18

Quality Control (QC)

The quality of chemical analyses is of utmost importance in environmental sciences. Chemical analyses are – as are any physical measurement(s) – confounded by uncertainty. Should the users of the analyses have to concern themselves with the quality of the data returned from the laboratory, or is that the laboratory's task? Today most laboratories are nationally accredited, i.e. they follow strict quality control procedures and frequently take part in "round-robin tests" (an analysis of the same samples is performed independently by several laboratories several times). Why should there still be any problems with analytical quality that applied geochemists and environmental scientists, the users of the chemical analyses, should be concerned with? In these times of accredited laboratories and certified analytical methods it is a temptation to take the analytical results and use them without any further concern. However, experience with large geochemical mapping projects during the last two decades has proven that external quality control of the results received from laboratories should, as ever, be an important concern.

For data analysis a good measure of the quality of the data is required; how reliable are they? Would the same results be obtained if the fieldwork was repeated and a second set of samples collected in the same general area as the first survey? Would the same analytical results be obtained if the same sample was re-analysed some weeks later in the same laboratory – or in a different laboratory? Can results of samples collected today be reliably compared with samples to be collected from the same area in ten years time?

To answer such questions, any project in applied geochemistry or environmental sciences should be carried out with its own built-in quality control procedures. It is not sufficient to rely on the accreditation of the laboratory and automatically accept the quality of the values received from a certified laboratory. A project's quality control procedure should include the following steps:

- collection and analysis of field duplicates (frequently, 5–10 per cent);
- randomisation of samples prior to analysis;
- insertion of international reference materials (sparsely, 1 per cent);
- insertion of project standard(s) (frequently, 10–20 per cent); and
- insertion of analytical duplicates of project samples (frequently, 10–20 per cent).

If it is desired to simultaneously quantitatively estimate both the field sampling and the analytical variability it is preferable to prepare the analytical duplicate from the field duplicate.

Statistical Data Analysis Explained   Clemens Reimann, Peter Filzmoser, Robert G. Garrett, Rudolf Dutter
© 2008 John Wiley & Sons, Ltd.

Although this chapter focuses predominantly on statistical methods for analytical quality control it must be mentioned that quality control of sampling and an adequate sampling protocol for the whole project is at least as important (e.g., Garrett, 1983b; Argyraki *et al.*, 1995; Ramsey, 1997, 2002, 2004a,b).

18.1 Randomised samples

Why should the samples be submitted in a random order to the laboratory? The reason is to spread any temporally systematic bias related to the measurement system(s) randomly over all the samples when they are returned to project order. Thus any time-dependent errors that may occur in the laboratory, such as a slow drift from lower to higher reporting levels, does not appear as a feature in the map.

When submitting randomised samples, the laboratory must be instructed to analyse the samples in the sequence of the submitted sample numbers – otherwise the laboratory may randomise the samples once more. Whether either randomised sample numbers are used during field sampling, or new randomised sample numbers are assigned to samples during sample preparation (easiest because standards and duplicates also need to be inserted so that it is not easy for a laboratory to detect the QC samples), is not important as long as the samples are randomised. For a large project it can be advantageous to wait until all samples are ready to be submitted as a single large batch. Submitting the samples as several batches can create serious problems when QC indicates that clear differences exist between the batches in terms of accuracy and/or precision (see below).

18.2 Trueness

Trueness is the closeness of agreement between the average value obtained from a large series of test results and an accepted reference value for a reference material. In this connection the term "bias" is important. Bias is the difference between the expectation of the test results and an accepted reference value. Bias is the total systematic error. Systematic error is that component of error which, in the course of a number of test results for the same characteristic, remains constant or varies in a predictable way. Systematic error must be contrasted to random error. Random error is the error, which in the course of a number of test results for the same characteristic, varies in an unpredictable way and thus cannot be corrected for. Bias may consist of one or more systematic error components.

It is impossible to absolutely determine the "true" composition of a reference material; it can be approached, but there will always be some uncertainty. The objective is to minimise this uncertainty. The data for international reference materials, analysed by many different laboratories, preferably employing a variety of different methods, are compiled and estimates made of the most likely "true" value. The standard deviation of the repeated analyses provides an estimate of the uncertainty associated with the most likely "true" value.

An issue in selecting reference materials is that they should be as similar as possible to the samples collected from the field and under study. The requirement is to match the matrix, the chemistry and mineralogy, of the reference materials as closely as possible to the matrix of the field study samples. Soil standards should be used in soil studies, vegetable materials should be used for moss or other plant materials, rocks for rock analyses, etc. A soil reference sample prepared from a calcareous soil should not be used for the analyses of soils with a low-carbonate siliceous matrix. Documentation of analytical "quality" for the project samples should involve results obtained from analysing one or more standard reference materials similar to the sample material, be it rock, soil, stream sediment, vegetation or water, analysed a number of times throughout the duration of the analytical project so that values for the MEAN and SD can be computed.

Results obtained when analysing an international reference material together with the project samples are usually summarised in a table showing the most important statistical measures, i.e. MEAN, SD, and Coefficient of Variation (CV, that is the same as Relative Standard Deviation, RSD) of the results and of the agreed certified value of the reference material. Table 18.1

Table 18.1 Results for the nine replicate analyses of international reference material CRM NIST 1547, Peach leaves, carried out while analysing the Kola moss samples; together with the MEAN and SD for the certified values. Results of CRM values in square brackets are not certified. All units mg/kg

Element	Certified value	Kola Project		
	MEAN ± SD	MEAN	SD	CV %
Al	249 ± 8	213	8	3.9
As	0.06 ± 0.018	0.098	0.007	6.7
B	29 ± 2	27.4	1.9	7.1
Ba	124 ± 4	126	3.2	2.6
Ca	15600 ± 200	16210	20	1.3
Cd	0.026 ± 0.003	0.021	0.003	13.2
Co	[0.07]	0.068	0.004	6.5
Cr	[1.0]	0.88	0.11	12.4
Cu	3.7 ± 0,4	3.50	0.20	5.7
Fe	218 ± 14	209	3.4	1.6
Hg	0.031 ± 0.007	0.035	0.012	33.2
K	24300 ± 300	24967	524	2.1
La	[9]	9.6	0.35	3.6
Mg	4320 ± 8	4318	100	2.3
Mn	98 ± 3	97.9	1.6	1.7
Mo	0.06 ± 0.008	0.050	0.008	16.5
Na	24 ± 2	21.8	4.9	22.4
Ni	0.69 ± 0.09	0.602	0.085	14.1
P	1370 ± 7	1451	23	1.6
Pb	0.87 ± 0.03	0.86	0.04	4.6
Rb	19.7 ± 1.2	19.1	0.5	2.6
S	[2000]	1630	23	1.4
Sb	[0.02]	0.023	0.008	34.7
Si	—	263	19	7.1
Sr	53 ± 4	57.9	1.7	2.9
Th	[0.05]	0.050	0.008	15.7
Tl	—	0.020	0.0022	11.0
U	[0.015]	0.011	0.0026	24.7
V	0.37 ± 0.03	0.35	0.02	6.0
Y	[0.2]	3.00	0.09	2.9
Zn	17.9 ± 0.4	18.1	0.5	2.7

contains the results for the control reference material (CRM) NIST 1547, Peach leaves, analysed nine times together with the Kola moss samples. Such a table can be used to convey information on the trueness of the received analytical results of a project. Other scientists, using the same reference material can now directly judge the comparability of the project data to their data. Furthermore, it will give a first impression as to the overall analytical quality achieved.

Due to the high price of international reference materials, they will usually not be used to monitor for accuracy (see below).

18.3   Accuracy

Accuracy is essentially the absence of bias. Note that analytical results can be highly accurate without being "true". International reference materials are expensive. In addition they are usually easy for a laboratory to identify when inserted among the project samples. Thus to detect any drift or other changes in analytical results over time, one or more project standards, control reference materials (CRMs), are required that are quite frequently inserted among the field survey samples – probably at least at an overall rate of 1 in 20. These samples should not be different from all other project samples, either in looks or in specific gravity (density), or in the amount submitted for analysis. A common procedure is to collect one or more large samples from the project area, prepare them as for all the field samples of that material, and split them down into the appropriate number of aliquots that can be inserted among the samples prior to submission for laboratory analysis. These samples are then used to monitor for consistency, lack of drift across the duration of the project. The number of CRMs used depends upon the geochemical complexity of the survey area. On one hand the CRMs should reflect the different mineralogical matrices, but there should not be so many CRMs that insufficient analyses are made to be able to monitor the analytical results adequately. Rarely should more than three CRMs be used for a sample material, and in some cases, e.g., mosses with a consistent matrix, only one would be needed.

Immediately after the analytical results are received from the laboratory, the control reference material results should be retrieved from the file. A table for these results can be prepared providing estimates of the repeatability of the measurements (see Section 18.4) at the concentration of the standard sample for each element (Table 18.2). Such a table can be sorted alphabetically for the elements (Table 18.2, left) or in increasing order of the CV to get a rapid overview of the overall quality (Table 18.2, right). The sorted table (Table 18.2, right) shows at one glance that the elements analysed by XRF (recorded as oxides) show an exceptional high repeatability and highlights the elements with possible analytical quality problems at the level of the CRM (poor repeatability could be due to proximity to the DL).

In addition, the data for each variable are plotted against the sample number. The sample number provides an estimate of "time of analysis" (again, the laboratory must be instructed to analyse the samples in the exact sequence of the submitted sample numbers). In such diagrams, also called "x-Charts", any trends (e.g., due to instrumental drift) or gross deviations (e.g., due to sample mix-ups) from the average analytical result for the project CRM become immediately visible. It is common practice for laboratories to insert their own control materials, as a result errors due to an instrument malfunction are usually "caught" prior to the data being returned to the project. However, if a problem has remained undetected by the laboratory, x-Charts will likely detect them.

Table 18.2 Average repeatability for selected elements calculated from analyses of the CRM used for the Kola Project C-horizon data ($n = 52$)

Element	Unit	DL	MEAN	SD	CV %	Element	Unit	DL	MEAN	SD	CV %
Ag	mg/kg	0.001	0.0075	0.0014	18	Al_2O_3	wt%	0.05	10	0.200	2.0
Al	mg/kg	10	4015	254	6.3	CaO	wt%	0.007	2.2	0.057	2.6
Al_2O_3	wt%	0.05	10	0.200	2.0	Fe_2O_3	wt%	0.02	3.3	0.098	3.0
As	mg/kg	0.1	0.22	0.078	35	Ni	mg/kg	1	16	0.775	4.9
Ba	mg/kg	0.5	35	2.42	7.0	Fe_INAA	mg/kg	100	24898	1278	5.1
Bi	mg/kg	0.005	0.013	0.0031	24	La	mg/kg	0.5	8.0	0.478	6.0
CaO	wt%	0.007	2.2	0.057	2.6	La_INAA	mg/kg	1	15	0.908	6.1
Cd	mg/kg	0.001	0.010	0.0016	15	Al	mg/kg	10	4015	254	6.3
Co	mg/kg	0.2	5.5	0.666	12	Fe	mg/kg	10	9528	628	6.6
Cr	mg/kg	0.5	42	3.51	8.3	Cu	mg/kg	0.5	12.6	0.84	6.7
Cu	mg/kg	0.5	12.6	0.84	6.7	Ba	mg/kg	0.5	35	2.42	7.0
Fe	mg/kg	10	9528	628	6.6	V	mg/kg	0.5	21	1.54	7.5
Fe_INAA	mg/kg	100	24898	1278	5.1	Zn	mg/kg	0.5	9.1	0.689	7.5
Fe_2O_3	wt%	0.02	3.3	0.098	3.0	Cr	mg/kg	0.5	42	3.51	8.3
La	mg/kg	0.5	8.0	0.478	6.0	Th_INAA	mg/kg	0.2	3.0	0.305	10
La_INAA	mg/kg	1	15	0.908	6.1	Sc_INAA	mg/kg	0.1	9.5	0.972	10
Ni	mg/kg	1	16	0.775	4.9	Co	mg/kg	0.2	5.5	0.666	12
Pb	mg/kg	0.2	0.78	0.096	12	Pb	mg/kg	0.2	0.78	0.096	12
S	mg/kg	5	13	8.07	63	Cd	mg/kg	0.001	0.010	0.0016	15
Sc_INAA	mg/kg	0.1	9.5	0.972	10	Ag	mg/kg	0.001	0.0075	0.0014	18
Se	mg/kg	0.01	0.034	0.030	89	Bi	mg/kg	0.005	0.013	0.0031	24
Te	mg/kg	0.003	0.0048	0.0034	72	As	mg/kg	0.1	0.22	0.078	35
Th_INAA	mg/kg	0.2	3.0	0.305	10	S	mg/kg	5	13	8.07	63
V	mg/kg	0.5	21	1.54	7.5	Te	mg/kg	0.003	0.0048	0.0034	72
Zn	mg/kg	0.5	9.1	0.689	7.5	Se	mg/kg	0.01	0.034	0.030	89

Figure 18.1 shows two such x-Charts constructed with the data obtained from the Kola Project soil CRM, a large soil sample collected within the project area in northern Norway. In such a plot it is desirable to have an indication of whether a sample falls within the range of "normal" variation around a central value, or is an outlier. This can be done by plotting lines for the MEAN, and multiples of the standard deviation (Figure 18.1). This graphical inspection can be used, for example, to decide whether the block of samples containing the CRM result is accepted from the laboratory or rejected and the block needs to be re-analysed. x-Charts will often show time trends or unusual breaks in the data, related to changes in the laboratory – examples displaying pure random variation (the ideal case) are rare. In the examples used in Figure 18.1 the x-Chart for Al_2O_3 indicates the existence of one outlier (ID 510) – this is possibly an indication that while loading the automatic sample changer of the XRF-machine, samples were inadvertently exchanged. Note also the tendency for higher results at IDs > 600. The plot for Sm shows a clear trend towards lower values over time (Figure 18.1). It appears that only for the very last analytical batches (indicated by the last two CRM results) the laboratory became aware of this trend and corrected the calibration. This plot illustrates why samples should be randomised: a time trend like this might otherwise lead to unusual patterns on a geochemical map. Randomisation ensures this drift is distributed as a random component across the map. The plot also

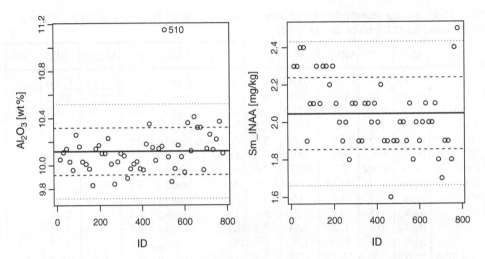

Figure 18.1 x-Chart showing the analytical results of the Kola Project CRM for Al$_2$O$_3$ as determined by XRF and for Sm determined by INAA. Dotted lines shown in the plot are MEAN $\pm$ 2 $\cdot$ SD, dashed lines are drawn at MEAN $\pm$ 1 $\cdot$ SD, and the solid line at the MEAN

indicates a too-severe rounding, discretisation, of the Sm results (discrete "lines" of results) by the laboratory.

18.4 Precision

Precision is the closeness of agreement between independent test results obtained under stipulated conditions. It depends only on the distribution of random errors and does not relate to the true value. Precision is usually quantitatively expressed in terms of imprecision and estimated through the standard deviation of the test results: a low standard deviation will indicate a high precision. The precision is usually adjusted for the mean and expressed as the CV, or equivalent RSD, both quoted as a percentage (e.g., Massart *et al.*, 1988). When referring to precision, the conditions of measurement must be specified. There is an important difference between repeatability conditions and reproducibility conditions. Repeatability conditions refer to situations where independent test results are obtained using the same method on identical test items (samples) in the same laboratory by the same operator using the same equipment within short intervals of time. Most references to precision found in literature will refer to repeatability conditions. Reproducibility conditions refer to situations where test results are obtained with the same method on identical test items (samples) in different laboratories with different operators using different equipment. It is also important to note that repeatability conditions involve repeated execution of the entire method from the point at which the material for analysis reaches the laboratory, and not just repeat instrumental determinations on prepared extracts. The latter give impressive estimates of precision – but have no relevance to the precision achieved when field samples are analysed in the laboratory as it does not take the natural inhomogeneity of the sample material, an important source of variability, into account.

18.4.1 Analytical duplicates

Precision is routinely estimated via the insertion of duplicates of real project samples, usually at a rate of 1 in 20 (or 1 in 10). It is, for example, straightforward to always reserve position "20" (20, 40, 60, ...) for a duplicate of one of the preceding 18 real samples (plus one project standard). Again the results for all duplicate pairs are retrieved once the laboratory delivers the results. For each pair the squared difference is calculated. The sum of these values divided by the number of samples is a measure of variability. To obtain the standard deviation the square root of this variability measure is taken. The resulting measure of precision as shown in Table 18.3 corresponds to a CV value, because the standard deviation is divided by the overall mean of the samples.

Table 18.3 shows the results of the analytical duplicates as received for the Kola Project C-horizon samples. Again it can be advantageous to show both a table sorted according to the alphabetical sequence of the elements (Table 18.3, left) and the same table sorted according to precision (Table 18.3, right). On average, precision is quite acceptable for most elements. Even the problematic elements in the CRM table (Table 18.2), Se and Te, show a better than expected precision in Table 18.3.

Table 18.3 Precision as calculated for selected elements from the 52 analytical duplicates from the analyses of Kola Project C-horizon soils

Element	Unit	DL	Precision %	Element	Unit	DL	Precision %
Ag	mg/kg	0.001	20	Al_2O_3	wt%	0.05	4
Al	mg/kg	10	5.5	Ni	mg/kg	1	5.2
Al_2O_3	wt%	0.05	4.0	Ba	mg/kg	0.5	5.3
As	mg/kg	0.1	28	Al	mg/kg	10	5.5
Ba	mg/kg	0.5	5.3	Zn	mg/kg	0.5	5.6
Bi	mg/kg	0.005	18	Cu	mg/kg	0.5	6.1
CaO	wt%	0.007	16	Co	mg/kg	0.2	6.7
Cd	mg/kg	0.001	11	Cr	mg/kg	0.5	7.2
Co	mg/kg	0.2	6.7	Fe_INAA	mg/kg	100	7.3
Cr	mg/kg	0.5	7.2	Sc_INAA	mg/kg	0.1	7.5
Cu	mg/kg	0.5	6.1	La_INAA	mg/kg	1	8.3
Fe	mg/kg	10	11	S	mg/kg	5	9.3
Fe_INAA	mg/kg	100	7.3	V	mg/kg	0.5	10
Fe_2O_3	wt%	0.02	15	Cd	mg/kg	0.001	11
La	mg/kg	0.5	18	Fe	mg/kg	10	11
La_INAA	mg/kg	1	8.3	Fe_2O_3	wt%	0.02	15
Ni	mg/kg	1	5.2	CaO	wt%	0.007	16
Pb	mg/kg	0.2	25	Bi	mg/kg	0.005	18
S	mg/kg	5	9.3	La	mg/kg	0.5	18
Sc_INAA	mg/kg	0.1	7.5	Se	mg/kg	0.01	18
Se	mg/kg	0.01	18	Th_INAA	mg/kg	0.2	18
Te	mg/kg	0.003	33	Ag	mg/kg	0.001	20
Th_INAA	mg/kg	0.2	18	Pb	mg/kg	0.2	25
V	mg/kg	0.5	10	As	mg/kg	0.1	28
Zn	mg/kg	0.5	5.6	Te	mg/kg	0.003	33

Again it is advantageous to represent these results in a graphical form. "Thompson and Howarth" plots (Thompson and Howarth, 1978) are a frequently used graphic for this purpose (Figure 18.2). Here the absolute difference between the two analyses, $|D1 - D2|$, is plotted against the mean of the duplicate results $(D1 + D2)/2$, and the overall analytical performance can be grasped at once. These graphs can also be used to estimate the practical detection limit (PDL – see discussion below) for an analytical procedure (the point where precision becomes worse than ±100 per cent) via regression analysis. It is possible to draw lines for any certain predefined precision (e.g., ±10 per cent) into these diagrams (in Figure 18.2 at $\pm$ 10 per cent and ±20 per cent). It is then directly recognisable if any, and if so how many, of the samples plot above the line(s). This can support the decision as to whether an analytical batch from the laboratory is accepted or rejected. Figure 18.2 shows that precision is excellent for potassium (K_2O) as analysed by XRF. Four samples that fall above the 20 per cent line are an exception. This is again a strong indication that a number of samples were exchanged during the analytical process. For Zn as analysed by INAA, precision is not so good. The fact that a number of duplicate pairs plot along a straight line to the left in the diagram is caused by a high detection limit of 50 mg/kg. When one of the duplicate samples returns a value above the detection limit and the other a result below the detection limit (set to 25 mg/kg) points on a line result.

In reality precision depends on concentration and follows a curve from poor precision near the detection limit to high precision in the optimal working range of the analytical technique back to poorer precision at the upper limit of concentrations that can be measured with this technique (for an example, see Fauth *et al.*, 1985). Thus there exists both a "lower" limit and an "upper" limit of detection, and fitting just one regression line into these plots is a simplification that has its limits when the duplicates cover a very wide concentration range.

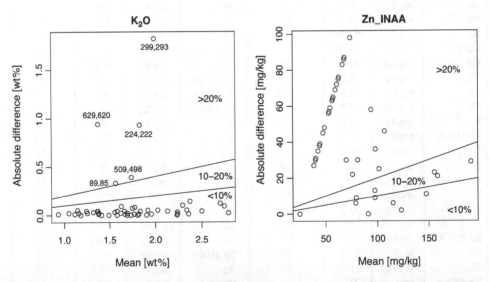

Figure 18.2 Thompson and Howarth plots for analytical duplicate results of K_2O (XRF) and Zn (INAA) determined in Kola Project C-horizon soils. Number pairs annotating samples in the left plot identify analytical duplicates with unusually poor precision for follow-up

18.4.2   Field duplicates

In a comprehensive quality control program field duplicates will have been collected at a
number of randomly selected sites (usually 5–10 per cent). These samples are used to estimate
the variation introduced by sampling and to answer the question of whether more or less
the same analytical results would be obtained if undertaking the survey a second time at
approximately the same sample sites. An estimate of the field variability is especially important
in monitoring programs, i.e. when the sampling exercise is to be repeated after a number of
years to detect any changes over time.

The precision of field duplicates can be estimated in the same way as was the precision
of analytical duplicates (Table 18.4), and even Thompson and Howarth plots could be drawn
for field duplicates (Figure 18.3). The precision determined from field duplicates includes
variability due to both sampling and analysis. Table 18.4 and the Thompson and Howarth
plots for the field duplicates (Figure 18.3) can be directly compared to the same results
from the analytical duplicates to get an initial impression of the relative magnitude of the
sampling error in relation to the analytical error. Again, sorting Table 18.4 according to
decreasing precision (right half) can aid interpretation and help to detect any problematic
elements.

Table 18.4   Combined sampling and analytical precision calculated for selected elements from
the 49 field duplicates of Kola Project C-horizon soils

Element	Unit	DL	Precision %	Element	Unit	DL	Precision %
Ag	mg/kg	0.001	57	Al_2O_3	wt%	0.05	4.2
Al	mg/kg	10	16	Fe_INAA	mg/kg	100	8.1
Al_2O_3	wt%	0.05	4.2	Fe	mg/kg	10	9.3
As	mg/kg	0.1	37	V	mg/kg	0.5	11
Ba	mg/kg	0.5	21	CaO	wt%	0.007	12
Bi	mg/kg	0.005	26	Fe_2O_3	wt%	0.02	13
CaO	wt%	0.007	12	Co	mg/kg	0.2	14
Cd	mg/kg	0.001	71	La_INAA	mg/kg	1	14
Co	mg/kg	0.2	14	Sc_INAA	mg/kg	0.1	14
Cr	mg/kg	0.5	23	Zn	mg/kg	0.5	14
Cu	mg/kg	0.5	30	Al	mg/kg	10	16
Fe	mg/kg	10	9.3	La	mg/kg	0.5	16
Fe_INAA	mg/kg	100	8.1	Th_INAA	mg/kg	0.2	17
Fe_2O_3	wt%	0.02	13	Ba	mg/kg	0.5	21
La	mg/kg	0.5	16	Cr	mg/kg	0.5	23
La_INAA	mg/kg	1	14	Bi	mg/kg	0.005	26
Ni	mg/kg	1	33	S	mg/kg	5	27
Pb	mg/kg	0.2	44	Cu	mg/kg	0.5	30
S	mg/kg	5	27	Ni	mg/kg	1	33
Sc_INAA	mg/kg	0.1	14	As	mg/kg	0.1	37
Se	mg/kg	0.01	39	Se	mg/kg	0.01	39
Te	mg/kg	0.003	170	Pb	mg/kg	0.2	44
Th_INAA	mg/kg	0.2	17	Ag	mg/kg	0.001	57
V	mg/kg	0.5	11	Cd	mg/kg	0.001	71
Zn	mg/kg	0.5	14	Te	mg/kg	0.003	170

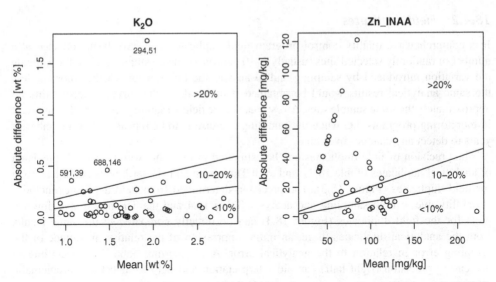

Figure 18.3 "Thompson and Howarth" plots for field duplicate results of K_2O (XRF) and Zn (INAA) determined in Kola Project C-horizon soils. Number pairs annotating samples in the left plot identify analytical duplicates with unusually poor precision for follow-up

18.5 Analysis of variance (ANOVA)

In a more formalised approach, results of all samples, field and analytical duplicates can be used to carry out an unbalanced Analysis of Variance (ANOVA) (see, e.g., Scheffé, 1999). By "unbalanced" is meant that unequal numbers of analyses occur at each level of the design. A "balanced" design for studying field and analytical variability would require that each sample collected at a field duplicate site would be split and analysed twice. In an unbalanced design only one of the field duplicate pairs is split and analysed twice. The same information can be extracted from both designs, however, the unbalanced design makes a more efficient use of resources. The results of the ANOVA include estimates of the proportion of the total variability due to each of sampling and analysis. Whether or not the analytical variability is significantly smaller than the sampling variability can be determined by a formal statistical test. However, this formal analysis can only be undertaken when the analytical duplicate is prepared from the field duplicate. This was not done for the Kola Project. In its place two independent sets of duplicates were obtained: one to estimate the combined sampling and analytical variability from field duplicates; and the second to estimate the analytical variability from a set of duplicates prepared from the field samples. The "at-site" sampling variability should be considerably larger than the analytical variability. However, this depends on how the field duplicates were collected. Were they just a split of the field sample (not a "true" field duplicate!); were they, for example, taken from opposite walls of the same soil pit; or do they represent a "true" duplicate, i.e. collected at a second site in the vicinity of the first site? For the Kola Project all field duplicates were taken at a second site about 100 m removed from the first site. This is a "worst case" scenario that will provide a very realistic picture of at-site variability.

An ANOVA of the field duplicate pairs decomposes the variability into two components: (1) the variability between the sites where duplicates were collected; and (2) the variability at the field duplicate site. In addition, a formal F-test can be undertaken to determine if the variability

Table 18.5 ANOVA table for the 49 field duplicate pairs for Cu in Kola Project C-horizon soils

Source	DF	MSS	F	p-value	Variance	Percentage of variation
Between sites	48	0.16536	16.5	<0.0001	0.077673	88.6
Within sites	49	0.01001			0.010009	11.4

at the sites is significantly different from the variability between the sites. This test uses what is called a "random effects model" which is somewhat different from the "fixed effects" one-way ANOVA model discussed in Section 9.7.1. When only two sources of variability are being considered the two methods, "fixed effects" and "random effects", are computationally identical except for the calculation of the proportions of the variability related to the two sources. If the variability between the sites is not significantly greater than the variability within the sites, preparing maps of the data may be misleading. Using the logarithmically transformed Cu data for C-horizon soils, the ANOVA table presented in Table 18.5 may be generated. A logarithmic transformation was used in order to meet the requirement of homogeneity of variance and to approach a normal distribution (see discussion in Section 16.1.1). Additionally an inspection of the data shows that the Cu data span more than one order of magnitude, a useful guide to whether a transformation is necessary.

The ANOVA (Table 18.5) indicates that some 89 per cent of the total variability is due to variations between the regional sites where field duplicates were collected, and only 11 per cent was due to variability at the field duplicates sites. The F-test and the associated p-value confirm that this partition is highly significant and that maps can be prepared with confidence.

The same approach can be applied to analytical duplicate pairs, see Table 18.6.

The between sites variability now comprises the variability between the sites chosen for analytical replication across the survey area, and the only remaining variability is due to analysis, i.e. weighing out and analysing a second aliquot of the prepared sample. As might be expected, virtually all the variability now lies between the regional sampling sites and only 0.4 per cent is due to analysis. The F-test and p-value demonstrate that this partition is highly significant.

As a formal staggered unbalanced design (Garrett, 1983b) was not employed in the Kola Project, an approximation has to be used in order to partition the variability between the three sources, regional, at sites, and analytical. This can be done because variances are additive. The first task is to determine if the regional variability estimates, between sites, as determined from the data used in both ANOVAs, are sufficiently similar to proceed. The ratio of the "between" variances, both measures of the variability across the survey area, from the two ANOVAs is calculated; it is 1.45, which corresponds to a p-value of 0.1. This is above the five per cent significance level, and it cannot be accepted that the two variances are equal.

Table 18.6 ANOVA table for the 52 analytical duplicate pairs for Cu in Kola Project C-horizon soils

Source	DF	MSS	F	p-value	Variance	Percentage of variation
Between sites	51	0.22600	454	<0.0001	0.11274	99.6
Between analyses	52	0.00051			0.00050838	0.4

However, some estimate of the relative variability between the sources is still desirable. Therefore, the field duplicate "between sites" variance was selected, as sites selected for duplicate sampling are likely more evenly distributed across the project area than those selected to monitor analytical variability. As variances are additive, the analytical, "between analyses", variance determined from the analytical duplicate pairs can be subtracted from the "within sites" variance to estimate that part of the "within sites" variance that is due to the "at sample sites" variance alone. These three variances may now be expressed in percentage form. Thus, for Cu the variability is partitioned: 88.6 per cent to regional, 10.8 per cent to variability at sampling sites, and 0.6 per cent to analytical variability. This is a satisfactory outcome for a regional survey.

The above approximation will sometimes lead to negative estimates for the "at sampling sites" variances; these are impossible. However, they are the outcome of not employing a proper unbalanced sampling design for estimating the variance components. By convention, negative estimates are set to zero, and the computation completed.

Table 18.7 presents the results of the above partitioning of the variability into variance components (expressed as percentages) for the major element oxides determined by XRF (left) and selected trace elements (right), and the p-value is for the F-test with the ratio of the two "between" variances determined from the field duplicates and analytical duplicates.

In most cases the p-values in Table 18.7 are greater than 0.05, indicating an inequality of the "between" variances for the field and analytical duplicates, raising some doubt over the

Table 18.7 Distributed percentage variabilities for the major element oxides (determined by XRF) and selected trace elements for Kola Project C-horizon soils. The p-value is for the F-test to determine if the variances at the "between" level are equal for the field and analytical duplicates. With the exception of pH all variables were log-transformed prior to the calculation

Element	Regional	Site	Analytical	p-value	Element	Regional	Site	Analytical	p-value
Al_2O_3	88.65	1.05	10.30	0.88	Ag	63.14	17.29	19.57	0.08
CaO	86.17	0	13.83	1	As	90.50	3.52	5.98	0.03
Fe_2O_3	80.11	0	19.89	0.05	Be	95.30	3.51	1.19	0.45
K_2O	80.32	0	19.68	0.8	Bi	79.65	7.56	12.80	0.11
MgO	86.86	2.40	10.75	0.54	Cd	80.25	17.47	2.28	0.48
MnO	66.73	0	33.27	0.15	Co	92.76	5.47	1.77	0.15
Na_2O	87.62	0	12.38	0.11	Cr	94.68	4.15	1.18	0.42
P_2O_5	91.37	4.20	4.43	0.94	Cr_INAA	90.44	8.12	1.44	0.82
SiO_2	82.56	0	17.44	0.88	Cu	88.59	10.83	0.58	0.10
TiO_2	82.53	0	17.47	0.04	Hg	49.99	10.20	39.81	0.10
					Mo	72.32	13.88	13.80	0.25
					Ni	92.02	7.08	0.90	0.25
					Pb	80.01	11.50	8.48	0.87
					pH	28.19	11.23	60.57	0.01
					S	72.73	23.96	3.31	0.07
					Sb	65.32	25.09	9.59	0.09
					Se	67.40	19.11	13.48	0.06
					Sr	95.91	1.52	2.57	0.86
					Th	82.05	9.92	8.03	0.30
					V	95.34	0.68	3.98	0.40
					Zn	93.26	5.59	1.15	0.26

validity of this *ad hoc* decomposition of the variances. The critical fact is that all the field duplicate ANOVAs, except Hg and pH, indicated that the combined sampling and analytical variability is significantly smaller than the "between sites" regional variability. That is the most important conclusion and supports the validity of the maps. As this critical test was passed, it was considered informative to proceed with the *ad hoc* estimation of variance components so as to be able to place the relative levels of "at site" and "analytical" variability in context with the "overall regional" variability.

The table (Table 18.7, left) indicates that the "at sampling sites" variability for the major elements is very small or zero relative to the analytical variability. This indicates that the soils at the 100 m scale are very homogeneous with respect to major elements. For all oxides, except Mn, in excess of 80 per cent of the variability is at a regional scale. The partitioning for the trace elements (Table 18.7, right) presents a more diverse story. Elements such as Ag, Hg, Sb and the pH all have high local variabilities, i.e. the "between sites" regional variability is less than 67 per cent. This reflects the difficulties in determining these elements at low levels and the problems of measuring pH in soil samples. Bi, Mo, S and Se all have "between sites" variabilities in the range of 67–80 per cent, indicating that maps of these elements are likely reliable. For all of the remaining elements in excess of 80 per cent of their variability is at the regional scale indicating that they may be mapped with confidence.

18.6 Using maps to assess data quality

One additional quality criterion will be the appearance of the map when the analytical results are mapped (see Chapter 5). Figure 18.4 shows two kriged surface maps for Ca: left map, Ca in the O-horizon; and right map, Ca in the C-horizon. The question to be asked is, do the maps contain any clear regional features or could they as easily represent random variability

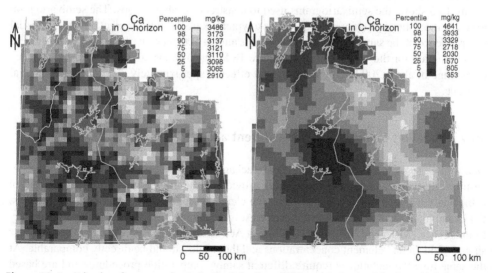

Figure 18.4 Kriged surface maps for Ca in the O-horizon (left) and Ca in the C-horizon (right) of Kola Project soils

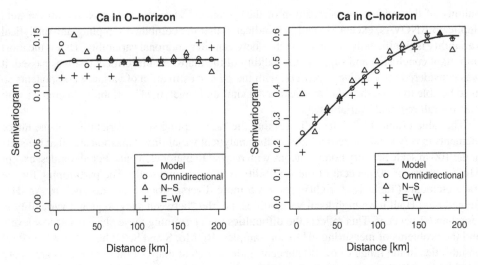

Figure 18.5 Semivariograms for Ca in the O-horizon (left) and C-horizon (right) Kola Project soils

due to analysis, sampling variability, and a lack of any significant regional trends? The map of Ca in the O-horizon is very noisy without any clear spatial structure, while the map of Ca in the C-horizon shows clear spatial structures (Figure 18.4, right). In the first case it could be assumed that the sample density was too low or (and) the sampling and analytical variability was too high relative to the regional variability to obtain a useful map. It is also possible that no clear regional structure for Ca in the O-horizon exists because its regional distribution is dominated by local small-scale effects.

Information about data quality, or better suitability for mapping, can also be directly derived from the semivariogram (see Section 5.7.2), if kriging was used as the interpolation method. Figure 18.5 shows the semivariograms used to construct the above maps. The semivariogram for Ca in the O-horizon (Figure 18.5, left) is a typical example of the nugget effect, i.e. spatial independence of the samples; the nugget effect is almost 100 per cent of the total variance. The semivariogram for the C-horizon data (Figure 18.5, right) shows a clear spatial dependency for a range of about 200 km, here the nugget effect is approximately 30 per cent of the total variance.

18.7 Variables analysed by two different analytical techniques

Due to the use of multi-element analytical packages, data for several elements may be determined by more than one technique. This provides an additional means for quality control if the analytical techniques are measuring the elements in the same mineralogical fractions. For example, in the Kola data set (C-horizon) several elements were analysed by both X-Ray Fluorescence (XRF) and Instrumental Neutron Activation Analysis (INAA). Both techniques should yield "total" element concentrations and the results should be directly comparable. At the same time the techniques require different sample preparation procedures and are based on completely different physical principles. Thus they provide an ideal opportunity for quality control investigations, covering the whole range of the data. Some elements may have been analysed by different techniques, giving some combination of "partial" and/or "total" results. In these instances a comparison can usually not be made (other than in a gross sense) for quality control purposes. However, the data can be used to gain a better understanding of modes of transport for trace elements and the forms in which they are sequestered in the sample materials.

Iron (Fe) is an example of an element that was determined by three different techniques: XRF, INAA and Inductively Coupled Plasma Atomic Emission Spectrometry (ICP-AES) following a partial aqua-regia extraction. In the environmental literature an aqua-regia extraction is often referred to as providing "total" element concentrations because it is a strong acid attack.

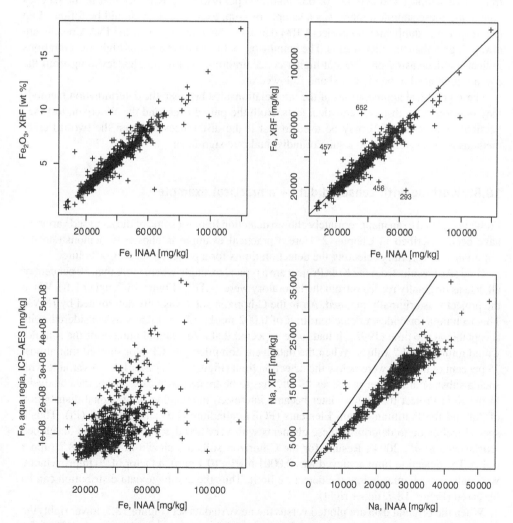

Figure 18.6 Fe in the Kola Project C-horizon soils as determined by three different techniques: XRF expressed as wt% Fe_2O_3 (upper left) and mg/kg Fe (upper right), INAA and in an aqua-regia extraction with ICP-AES finish (lower left); and Na as determined by XRF and INAA in Kola Project C-horizon soils (lower right)

However, elements bound in the lattices of many silicate and oxide minerals are not freed by an aqua-regia extraction and the results obtained are thus often far from "total".

When plotting the results received from the laboratory (XRF results of major elements are routinely reported in wt% of the oxide – for Fe this is usually Fe_2O_3) a generally close relationship between the XRF and the INAA results is visible (Figure 18.6) – with some samples deviating from this general trend. To obtain a better direct comparison, the XRF results should be re-expressed in mg/kg Fe. The conversion factor from wt% Fe_2O_3 to mg/kg Fe is 6994 (conversion factors for all elements are, for example, given in Reimann and de Caritat, 1998b). Following the conversion XRF and INAA results can be directly compared and a 1:1 line can be drawn in the diagram (Figure 18.6, upper right). Pairs of samples that deviate from the general trend (like samples 456 and 457 or 652 and 293) point to the possibility that in the batches containing these samples, there was a sample mix-up, exchange, that should be followed up (i.e. inspect the duplicate and project CRM data). For Na it appears that the INAA results are always higher than the XRF results. The fanning out of the samples towards high concentrations indicates a decreasing precision at higher concentrations. Again samples deviating from the overall trend need to be identified and followed up.

These graphical appreciations of the interrelationships between the determination methods may be formalised through statistical tests. Both the paired t-test and the Wilcoxon test (see Sections 9.5.1 and 9.5.2) may be used to test if the differences between the two different methods of measurement on a suite of individuals are significant.

18.8 Working with censored data – a practical example

Problems related to an inappropriately chosen detection limit for some of the analysed variables have been described in Chapter 2. Here a practical example is chosen to demonstrate the importance of carefully selecting the detection limits for a project's success or failure.

Gold (Au) results from the Kola Project are a typical example where more than 50 per cent of all data as originally received from the laboratory were "<DL" (Figure 18.7, upper left). When the project was originally planned, Au in the C-horizon soils was only determined by INAA. This technique provides a detection limit of 0.002 mg/kg (2 ppb) that was considered quite acceptable at the time (1995). It had to be expected that a substantial number of the samples would return "<DL" values. When the data were compiled, the CP-plot showed that almost 75 per cent of the data were below the detection limit (Figure 18.7, lower left). A variable with such a substantial number of values "<DL" has to be treated with care during data analysis. In the Kola Project the authors later became interested in studying the regional distribution of Au and the Platinum Group Elements (PGE) palladium (Pd) and platinum (Pt). Thus a special technique to determine these elements with very low detection limits was developed (Niskavaara et al., 2004). Results for the C-horizon soils are shown in Figure 18.7 (upper right). The detection limit achieved was 0.0001 mg/kg (0.1 ppb), a factor of 20 improvement, with only two samples below the detection limit. Thus the complete data distribution can be observed (Figure 18.7, upper right).

When the original data are plotted versus the new Au data set (Figure 18.7, lower right) the major problem caused by the censored data in the original data set can be observed. Even for that part of the data where real values are available the expected 1:1 relationship is not present. This is in part due to poor analytical data quality close to the detection limit (original Au data), and in part caused by high natural sample variability due to the presence or absence of high

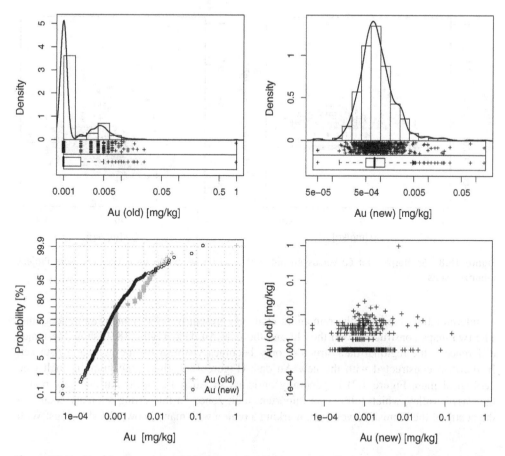

Figure 18.7 Combination plots of histogram, density trace, one-dimensional scatterplot and Tukey boxplot for Au old (upper left) and Au new (upper right) in the Kola Project C-horizon soils. Combined CP-plot for Au old and Au new (lower left), and scatterplot for Au new versus Au old (lower right)

specific gravity micro-nuggets in the aliquots weighed out for analysis (Clifton *et al.*, 1969; Stanley, 1998).

The problems caused by censored data become even more obvious in a scatterplot. When plotting Cu versus Au almost all the structure visible in the diagram will be hidden in that part of the data that is below the detection limit of the original analyses (Figure 18.8, left). The original data do not reflect the relation between Au and Cu that is so apparent in the new Au data (Figure 18.8, right). A clear message from this plot is that variables with a high proportion of censored data should not be used for multivariate data analysis. A typical example would be correlation-based methods, where the results will be distorted and interpretation severely hindered if variables with a high proportion of censored data are included. Such variables can be used to document the data distribution and for mapping (though with care due to possible data quality problems). When highly censored data exist for elements that are characteristically high for some particular process, such as the formation of a mineral deposit type or industrial contamination, their presence in measurable amounts can be used to confirm

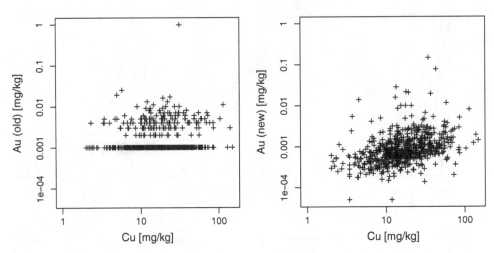

Figure 18.8 Scatterplot for Cu versus Au old (left) and Cu versus Au new (right) in Kola Project C-horizon soils

an interpretation, based on the remaining elements present in reliably measurable amounts. The two maps constructed with the old and the new Au data (Figure 18.9) show the major difference – the regional data structure only becomes visible with the complete data set. In the map constructed with the new Au data (Figure 18.9, right), the granulite belt (see geological map, Figure 1.2) is generally enriched in Au, and a linear Au anomaly marks the Pasvik valley, which follows an important fault zone. An interesting anomaly occurs in the centre of the Finnish survey area, marking a region with many known Au showings. With

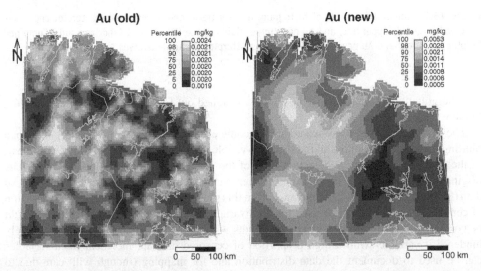

Figure 18.9 Distribution maps of Au determined with the old (left) and new (right) method. For Au determined with the old method about 75 per cent of all values were below the detection limit (Figure 18.7)

knowledge of these features it may be possible to detect them in the map for the original Au determinations. However, the original map by itself is so noisy that no clear conclusions about the processes determining the regional distribution of Au can be deduced.

18.9 Summary

Quality control should be an integral part of any project in applied geochemistry and environmental sciences. The design of a QC protocol needs to be an integral part of project planning. International reference materials, at least one project standard, field and analytical duplicates all need to be inserted in any batch of samples to be analysed. A contract with the analytical laboratory should clearly spell out the consequences if quality problems are detected. It is neither sufficient nor prudent to rely on the quality certification of any laboratory. Ultimately the survey activity is the responsibility of the persons organising it, and it is their responsibility to independently check the quality of all aspects of the work from planning, through field and laboratory activities to final report preparation.

19

Introduction to R and Structure of the DAS+R Graphical User Interface

19.1   R

R is freely available in source code form under the terms of the Free Software Foundation's GNU General Public License. It is a different implementation of the S language developed at Bell Laboratories (Becker *et al.*, 1988). R was initially written by Ross Ihaka and Robert Gentleman (University of Auckland) around 1996 (Ihaka and Gentleman, 1996). It is mainly used for statistical data analysis. It functions on various platforms like Linux, Windows and MacOS.

R consists of the R base system and contributed packages. At present there exist about 1000 contributed packages. Anybody can contribute a new package to the R core team via the R homepage (http://www.r-project.org). This is also the site to search for R documentation.

R can be freely downloaded from the CRAN-server at http://cran.r-project.org (or just search on the web for "R"). A download will install the base system. For the functionality needed to produce the graphics and tables in this book additional packages are needed. These additional packages are best installed in a session within the R base system.

Becker *et al.* (1988), Chambers and Hastie (1992), Chambers (1998), Dalgaard (2002), Fox (2002), Venables and Ripley (2002), Everitt (2005), and Crawley (2007) provide introductory texts to R (or S) and statistical analysis based on R (S). The R homepage (http://www.r-project.org) provides online manuals and Wiki pages for an introduction to R.

19.1.1   Installing R

R is installed via web-download from http://cran.r-project.org. Just follow the instructions on the screen. There are slight differences depending on the chosen operating system. However, in any case the R base system needs to be installed first. Under Windows the base system is installed on your computer via downloading and executing the ".exe" file.

Statistical Data Analysis Explained   Clemens Reimann, Peter Filzmoser, Robert G. Garrett, Rudolf Dutter
© 2008 John Wiley & Sons, Ltd.

19.1.2   Getting started

After installing R the program can be executed by double-clicking on the appropriate icon (Windows) or typing R into the console (Linux, etc.). Windows will open a window called RGui, containing a window called "R Console", a simple graphical user interface. Other platforms will only show a prompt sign (e.g., ">") and wait for R commands (the prompt also appears under Windows in the "R Console" window).

For working under Windows it is an advantage to change the standard setup. Click "edit" "Gui preferences" and then "SDI" and "multiple windows" in the mask and "save". R needs to be shut down and started again for these settings to become active. When re-starting R, only the "R Console" window will appear on the screen.

Under Windows the working directory can either be defined in the "R Console" window ("File", "Change dir..." and entering a directory, e.g., C:/StatDA) or via directly entering the command setwd("C:/StatDa"). R will then save all output to the directory "C:\StatDA". This directory should then also be the default directory for providing all data and background information (e.g. maps) files.

19.1.3   Loading data

Any data set that has the required format for R (see Chapter 2, Figure 2.1) can now be loaded. Generally the data set should be available as a simple ".txt" or ".csv" file. The command to load a ".txt" file, where the first row contains the variable names is

```
> dat = read.table("filename.txt",header=TRUE)
```

The equivalent command for reading a ".csv" file is

```
> dat = read.csv("filename.csv")
```

for files where the field (column) separator is a comma, "," and the decimal point is a ".", and

```
> dat = read.csv2("filename.csv")
```

for files where the field (column) separator is a semicolon ";" and the decimal point is a ",".

"dat" is the object in R which contains the data. It is possible to use any name for "dat", e.g., "chorizon" (i.e. the filename can be used as the object name as well). The first character of the name must be a letter and special signs should generally be avoided for object names (see Chapter 2).

For demonstrating the basic functionality of R, and the R command structure, the example data that have been used throughout the book will be loaded. The Kola data sets are included in the contributed R package "StatDA" and can be downloaded from the CRAN-server. This is done during an active R session via entering:

```
> install.packages("StatDA")
```

The package StatDA is activated via entering

```
> library(StatDA)
```

and the Kola C-horizon data are loaded via entering

```
> data(chorizon)
```

This command creates an R object called "chorizon" containing 605 rows (observations) and 111 columns (variables).

19.1.4 Generating and saving plots in R

It is now possible to directly "look" at the data via entering the necessary R commands. If the task is to study the variable Ba of the Kola C-horizon data set this variable can be extracted separately:

```
> Ba=chorizon[1:605,"Ba"]
```

Ba is now a new object, containing the 605 measured concentrations of the variable Ba in the Kola C-horizon data set. The object chorizon is a data matrix with 605 rows and 111 columns. R needs to be told which rows and which columns define the object Ba. The square brackets are needed to select rows and columns of the matrix, separated by a comma. In the above example all 605 rows are needed (1:605) and it is planned to use the variable with the name "Ba". Because all rows of the matrix are used it would be possible to avoid the "1:605" in front of the comma, giving the abbreviated form of above command:

```
> Ba=chorizon[,"Ba"]
```

For plotting the histogram of the variable Ba the following simple command is now needed:

```
> hist(Ba)
```

Figure 19.1 left shows the resulting histogram (compare Figure 3.3 upper left). The histogram does not really show the data structure due to the strongly skewed data. Thus a histogram of the log-transformed data will probably give a better impression of the data distribution. The command

```
> hist(log10(Ba))
```

first performs a log-transformation to the base 10 of the Ba data and then plots the histogram (Figure 19.1 right – compare Figure 3.1 lower right). Thus the x-axis is now scaled according to the log-transformed data and not according to the original data scale. To construct a scale that relates back to the original data is possible (see, e.g., R-script for Figure 3.1 lower right) but requires advanced knowledge of the R graphics parameters.

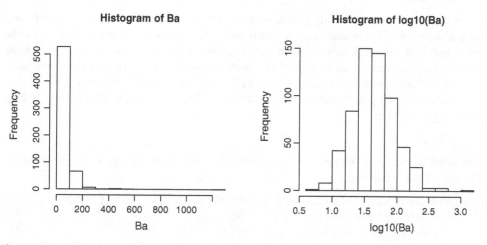

Figure 19.1 Histogram of the variable Ba in the Kola C-horizon data set, original data (left) and log-transformed data (right)

The plots are presented in the R Graphics Device providing the space for one plot only. Thus the histogram for the log-transformed values overplots the histogram for the untransformed values.

Each graphic output of R can be saved to a file (see Section 19.3). To save the histogram of Ba the required commands are

```
> pdf("histBa.pdf",width=6,height=6)
> hist(Ba)
> dev.off()
```

telling R that a .pdf file with the name "histBa.pdf" will be generated, using the R standard width and height of 6 inches (15.2 cm) to plot the histogram and then to close the graphics file (dev.off = device off).

It is often interesting to directly compare two (or more) graphics side by side. For the this purpose the R graphics device is split into rows and columns, just like a table. For the above example (Figure 19.1) two plots will be arranged side by side. Thus the "table" will contain 1 row and 2 columns. The simple R command then is:

```
> par(mfrow=c(1,2))
```

The command "par" handles many graphical parameters in R. "mfrow" is one of these parameters, determining the number of rows (first number) and the number of columns (second number) of the graphic device.

The complete sequence of commands for producing Figure 19.1 is thus:

```
> library(StatDA)
> data(chorizon)
> Ba=chorizon[,"Ba"]
> pdf("fig-19-1.pdf",width=9,height=4.5)
> par(mfrow=c(1,2))
> hist(Ba)
> hist(log10(Ba))
> dev.off()
```

The first graphic is automatically plotted into the left frame, the second into the right frame. Such a sequence of commands is called an R-script. Because it is tedious to write the same commands time and again such scripts are best saved in a script library and the same histograms for other variables can simply be plotted via editing the variable name (e.g., replace "Ba" with "Al_XRF"). More details on the use of R-script files will be provided in Section 19.3.

There exist many possibilities to change the basic plot. These can be found in the R help pages, e.g., by typing:

```
> ?hist
```

A detailed description of the function hist with all possibilities to change the plot will appear. The core of the function is:

```
hist(x,breaks="Sturges",freq=NULL,probability=!freq,
    include.lowest=TRUE,right=TRUE,
    density=NULL,angle=45,col=NULL,border=NULL,
    main=paste("Histogram of",xname),
    xlim=range(breaks),ylim=NULL,
```

```
        xlab=xname,ylab,
        axes=TRUE,plot=TRUE,labels=FALSE,
        nclass=NULL,...)
```

In the above example only the first argument x of the function hist was used for providing the data (x=Ba) and for all other arguments the R default was used. Each argument can now be used to change the plot.

For example, the plot title can be deleted using the argument main:

```
> hist(Ba,main="")
```

main expects a character string, which is defined using quotation marks. If there is nothing between the quotation marks, an empty title will be plotted. Of course it would also be possible to include any desired title between the quotation marks.

To additionally change the x-axis text (the label of the x-axis) the command is:

```
> hist(Ba,main="",xlab="Ba in C-horizon [mg/kg]")
```

To change the y-scale from absolute frequencies to relative frequencies the command is:

```
> hist(Ba,main="",xlab="Ba in C-horizon [mg/kg]",freq=FALSE)
```

The freq=FALSE command will automatically change the y-axis label from "Frequency" to "Density". This example demonstrates that it is necessary to read the complete help file to understand which parameters will cause which changes. In many cases the examples at the end of the help page will be very useful to understand the command structure. These examples can be executed in R using the command:

```
> example(hist)
```

When entering the command, here several graphics will flash up on the screen until the last one is displayed in the graphics window. In cases where there are several examples provided it is thus needed to first enter the command:

```
> par(ask=TRUE)
> example(hist)
```

The command par(ask=TRUE) requires that the user enters a <return> before an example graphic is replaced.

19.1.5   Scatterplots

For producing a scatterplot two variables have to be selected (if only one variable is selected, R will automatically plot the selected variable against the number of the row in the data matrix, the index).

```
> library(StatDA)
> data(moss)
> Cu=moss[,"Cu"]
> Ni=moss[,"Ni"]
```

These commands select the variables Cu and Ni from the Kola moss data set.

To plot a scatterplot the command is:

```
> plot(Cu,Ni)
```

The first variable is used for plotting the *x*-axis, the second is used for the *y*-axis. The resulting plot is already quite close to the plot shown in Figure 6.1, left. To change the plot symbols and the axis labels, the following commands are entered:

```
> plot(Cu,Ni,xlab="Cu in Moss [mg/kg]",
ylab="Ni in Moss [mg/kg]",pch=3,cex=0.7,cex.lab=1.2)
```

This command delivers exactly the graphic shown in Figure 6.1, left. The axis titles are changed using the arguments xlab and ylab (see above, histogram). The plot symbol is changed using the argument pch (plot character). To see the standard R symbol table the command

```
> example(points)
```

can be entered. Here we have replaced the standard circle (1) by a plus (3).

To change the size of the symbols, the argument cex (character expansion factor) is used. The R default size is cex=1, here the size of the symbol was now reduced to 0.7. To change the size of the two axis titles the argument cex.lab (character expansion for *x*- and *y*-axis label) was used and the text size is increased by 20 per cent.

The above two examples are shown in Figure 19.2. The complete R-script for Figure 19.2 is:

```
> library(StatDA)
> data(moss)
> Cu=moss[,"Cu"]
> Ni=moss[,"Ni"]
> pdf("fig-19-2.pdf",width=9,height=4.5)
> par(mfrow=c(1,2))
> plot(Cu,Ni)
> plot(Cu,Ni,xlab="Cu in Moss [mg/kg]",
ylab="Ni in Moss [mg/kg]",pch=3,cex=0.7,cex.lab=1.2)
> dev.off()
```

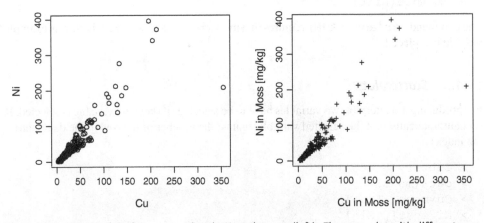

Figure 19.2 Scatterplot of Cu versus Ni, Kola Moss data set (left). The same plot with different axes labels and symbols (right)

The structure is thus very similar to the above example for plotting the histogram and it was possible to simply copy and paste substantial parts of the R-script.

19.2 R-scripts

The R-scripts used to produce all graphics, tables, and statistical tests shown in this book are provided at the following website:

    http://www.statistik.tuwien.ac.at/StatDA/R-scripts/

The R-scripts used throughout the book contain our own functions that are not included in the R base system. To be able to use the full functionality of these scripts it is necessary to install the following contributed package from the CRAN-server:

```
> install.packages("StatDA")
```

When this package is installed, each script file will automatically work with the provided Kola example data sets. With some basic knowledge of R it is possible to edit these files to fit the reader's own data.

For working with the examples provided in the book it is also necessary to create a personal directory called, e.g., C:\StatDA\data and to download the provided files containing all additional files needed to create some of the figures, maps or tables in the book from:

    http://www.statistik.tuwien.ac.at/StatDA/R-scripts/data/

For example, for plotting the maps shown in Figure 5.1, the R-script looks like the following:

```
library(StatDA)
data(ohorizon)
X=ohorizon[,"XCOO"]
Y=ohorizon[,"YCOO"]
pdf("fig-5-1.pdf",width=8.2,height=8.6)
```

The width and height of the resulting plot need to be defined. In the example the plot (the pdf file) is scaled to 8.2 inches × 8.6 inches (20.8 cm × 21.8 cm). In principle the size of the graphic does not really matter because it can be resized in a word processing environment. However, proportions and text size must be chosen to fit the final size.

```
par(mfrow=c(2,2),mar=c(1.5,1.5,1.5,1.5))
```

The parameter mfrow divides the plot into a table of two rows and two columns (for plotting four graphics). The parameter mar determines the width of the figure margins on the four sides of each graph. This parameter is automatically valid for all four graphs if not changed again during the session.

```
# library(pixmap)
```

# denotes that the rest of the command line is not executed. Here the library(pixmap) was automatically loaded when loading the library(StatDA). If working without the library StatDA this command would need to be executed in R (delete the "#").

```
m=read.pnm("C:/StatDA/data/kolageology.pnm")
```

```
plot(m[1:500,1:494])
```

For plotting the geological background map a bitmap file is needed. The `read.pnm` command requires a file with the extension ".pnm". Under Linux it is very easy to convert a ".jpg" file to a ".pnm" file, e.g., via executing the command "`convert kolageology.jpg ko-lageology.pnm`". Under Windows it is necessary to find a graphics program that can export bitmaps in ".pnm" format. The R object m now contains the information of the bitmap file in the form of a data matrix. This information can be directly plotted using the `plot` command. The original bitmap file contains not only the map but also the legend for the map (see "kolageology.jpg"). For plotting the geology in Figure 5.1, upper left, the legend is not needed and cut from the plot via specifying the required rows and columns of the matrix – to find the exact rows and columns of the matrix m will need several trials. The result is the plot in Figure 5.1, upper left.

```
plot(X,Y,frame.plot=FALSE,xaxt="n",yaxt="n",xlab="",ylab="",
type="n",asp=TRUE)
```

This command is supposed to open an empty plot that is scaled according to the x- and y-coordinates of the data file. `frame.plot=FALSE` results in no frame for the plot, `xaxt="n",yaxt="n",xlab="",ylab=""` avoids plotting axes and labels for the axes, and `type="n"` ("n" for "no/none") avoids plotting symbols for each x, y-coordinate. The final argument `asp=TRUE` guarantees the correct aspect ratio of the x- and y-coordinates, independent of width and height of the plot.

```
data(kola.background)
```

This command loads the data set `kola.background`, containing the x- and y-coordinates of the Kola background map shown in Figure 5.1, upper right. A background map file consists of a matrix with two columns (x- and y-coordinates) of the polygons shown in the map. When one of the polygons is finished the matrix must contain a row with two "NA" values (not available). For finishing a background file, two rows with "NA" are needed. In case of the Kola background map there exist four different layers: the project boundary, the coastline, the international country borders and shorelines of some major lakes. Each of these layers exists originally in its own .csv file. These four separate .csv files need to be read into R, e.g.,

```
> data.boundary=
read.csv("C:/StatDA/data/kola-background-boundary.csv")
> data.coast=
read.csv("C:/StatDA/data/kola-background-coast.csv")
> data.borders=
read.csv("C:/StatDA/data/kola-background-borders.csv")
> data.lakes=
read.csv("C:/StatDA/data/kola-background-lakes.csv")
 > kola.background=list(boundary=data.boundary,
coast=data.coast,borders=data.borders,lakes=data.lakes)
```

to produce the "kola.background" file that is used in the R-scripts for all Kola maps. Such a background file needs then to be saved:

```
> save(file="C:/StatDA/data/kola-background.Rdata",
"kola.background")
```

In a new R session the file can then be loaded with:

```
> load(file="C:/StatDA/data/kola-background.Rdata")
```

In the special case of the Kola background data the file is included in the package StatDA and could thus be loaded with the above short command data(kola.background). Such a map for a background file is usually digitalised using a geographical information system like ArcInfo™. It is easiest to use the same coordinates for the map as are used in the data file.

```
plotbg(map="kola.background",map.col=c("gray","gray",
"gray","gray"),map.lwd=c(2,1,2,1),add.plot=TRUE)
```

The command plotbg plots the background map into the empty plot generated above. The arguments map.col and map.lwd allow specifying the colour and line width for each of the four layers in the background map.

```
> scalebar(761309,7373050,761309+100000,7363050,
shifttext=-0.5,shiftkm=4e4,sizetext=0.8)
```

The command scalebar plots the map scale. To place the map scale in the map the coordinates for the map scale need to be specified (x lower left, y lower left, x upper right, y upper right). shifttext and shiftkm allow to move the text in relation to the bar. At present the function scalebar will always plot a 100 km bar as used for the Kola map. To change the bar one can either change the function or execute the commands defined in the function with edited parameters. To see the definition of the function enter:

```
> scalebar
```

```
> Northarrow(362602,7818750,362602,7878750,362602,7838750,
Alength=0.15, Aangle=15,Alwd=1.3,Tcex=1.6)
```

Accordingly the function Northarrow plots the north arrow onto the map. All parameters starting with capital A refer to the arrow (length, angle and width). The parameter Tcex determines the size of the "N" in the arrow.

This command finishes Figure 5.1 upper right. To plot the map in Figure 5.1, lower left, the following commands are needed:

```
plot(m[1:500,1:494])
Mg=ohorizon[,"Mg"]
bubbleFIN(X/1100-300,Y/1100-6690,Mg,radi=8,S=9,s=2,
plottitle="",leg=FALSE)
```

The command bubbleFIN plots a growing dot map of the selected variable (Mg). Because in this case a general imported .jpg (.pnm) file was used as the background, the x- and y-coordinates need to be transformed to fit the background map. This is done via the first two arguments and needs to be tried out until both maps fit on top of one another. Then the size of the growing dots needs to be specified; S and s determine maximum and minimum size of the dots and `radi` is a multiplication factor that determines the size of the dots in relation to the map scale. In this specific example no legend will be plotted (`leg=FALSE`).

Finally, to plot the map in Figure 5.1, lower right, the following commands are needed:

```
plot(X,Y,frame.plot=FALSE,xaxt="n",yaxt="n",xlab="",ylab="",
type="n",asp=TRUE)
plot.bg(map="kola.background",map.col=c("gray","gray",
"gray","gray"),   map.lwd=c(2,1,2,1),add.plot=TRUE)
bubbleFIN(X,Y,Mg,radi=10000,S=9,s=2,plottitle="",
legendtitle="Mg [mg/kg]",  text.cex=0.63,legtitle.cex=0.80)
```

The `text.cex` argument determines the size of the numbers in the legend, `legtitle.cex` determines the size of the legend title.

```
scalebar(761309,7373050,761309+100000,7363050,
shifttext=-0.5,  shiftkm=4e4,sizetext=0.8)
Northarrow(362602,7818750,362602,7878750,362602,7838750,
Alength=0.15, Aangle=15,Alwd=1.3,Tcex=1.6)
dev.off()
```

This command finishes plotting Figure 5.1, the file is saved as *fig-5-1.pdf* in the folder where R is running. It is possible to specify the folder where the plot should be stored before starting the whole R-script via entering the command:

```
> setwd("C:/StatDA/plots")
```

It is of course required that the directory "C:\StatDA\plots" exists on your system.

The R-script can be copied and pasted into the R Console and will then be executed, resulting in the figure. Alternatively the following command can be entered:

```
> source("fig-5-1.R")
```

This command requires that the R file is in the same directory where the working directory actually is, otherwise the full path name must be provided. These ".R" files can be opened and edited in the notepad or in any text editor.

However, the best choice for Windows users is probably to open the script in a running R session:

```
click "File" – "Open script ..."
```

and select the appropriate ".R" file. This file can then be edited, saved and executed within the running R session.

Any other figure, table or statistical result presented in the book can be generated using the above procedure. An overview of all scripts is provided at the following web site:

```
http://www.statistik.tuwien.ac.at/StatDA/R-scripts/
overview/
```

Here the scripts are sorted according to the table of contents of the book. Via clicking at the subsection in the table of contents all outputs and the required R-scripts are shown.

19.3 A brief overview of relevant R commands

Help

`start.help()` starts the help system.

`help(function)` or `?function` provides help to a specific function (e.g., `help(hist)` or `?hist`).

`help.search("topic")` (fuzzy) search for the desired topic (e.g., `help.search("map")`) in all installed packages.

Reading and loading data, functions and tables

`read.table("MyFile.txt", header=TRUE)` reads data saved as text file.

`read.csv("MyFile.csv")` or `read.csv2("MyFile.csv")` reads data saved as .csv file.

`load("MyFile.RData")` loads an R binary file, see `save`.

`source("text")` loads a text file (e.g., R-script), see `dump`.

`library(StatDA); data(chorizon)` loads data that are available in a package, here the Kola C-horizon data from library `StatDA`.

Saving data

`write.table(dataset,"MyFile.txt")` save R object data set into a text file.

`write.csv, write.csv2` like above, but save in .csv format.

`save(x,"MyFile.RData")` saves a more general R object x into a binary file.

`dump("Robject","text")` saves an R object (e.g., a function) as text file.

Mathematical and logical operators, comparisons

+		addition
−		subtraction
*		multiplication
/		division
^		power
%*%		matrix multiplication
==		equal
!=		unequal
>	<	larger, less
>=	<=	larger (less) or equal
!		negation
&&		logical AND
||		logical OR
&	|	elementwise logical AND (OR).

Mathematical functions (here applied to an object x)

sqrt(x)	square root
min(x), max(x)	minimum, maximum
abs(x)	absolute value
sum(x)	sum of all values
log(x), log10(x)	natural logarithm, base 10 logarithm
exp(x)	exponential function
sin(x), cos(x), tan(x)	trigonometric functions
eigen(x)	eigenvectors and eigenvalues of x.

Statistical functions

mean(x), median(x)	mean, median
sd(x), mad(x)	standard deviation, median absolute deviation
var(x), cov(x)	variance and covariance of the elements of x
var(x,y), cov(x,y)	covariance between x and y
cor(x)	correlation matrix of x
cor(x,y)	correlation between x and y
scale(x)	scaling and centring of a matrix
quantile(x,probs)	quantile of x to probabilities probs
dnorm, pnorm, qnorm, rnorm	density, distribution function, quantile function, random generation for the normal distribution.

Density, distribution function, quantile function and random generation is also available for various other distributions. Instead of dnorm the functions are named dt, dunif, dchisq, dexp, dbinom, dpois, dlnorm, etc.

Data types

logical	logical value, either TRUE (T) or FALSE (F)
integer	integer (number without decimals) value (e.g., 37)
double	a real numeric value (e.g., 3.58)
character	character expression, e.g., "abc"

with the following special substitutes:

Inf	infinity
-Inf	minus infinity
NULL	null (empty) object
NA	not available (missing value)
NaN	not a number (not defined).

Conversion of x to a different data type (if appropriate)

as.logical(x), as.numeric(x), as.integer(x), as.double(x), as.character(x)

Checking for the data type

is.logical(x), is.numeric(x), is.integer(x), is.double(x), is.character(x) will check whether x is of the specified data type.
is.na(x) will check whether x contains missing values.

Data structures

`vector`	consists of a desired number of elements which are all of the same data type, e.g., logical, numeric, or character.
`c(,,,)`	generates a vector (a combination of single elements separated by a comma).
`seq(from,to,by)`	creates a sequence of numbers.
`rep(x,n)`	replicates the value x, n times.
`factor`	contains information about categorical data; it contains only a few different levels (which can be characters, numbers, or combinations thereof).
`matrix`	matrix of any dimension; all elements must consist of the same data type.
`data.frame`	a data frame can combine different data types, e.g., a matrix where factors, numeric, and character variables can occur in different columns.
`list`	is a general data object combining vectors, factors, matrices, data frames and lists.

Extracting data

Extracting data from a vector x or factor x:

`x[k]`	select k-th element of x
`x[-k]`	select all except the k-th element of x
`x[1:k]`	select the first k elements of x
`x[-(1:k)]`	select all except the first k elements of x
`x[c(3,5,7)]`	select specific elements from x
`x[x>5]`	select those elements from x where x is larger than 5 (only if x is numeric)
`x[x=="abc"]`	select the group abc from x (only if x is a factor).

Extracting data from a matrix x or data frame x:

`x[i,j]`	select element in i-th row and j-th column of x
`x[i,]`	select row i of x
`x[,j]`	select column j of x
`x[,(1:k)]`	select the first k columns of x
`x[,"col"]`	select column with name col from x
`x$col`	select column with name col from x (only if x is a data frame)
`x[x=="abc"]`	select the group abc from x (here x is a factor).

Extracting data from a list x:

`x$listk`	select list element with name listk from x
`x[[k]]`	select k-th list element from x.

Working with matrices or data frames

`rbind, cbind`	combine rows (columns)
`t(x)`	matrix transpose of x
`solve(x)`	matrix inverse of x
`apply(x,1,function)`	apply a function (e.g., mean) to all rows of x
`apply(x,2,function)`	apply a function (e.g., mean) to all columns of x
`dim(x)`	provides number of rows and columns of x.

Graphics

Functions relevant for graphics:

plot(x)	plot the values of a column (vector) x versus the index
plot(x,y)	plot the elements in the column (vector) x against those in y
points(x,y)	add points to a plot
lines(x,y)	add lines to a plot
text(x,y,"text")	place the text text at the location specified by x and y
legend(x,y,leg)	place legend given by vector leg at location x and y.

Relevant plot parameters:

main="title"	add title title to the plot
xlab,ylab	character for *x*- (or *y*-) axis label
xlim,ylim	vector with minimum and maximum for *x*- (or *y*-) axis
pch	number (e.g., pch=3) of the plot symbol (plot character)
col	name (e.g., col="red") or number (e.g., col=2) of the symbol colour
cex	scaling for text size and symbol size.

Statistical graphics:

hist(x)	plot a histogram of the frequencies of x
plot(density(x))	plot a density function of x
boxplot(x)	boxplot of x
qqnorm(x)	QQ-plot of x.

Saving graphic output

First the graphics device has to be opened. Depending on the desired file format, this is done by the commands postscript, pdf, png, jpeg. In the arguments of these functions the file name has to be provided, and optionally the size of the graphic (width and height). Then the plot is generated. Finally, the graphics device needs to be closed with the command dev.off(). The command sequence

```
pdf(file="figx.pdf",width=6,height=6)
hist(x)
dev.off()
```

generates the file *figx.pdf* with the size 6 inches × 6 inches (15.2 cm × 15.2 cm) containing a histogram of x. Under Windows it is possible to directly save the graphics displayed in the R Graphics Device in the desired format via clicking at "File" and "Save as".

Terminating an R session

q() quits (ends) the R session. The user is asked whether data objects should be saved. If this is desired, all newly created objects (data objects, functions, etc.) will be saved in a special binary format in the current working directory, and will be available in the next R session. If terminating without saving, all newly generated objects are lost, and only objects saved in previous R sessions will be available.

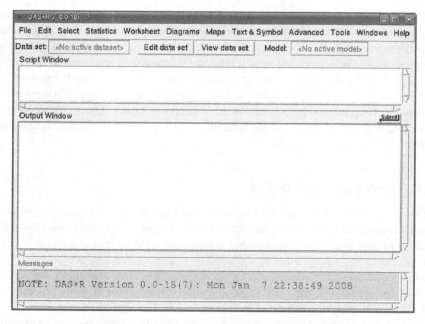

Figure 19.3 Screen shot of the DAS+R main menu window

19.4 DAS+R

DAS+R provides a graphical user interface (GUI) to R, based on the R Commander of Fox (2005). Much of the system analysis for the functionality of the GUI was copied from the old DAS system (Dutter *et al.*, 1990). DAS+R allows storing and using more information about the data than R alone (see Chapter 2) during data analysis. It runs in its own window while the R Console is active. It can also be used to build or edit R-scripts.

DAS+R is presently available at the following website:

```
http://www.statistik.tuwien.ac.at/StatDA/DASplusR/
```

It needs to be downloaded and then installed as a package in a running R session (in Windows: click at "packages"–"install packages from local .zip files ..."and select the .zip file). The general command is:

```
> install.packages(repos=NULL,"DASplusR.zip")
```

Because DAS+R is still under development it is advisable to check the web site regularly for updates. Manuals and help files are available at the same web site.

The program is started in a running R session via loading the already installed package DASplusR:

```
> library(DASplusR)
```

(note that a "+" is not allowed in object names in R). Two windows will open:

- the R Console window and
- the DAS+R window.

The DAS+R window consists of a menu bar and a tool bar on top of the window and the window which is subdivided into three sub-windows (the *script window*, the *output window* and the *messages window*). Any run time errors and other messages will be displayed in the messages window. The script window will show the R commands that result from the clicked selections. It is also possible to use the script window for directly entering R commands. The output window shows the complete submitted command and any text output that is created from this command. Graphical output appears in the R Graphics Device. Due to the many stacked windows that will occur when working with the system a large screen (or a combination of two screens) clearly has advantages.

19.4.1   Loading data into DAS+R

A session always starts with either entering a new data set, loading a .txt or .csv file containing a new data set or activating an existing data set. For activating an existing data set such as the Kola C-horizon example data, which are included in the package DAS+R, it is necessary to click at "File" – "Open" – "from Packages" – "Data Set from an Attached Pack-age", select "DASplusR" (double-click), "Kola95_Chor" (double click) and "OK". The now active data set Kola95_Chor will show up in blue in the tool bar. It is now possible to start a DAS+R session analysing this data set.

To load a new data set that exists as a .txt or .csv file (for file requirements see Chapter 2) it is necessary to click the "File" – "Open" – "Text File" buttons and follow the instructions on the screen. For directly typing a data set into the system the "New" button is clicked and an empty table for entering data will appear on the screen.

19.4.2   Plotting diagrams

Once a data set is indicated in blue print on the "Data set" button as active, DAS+R can be used to analyse the data set. Maps can of course only be plotted if coordinates exist in the data set.

For example, to plot a histogram of any of the variables in the data set use the pull down menu:

- Diagrams
- Histogram
- select the variable from the provided list (e.g., Al_XRF) and click
- OK

and a combination plot of histogram, density trace, one-dimensional scatterplot and boxplot (see Chapter 3) will be produced in the R Graphics window. The graph can be saved to a bitmap (e.g., .jpg format), a pdf file or in ps- or eps-format using the "Save Graph to File" button under "Diagrams" (note that the save buttons under "File" will save data and/or command files but not the graphical output!). Density trace, one-dimensional scatterplot and boxplot can be switched off. It is also possible to change the number of bins used for plotting the histogram and other features as outlined in the screen menu.

All other plots function accordingly. Some (e.g., the Xyplot) have numerous additional functions. Changes to the axes are made under "Details" (separately for each axis) and additional functions for a plot (e.g., the correlation coefficient, a regression line, or the identification of samples) can be found under "Additionals".

19.4.3 Tables

It will also be interesting to look at a number of summary tables. These tables will also provide a fast routine for checking that everything went OK when importing the data. DAS+R provides a number of easy functions to produce a variety of data tables.

"Status short" and "Status long" provide an overview of the current data set.

An overview of many statistical parameters can be produced under:

"Statistics" – "Summaries" – "DAS+R (standard)"

where a mask for a selection of statistical parameters appears. It is possible to choose a selection of parameters (or all) to produce a statistical summary table of the data set (e.g., minimum – MEDIAN – maximum value of each variable). The resulting file can then be saved in .csv format and edited in Excel^{TM}.

19.4.4 Working with "worksheets"

The standard R Graphics Device allows only the plotting of one graphic at a time. It is, however, often desirable to combine several graphics in one plot (e.g., two or four as often used throughout the book). Such a worksheet can be constructed in the Worksheet menu.

When clicking "Worksheet" – "New" the interactive worksheet menu asks for the size of the output file in cm. For example, a worksheet of the size 25 × 18 cm can be constructed. It is then possible to divide this sheet interactively into several plotting frames of the same or different size. Frames can even be stacked. Small frames can also be inserted into larger frames (e.g., for a legend) or to construct overlapping frames. Before a graphic (or text) is generated it is necessary to activate the desired frame via a mouse click (colour red). Each worksheet design can be saved (without any graphics) under "Worksheet" – "Save Layout", using a unique file name (e.g., A44lands.Rws) and loaded again for use in later sessions ("Worksheet" – "Open"). The graphics are not in the Worksheet menu window but plotted in the corresponding positions in the R Graphics Device. The worksheet design allows overplotting of any activated frame without destroying the graphics in other frames.

The final design with graphics is stored the usual way (e.g., "Diagrams" – "Save Graph to File" or "Worksheet" – "Snapshot").

19.4.5 Groups and subsets

Factor variables (see Section 2.1) can directly define groups. Each sample belongs to a certain group. Groups can be used for creating data subsets for graphics, tabulations or statistical computations. For example the variables "COUN" (country of origin of the samples) and "ASP" in the Kola C-horizon data set contain information about the country of origin of the samples (Finland or Norway or Russia) and the aspect of the sample site. This information can be directly used to assign specific plot symbols and/or colours to these "groups". It is also possible to directly compare groups in boxplots. Groups can be used for density traces,

ECDF-plots, boxplots, in scatterplots, ternary diagrams, for mapping and multivariate data analysis. New groups can also be defined during an interactive session in DAS+R.

Subsets can be defined using a multitude of mathematical operators, via editing indicator vectors or graphically in various diagrams (e.g., scatterplots, ternary diagrams, maps, the biplot).

19.4.6 Mapping

Mapping requires that the samples have $x-$ and $y-$coordinates. On first use the variables that identify the coordinates have to be specified. Maps can be produced with or without geographical (topographical) background. All maps are produced through the pull down menu button "Maps". The "Maps" menus allow the variables containing the $x-$ and $y-$coordinates to be selected, the variable to be mapped, mapping method and levels for classes and the symbol set for mapping (where appropriate). When clicking "OK" in the map menu the R Graphics Device will open. In addition an "Add Annotations" window will appear allowing the placement of a legend, north arrow and a scale bar interactively in the map. Following these steps it is necessary to click "OK" and the final map will appear on the R Graphics Device. Just as any other graphic it can be saved to a bitmap (e.g., .jpg format), a pdf file or in ps- or eps-format using the "Save graph to file" button under "Maps".

For mapping it is often desired to include background information files in a map. DAS+R is able to load such a background file and plot a geochemical map on top of it. It is possible to either use postscript files containing polygon information or .pnm files (bitmaps) as background. This permits the use of scanned maps as background, that were transformed (initially saved) into postscript (.ps or .eps) or .pnm format. In the majority of cases the coordinates of the data and the coordinates of the background map will not be the same. Thus it is required to fit the coordinate system of the background map such that the sample points plot at the right locations. This can be done in DAS+R in the "Maps" menu using the "Background" – "Convert" buttons. A menu will appear where the file containing the background information can be selected. R transfers such background files into an R object. This is a time and memory intensive procedure. The background files should thus not be too large (several Mb). For a screen presentation a file size of 1 Mb will usually be sufficient.

When the background file is converted it will appear in the R Graphics Device window. By selecting several sample sites from the geochemical map as reference points and their geographic locations on the background map the coordinates of the background map may be iteratively transformed to be conformable with those of the geochemical map. A number of transformation methods for the background map are available. This step requires experience and is best learned via trial and error. Once background map and sample locations fit together the transformed background map can be saved as a ".Rmap" file. For future application the transformed background map can then be loaded under "Maps" – "Background" – "Load".

19.5 Summary

R is a free open source software tool for statistical data analysis, which can be downloaded from http://www.r-project.org. Due to many contributed packages it follows the latest developments in statistics very closely. R is basically a command language. To master the command language needs frequent use of R. Commands to construct a figure, map or table can, however, be summarised in "R-scripts". Such scripts can be easily adopted to other data sets. Scripts for all graphics and tables in this book are provided at `http://www.statistik.tuwien.ac.at/StatDA/R-scripts/`. However, many users may prefer a graphical user interface to editing R-scripts. DAS+R (`http://www.statistik.tuwien.ac.at/StatDA/DASplusR/`) provides such a user interface. DAS+R is being developed for this book and is able to cover many of the applications described.

References

Abramowitz M. and Stegun I.A., editors. *Handbook of Mathematical Functions with Formulas, Graphs and Mathematical Tables*. Dover, New York, 1965. 1046 pp.

Afifi A.A. and Azen S.P. *Statistical Analysis: A Computer Oriented Approach*. Academic Press, New York, 1979. 442 pp.

Aitchison J. *The Statistical Analysis of Compositional Data*. Chapman and Hall, London, UK, 1986. 416 pp.

Aitchison J. *The Statistical Analysis of Compositional Data. Reprint*. Blackburn Press, Caldwell, NJ, USA, 2003. 416 pp.

Akima H. Rectangular-grid-data surface fitting that has the accuracy of a bicubic polynomial. *ACM Transactions on Mathematical Software*, 22:357–361, 1996.

AMAP. Assessment Report: Arctic Pollution Issues. Arctic Monitoring and Assessment Programme (AMAP), 1998. 859 pp.

AMC. What should be done with results below the detection limit? Mentioning the unmentionable. Analytical methods committee technical brief no. 5, April 2001. Royal Society of Chemistry, London, 2001, Two unnumbered pages.

Ansari A.R. and Bradley R.A. Rank-sum tests for dispersion. *Annals of Mathematical Statistics*, 31:1174–1189, 1960.

Argyraki A., Ramsey M.H. and Thompson M. Proficiency testing in sampling: pilot study on contaminated land. *Analyst*, 120:2799–2803, 1995.

Aubrey K.V. Frequency distribution of the concentrations of elements in rocks. *Nature*, 174:141–142, 1954.

Aubrey K.V. Frequency distributions of elements in igneous rocks. *Geochimica et Cosmochimica Acta*, 9:83–90, 1956.

Äyräs M. and Kashulina G. Regional patterns of element contents in the organic horizon of podzols in the central part of the Barents region (Finland, Norway and Russia) with special reference to heavy metals (Co, Cr, Cu, Fe, Ni, Pb, V and Zn) and sulphur as indicators of airborne pollution. *Journal of Geochemical Exploration*, 68:127–144, 2000.

Äyräs M. and Reimann C. Joint ecogeochemical mapping and monitoring in the scale of 1:1 mill. in the west Murmansk region and contiguous areas of Finland and Norway–1994–1996. Field manual. Technical Report 95.111, Norges Geologiske Undersøkelse, Trondheim, Norway, 1995. 33 pp.

Äyräs M., Niskavaara H., Bogatyrev I., Chekushin V., Pavlov V., de Caritat P., Halleraker J.H., Finne T.E., Kashulina G. and Reimann C. Regional patterns of heavy metals (Co, Cr, Cu, Fe, Ni, Pb, V and Zn) and sulphur in terrestrial moss samples as indication of airborne pollution in a 188,000 km^2 area in northern Finland, Norway and Russia. *Journal of Geochemical Exploration*, 58:269–281, 1997a.

Äyräs M., Pavlov V. and Reimann C. Comparison of sulphur and heavy metal contents and their regional distribution in humus and moss samples from the vicinity of Nikel and Zapoljarnij, Kola Peninsula, Russia. *Water, Air and Soil Pollution*, 98:361–380, 1997b.

Bandemer H. and Näther W. *Fuzzy Data Analysis*. Kluwer Academic Publication, Dordrecht, The Netherlands, 1992. 360 pp.

Barnett V. and Lewis T. *Outliers in Statistical Data*. John Wiley & Sons, Inc., New York, 3rd edition, 1994. 604 pp.

Barnsley M.F. and Rising H. *Fractals Everywhere*. Academic Press Professional, Boston, MA, USA, 1993. 531 pp.

Bartlett M.S. Properties of sufficiency and statistical tests. *Proceedings of the Royal Statistical Society, Series A*, 160:268–282, 1937.

Bartlett M.S. The use of transformations. *Biometrics*, 3:39–52, 1947.

Bartlett M.S. Tests of significance in factor analysis. *British Journal of Psychology (Statistics Section)*, 3:77–85, 1950.

Basilevsky A. *Statistical Factor Analysis and Related Methods. Theory and Applications*. John Wiley & Sons, Inc., New York, USA, 1994. 737 pp.

Becker R.A., Chambers J.M. and Wilks A.R. *The New S Language: A Programming Environment for Data Analysis and Graphics*. Chapman and Hall, London, 1988. 702 pp.

Berkson J. Application of the logistic function to bioassay. *Journal of the American Statistical Association*, 39:357–365, 1944.

Bezdek J.C. *Pattern Recognition with Fuzzy Objective Function Algoritms*. Plenum Press, New York, 1981. 256 pp.

Björklund A. and Gustavsson N. Visualization of geochemical data on maps: New options. *Journal of Geochemical Exploration*, 29:89–103, 1987.

Böhm P., Wolterbeek H., Verburg T. and Musilek L. The use of tree bark for environmental pollution monitoring in the Czech Republic. *Environmental Pollution*, 102:243–250, 1998.

Bølviken B., Bergstrøm J., Bjørklund A., Kontio M., Lehmuspelto P., Lindholm T., Magnusson J., Ottesen R.T., Steenfelt A. and Volden T. Geochemical Atlas of Northern Fennoscandia. Scale 1:4,000,000. Technical report, Geological Surveys of Finland, Norway and Sweden, Helsinki (Finland), Trondheim (Norway) and Stockholm (Sweden), 1986. 19 pp. and 155 maps.

Bonham-Carter G.F., Henderson P.J., Kliza D.A. and Kettles I.M. Comparisons of metal distributions in snow, peat, lakes and humus around a Cu smelter in western Québec, Canada. *Geochemistry: Exploration, Environment, Analysis*, 6:215–228, 2006.

Borggaard O.K. Composition, properties and development of Nordic soils. In O.M. Saether and P. de Caritat, editors, *Geochemical Processes, Weathering and Ground-Water Recharge in Catchments*, pages 21–75. Balkema, Rotterdam, The Netherlands, 1997.

Box G.E.P. and Cox D.R. An analysis of transformations. *Journal of the Royal Statistical Society, Series B*, 26:211–252, 1964.

Box G.E.P., Hunter W.G. and Hunter J.S. *Statistics for Experimenters: An Introduction to Design, Data Analysis and Model Building*. John Wiley & Sons, Inc., New York, 1978. 653 pp.

Boyd R., Niskavaara H., Kontas E., Chekushin V., Pavlov V., Often M. and Reimann C. Anthropogenic noble-metal enrichment of topsoil in the Monchegorsk area, Kola Peninsula, northwest Russia. *Journal of Geochemical Exploration*, 58:283–289, 1997.

Boyd R., Barnes S.-J., de Caritat P., Chekushin V.A., Melezhik V., Reimann C. and Zientek M.A. Assessment of heavy metal emissions from the copper-nickel industry on the Kola Peninsula and in other parts of Russia. Technical Report 98.138, Norges Geologiske Undersøkelse, Trondheim, Norway, 1998. 18 pp.

Breiman L. Bagging predictors. *Machine Learning*, 24:123–140, 1996.

Breiman L. Random forests. *Machine Learning*, 45:5–32, 2001.

British Geological Survey. *Regional Geochemistry of Southern Scotland and Part of Northern England*. Keyworth, Nottingham, UK, 1993. 96 pp.

Brown M.B. and Forsythe A.B. Robust tests of the equality of variance. *Journal of the American Statistical Association*, 69:364–367, 1974.

Buccianti A., Mateu-Figueros G. and Pawlowsky-Glahn V., editors. *Compositional Data Analysis in the Geosciences: From Theory to Practice*. Geological Society Publishing House, Bath, UK, 2006. 224 pp.

Butler J.C. Principal components analysis using the hypothetical closed array. *Mathematical Geology*, 8:25–36, 1976.

Calinski T. and Harabasz J. A dendrite method for cluster analysis. *Communications in Statistics*, 3:1–27, 1974.

Campbell N.A. Some aspects of allocation and discrimination. In W.W. Howells and G.N. van Vark, editors, *Multivariate Statistical Methods in Physical Anthropology*, pages 177–192. Reidel, Dordrecht, The Netherlands, 1984.

Cannings C. and Edwards A.W.F. Natural selection and the de Finetti diagram. *Annals of Human Genetics*, 31:421–428, 1968.

Carral E., Puente X., Villares R. and Carballeira A. Background heavy metal levels in estuarine sediments and organisms in Galicia (NW Spain) as determined by modal analysis. *The Science of the Total Environment*, 172:175–188, 1995.

Carroll J.B. An analytic solution for approximating simple structure in factor analysis. *Psychometrika*, 18:23–38, 1953.

Cattell R.B. The scree test for the number of factors. *Multivariate Behaviour Research*, 1:245–276, 1966.

Chambers J.M. *Programming with Data*. Springer, New York, 1998. 469 pp.

Chambers J.M. and Hastie T.J., editors. *Statistical Models in S*. Wadsworth and Brooks/Cole, Pacific Grove, California, 1992. 608 pp.

Chambers J.M., Cleveland W.S., Kleiner B. and Tukey P.A. *Graphical Methods for Data Analysis*. Duxbury Press, Boston, Massachussetts, 1983. 395 pp.

Chapman P. and Wang F. Issues in ecological risk assessments of inorganic metals and metalloids. *Human and Ecological Risk Assessment*, 6:965–988, 2000.

Chayes F. The lognormal distribution of elements: a discussion. *Geochimica et Cosmochimica Acta*, 6:119–121, 1954.

Chekushin V., Bogatyrev I., de Caritat P., Niskavaara H. and Reimann C. Annual atmospheric deposition of 16 elements in eight catchments of the central Barents region. *The Science of the Total Environment*, 220:95–114, 1998.

Cheng Q. and Agterberg F.P. Multifractal modeling and spatial point processes. *Mathematical Geology*, 27:831–845, 1995.

Cheng Q., Agterberg F.P. and Ballantyne S.B. The separation of geochemical anomalies from background by fractal methods. *Journal of Geochemical Exploration*, 51:109–130, 1994.

Chernoff H. The use of faces to represent points in k-dimensional space graphically. *Journal of the American Statistical Association*, 68:361–368, 1973.

Chester R. and Stoner J.H. Pb in particulates from the lower atmosphere of the eastern Atlantic. *Nature*, 245:27–28, 1973.

Chork C.Y. Unmasking multivariate anomalous observations in exploration geochemical data from sheeted-vein tin mineralisation near Emmaville, N.W.S., Australia. *Journal of Geochemical Exploration*, 37:205–223, 1990.

Chork C.Y. and Govett G.J.S. Comparison of interpretations of geochemical soil data by some multivariate statistical methods, Key Anacon, N.B., Canada. *Journal of Geochemical Exploration*, 23:213–242, 1985.

Chork C.Y. and Salminen R. Interpreting geochemical data from Outokumpu, Finland: an MVE-robust factor analysis. *Journal of Geochemical Exploration*, 48:1–20, 1993.

Cleveland W.S. *Visualizing Data*. Hobart Press, Summit, New Jersey, 1993. 360 pp.

Cleveland W.S. *The Elements of Graphing Data. Revised edition*. Hobart Press, Summit, New Jersey, 1994. 297 pp.

Cleveland W.S., Grosse E. and Shyu W.M. Local regression models. In J.M. Chambers and T.J. Hastie, editors, *Statistical Models in S*, pages 309–376. Wadsworth and Brooks/Cole, Pacific Grove, California, 1992.

Clifton H.E., Hunter R.E., Swanson F.J. and Phillips R.L. Sample size and meaningful gold analysis. Technical Report 625-C, United States Geological Survey, 1969. 17 pp.

Conover W.J. *Practical Nonparametric Statistics*. John Wiley & Sons, Inc., New York, 3rd edition, 1998. 584 pp.

Conover W.J. and Iman R.L. Rank transformations as a bridge between parametric and nonparametric statistics. *American Statistician*, 35:124–129, 1982.

Conover W.J., Johnson M.E. and Johnson M.M. A comparative study of tests for homogeneity of variances, with applications to the outer continental shelf bidding data. *Technometrics*, 23:351–361, 1981.

Cook R.D. and Weissberg S. *Residuals and Influence in Regression*. Chapman and Hall, New York, 1982. 230 pp.

Crawley M.J. *The R Book*. John Wiley & Sons, Inc., New York, 2007. 950 pp.

Cressie N. *Statistics for Spatial Data*. John Wiley & Sons, Inc., New York, 1993. 900 pp.

Croux C. and Dehon C. Robust linear discriminant analysis using S-estimators. *The Canadian Journal of Statistics*, 29:473–492, 2001.

Čurlík J. and Šefčík P. Geochemical Atlas of the Slovak Republic Part V: Soils, 1999. 99 pp. and 83 maps.

Dalgaard P. *Introductory Statistics with R*. Springer, New York, 2002. 267 pp.

Davis J.C. *Statistics and Data Analysis in Geology*. John Wiley & Sons, Inc., New York, 1973. 550 pp.

Davis J.C. *Statistics and Data Analysis in Geology*. John Wiley & Sons, Inc., New York, 3rd edition, 2002. 638 pp.

de Caritat P., Reimann C., Äyräs M., Niskavaara H., Chekushin V.A. and Pavlov V.A. Stream water geochemistry from selected catchments on the Kola Peninsula (NW Russia) and in neighbouring areas of Finland and Norway: 1. Element levels and sources. *Aquatic Geochemistry*, 2:149–168, 1996a.

de Caritat P., Reimann C., Äyräs M., Niskavaara H., Chekushin V.A. and Pavlov V.A. Stream water geochemistry from selected catchments on the Kola Peninsula (NW Russia) and in neighbouring areas of Finland and Norway: 2. Time series. *Aquatic Geochemistry*, 2:169–184, 1996b.

de Caritat P., Krouse H.R. and Hutcheon I. Sulphur isotope composition of stream water, moss and humus from eight Arctic catchments in the Kola Peninsula region (NW Russia, N Finland, NE Norway. *Water, Air and Soil Pollution*, 94:191–208, 1997a.

de Caritat P., Reimann C., Chekushin V., Bogatyrev I., Niskavaara H. and Braun J. Mass balance between emission and deposition of trace metals and sulphur. *Environmental Science and Technology*, 31:2966–2972, 1997b.

de Caritat P., Äyräs M., Niskavaara H., Chekushin V., Bogatyrev I. and Reimann C. Snow composition in eight catchments in the central Barents Euro-Arctic Region. *Atmospheric Environment*, 32:2609–2626, 1998a.

de Caritat P., Danilova S., Jæger Ø., Reimann C. and Storrø G. Groundwater composition near the nickel-copper smelting industry on the Kola Peninsula, central Barents Region (NW Russia and NE Norway). *Journal of Hydrology*, 208:92–107, 1998b.

de Caritat P., Reimann C., Bogatyrev I., Chekushin V., Finne T.E., Halleraker J.H., Kashulina G., Niskavaara H., Pavlov V. and Äyräs M. Regional distribution of Al, B, Ba, Ca, K, La, Mg, Mn, Na, P, Rb, Si, Sr, Th, U and Y in terrestrial moss within a 188,000 km^2 area of the central Barents region: influence of geology, seaspray and human activity. *Applied Geochemistry*, 16:137–159, 2001.

Dempster A.P., Laird N.M. and Rubin D.B. Maximum likelihood from incomplete data via the EM algorithm (with discussion). *Journal of the Royal Statistical Society, Series B*, 39:1–38, 1977.

Draper N.R. and Smith H. *Applied Regression Analysis*. John Wiley & Sons, Inc., New York, 3rd edition, 1998. 706 pp.

Dumitrescu D., Lazzerini B. and Jain L.C. *Fuzzy Sets and Their Application to Clustering and Training.* CRC Press, Boca Raton, Florida, 2000. 622 pp.

Dunn J.C. A fuzzy relative of the ISODATA process and its use in detecting compact well-separated clusters. *Journal of Cybernetics*, 3:32–57, 1973.

Dutter R., Leitner T., Reimann C. and Wurzer F. *DAS: Data Analysis System, Numerical and Graphical Statistical Analysis, Mapping of Regionalized (e.g. Geochemical) Data on Personal Computers. Preliminary Handbook.* Vienna University of Technology, Vienna, Austria, 1990. 236 pp.

Dutter R., Filzmoser P., Gather U. and Rousseeuw P.J., editors. *Developments in Robust Statistics. International Conference on Robust Statistics 2001.* Physika-Verlag, Heidelberg, Germany, 2003. 431 pp.

Eastment H.T. and Krzanowski W.J. Cross-validatory choice of the number of components from a principal component analysis. *Technometrics*, 24:73–77, 1982.

Efron B. Bootstrap methods: another look at the jackknife. *Annals of Statistics*, 7:1–26, 1979.

Egozcue J.J., Pawlowsky-Glahn V., Mateu-Figueros F. and Barceló-Vidal C. Isometric logratio transformations for compositional data analysis. *Mathematical Geology*, 35:279–300, 2003.

Everitt B.S. *Cluster Analysis.* Heinemann Educational, London, 1974. 248 pp.

Everitt B.S. *An R and S-Plus Companion to Multivariate Analysis.* Springer, New York, 2005. 221 pp.

Everitt B.S. and Dunn G. *Applied Multivariate Data Analysis.* Oxford University Press, New York, 2nd edition, 2001. 342 pp.

Fauth H., Hindel R., Siewers U. and Zinner J. *Geochemischer Atlas Bundesrepublik Deutschland. Verteilung von Schwermetallen in Wässern und Bachsedimenten. Bundesanstalt für Geowisseneschaften und Rohstoffe, Hannover.* Schweizerbart'sche Verlagsbuchandlung, Stuttgart, 1985. 79 pp.

Filzmoser P. Robust principal component and factor analysis in the geostatistical treatment of environmental data. *Environmetrics*, 10:363–375, 1999.

Filzmoser, P. Garrett R.G. and Reimann C. Multivariate outlier detection in exploration geochemistry. *Computers & Geosciences*, 31:579–587, 2005.

Fisher R.A. The use of multiple measurements in taxonomic problems. *Annals of Eugenics*, 7 (Part II):179–188, 1936.

Fox J. *Applied Regression Analysis, Linear Models and Related Methods.* Sage Publications, Thousand Oaks, CA, USA, 1997. 619 pp.

Fox J. *An R and S-Plus Companion to Applied Regression.* Sage Publications, Thousand Oaks, CA, USA, 2002. 312 pp.

Fox J. The R commander: a basic-statistics graphical interface to R. *Journal of Statistical Software*, 14:1–42, 2005.

Fraley C. and Raftery A.E. How many clusters? Which clustering method? Answers via model-based cluster analysis. *The Computer Journal*, 41:578–588, 1998.

Fraley C. and Raftery A.E. Model-based clustering, discriminant analysis and density estimation. *Journal of the American Statistical Association*, 97:611–631, 2002.

Friedman J.H. and Rafsky L.C. Graphics for the multivariate two-sample problem. *Journal of the American Statistical Association*, 76:277–291, 1981.

Friedman M. The use of ranks to avoid the assumption of normality implicit in the analysis of variance. *Journal of the American Statistical Association*, 32:675–701, 1937.

Furnival G.M. and Wilson R.W. Regressions by leaps and bounds. *Technometrics*, 16:499–512, 1974.

Gabriel K.R. The biplot graphic display of matrices with application to principal components analysis. *Biometrika*, 58:453–467, 1971.

Galton F. *Natural Inheritance.* Macmillan and Company, London, 1889. 266 pp.

Galton F. Kinship and correlation. *North American Review*, 150:419–431, 1890.

Garrett R.G. Regional geochemical study of Cretaceous acidic rocks in the northern Canadian Cordillera as a tool for broad mineral exploration. In M.J. Jones, editor, *Proceedings of the 4th International*

Geochemical Exploration Symposium, Geochemical Exploration 1972, pages 203–219. Institution of Mining and Metallurgy, London, 1972.

Garrett R.G. Copper and zinc in Proterozoic Acid volcanics as a guide to exploration in the Bear Province. In I.L. Elliott and W.K. Fletcher, editors, *Proceedings of the 5th International Geochemical Exploration Symposium, Geochemical Exploration 1974*, pages 371–388. Elsevier, Amsterdam, The Netherlands, 1975.

Garrett R.G. Opportunities for the 80's. *Mathematical Geology*, 15:389–402, 1983a.

Garrett R.G. Sampling methodology. In R.J. Howarth, editor, *Handbook of Exploration Geochemistry. Statistics and Data Analysis in Geochemical Prospecting*, Volume 2, Chapter 4, pages 83–113. Elsevier Scientific Publishing Company, Amsterdam, The Netherlands, 1983b.

Garrett R.G. Ideas–an interactive computer graphics tool to assist the exploration geochemist. Technical Report Paper 88-1F, Current Research Part F, Geological Survey of Canada, 1988. 13 pp.

Garrett R.G. The chi-square plot: A tool for multivariate outlier recognition. *Journal of Geochemical Exploration*, 32:319–341, 1989.

Garrett R.G. A robust multivariate allocation procedure with applications to geochemical data. In F.P. Agterberg and G.F. Bonham-Carter, editors, *Statistical Applications in the Earth Sciences*, pages 309–318. Geological Survey of Canada Paper 89-9, Canada, 1990.

Garrett R.G. Another cry from the heart. *Explore – The Association of Exploration Geochemists Newsletter*, 81:9–14, 1993.

Garrett R.G. and Grunsky E.C. Weighted sums – knowledge based empirical indices for use in exploration geochemistry. *Geochemistry: Exploration, Environment and Analysis*, 1:135–141, 2001.

Garrett R.G. and Nichol I. Factor analysis as an aid to the interpretation of stream sediment data. In F.C. Canney, editor, *Proceedings of the International Geochemical Exploration Symposium*, Volume 64, pages 245–264, 1969.

Garrett R.G., MacLaurin A.I., Gawalko E.J., Tkachuk R. and Hall G.E.M. A prediction model for estimating the cadmium content of durum wheat from soil chemistry. *Journal of Geochemical Exploration*, 64:101–110, 1998.

Gath I. and Geva A. Unsupervised optimal fuzzy clustering. *IEEE Transactions on Pattern Analysis and Machine Intelligence*, 11:773–781, 1989.

Gauss C.F. *Theoria Motus Corporum Coelestium*. Perthes et Besser, Hamburg, Germany, 1809. Reprinted 2004 by Dover, New York, 376 pp.

Gnanadesikan R. *Methods for Statistical Data Analysis of Multivariate Observations*. John Wiley & Sons, Inc., New York, 1977. 311 pp.

Goldschmidt V.M. The principles of distribution of chemical elements in minerals and rocks. *Journal of the Royal Chemical Society*, pages 655–673, 1937.

Gordon A.D. *Classification*. Chapman & Hall/CRC, Boca Raton, Florida, 2nd edition, 1999. 256 pp.

Graf U. and Henning H.J. Zum Ausreißerproblem. *Mitteilungsblatt Mathematischer Statistik*, 4:1–10, 1952.

Gregurek D., Reimann C. and Stumpfl E.F. Mineralogical fingerprints of industrial emissions–an example from Ni mining and smelting on the Kola Peninsula, NW Russia. *The Science of the Total Environment*, 221:189–200, 1998a.

Gregurek D., Reimann C. and Stumpfl E.F. Trace elements and precious metals in snow samples from the immediate vicinity of nickel processing plants, Kola Peninsula, NW Russia. *Environmental Pollution*, 102:221–232, 1998b.

Gregurek D., Melcher F., Niskavaara H., Pavlov V.A., Reimann C. and Stumpfl E.F. Platinum-Group Elements (Rh, Pd, Pt) and Au distribution in snow samples from the Kola Peninsula, NW Russia. *Atmospheric Environment*, 33:3291–3299, 1999a.

Gregurek D., Melcher F., Pavlov V., Reimann C. and Stumpfl E.F. Mineralogy and mineral chemistry of snow filter residues in the vicinity of the nickel-copper processing industry, Kola Peninsula, NW Russia. *Mineralogy and Petrology*, 65:87–111, 1999b.

Gustafson D.E. and Kessel W. Fuzzy clustering with a fuzzy covariance matrix. *Proceedings of the IEEE CDC*, 2:761–766, 1979.

Gustavsson N., Lampio E. and Tarvainen T. Visualization of geochemical data on maps at the Geological Survey of Finland. *Journal of Geochemical Exploration*, 59:197–207, 1997.

Gustavsson N., Bølviken B., Smith D.B. and Severson R.C. Geochemical landscapes of the conterminous united states – new map presentation for 22 elements. Technical Report 1648, Geological U.S. Survey Professional Paper, Geological U.S. Survey, Denver, 2001. 38 pp.

Haldiki M., Batistakis Y. and Vazirgiannis M. Cluster validity methods. *SIGMOD Record*, 31:40–45, 2002.

Halleraker J.H., Reimann C., de Caritat P., Finne T.E., Kashulina G., Niskavaara H. and Bogatyrev I. Reliability of moss (*Hylocomium splendens and Pleurozium schreberi*) as bioindicator of atmospheric chemistry in the Barents region: interspecies and field duplicate variability. *The Science of the Total Environment*, 218:123–139, 1998.

Hampel F.R., Ronchetti E.M., Rousseeuw P.J. and Stahel W.A. *Robust Statistics. The Approach Based on Influence Functions*. John Wiley & Sons, Inc., New York, 1986. 502 pp.

Harman H.H. *Modern Factor Analysis*. University of Chicago Press, Chicago, Illinois, USA 3rd edition, 1976. 487 pp.

Hartigan J. *Clustering Algorithms*. John Wiley & Sons, Inc., New York, 1975. 351 pp.

Hartigan J.A. and Wong M.A. Algorithm AS136: A k-means clustering algorithm. *Applied Statistics*, 28:100–108, 1979.

Hastie T., Tibshirani R. and Friedman J. *The Elements of Statistical Learning*. Springer, New York, 2001. 533 pp.

Hawkes H.E. and Webb J.S. *Geochemistry in Mineral Exploration*. Harper, New York, 1962. 415 pp.

He X. and Fung W.K. High breakdown estimation for multiple populations with applications to discriminant analysis. *Journal of Multivariate Analysis*, 72:151–162, 2000.

Helsel D.R. *Nondetects and Data Analysis. Statistics for Censored Environmental Data*. John Wiley & Sons, Inc., New York, 2005. 250 pp.

Helsel D.R. and Hirsch R.M. *Statistical Methods in Water Resources*, volume 49 of *Studies in Environmental Science*. Elsevier, Amsterdam, The Netherlands, 1992. 522 pp.

Hendrickson A.E. and White P.O. PROMAX: A quick method for rotation to oblique simple structure. *British Journal of Statistical Psychology*, 17:65–70, 1964.

Henriksen A., Skjelkvåle B.L., Lien L., Traaen T.S., Mannio J., Forsius M., Kämäri J., Mäkinen I., Berntell A., Wiederholm T., Wilander A., Moiseenko T., Lozovik P., Filatov N., Niinioja R., Harriman R. and Jensen J.P. Regional lake surveys in Finland – Norway – Sweden – Northern Kola – Russian Karelia – Scotland – Wales 1995. Coordination and design. Technical Report O-95001, Norwegian Institute for Water Research, Oslo, Norway, 1996. 30 pp.

Henriksen A., Traaen T.S., Mannio J., Skjelkvåle B.L., Wilander A., Fjeld E., Moiseenko T. and Vuorenmaa J. Regional lake surveys in the Barents region of Finland – Norway – Sweden and Russian Kola 1995 – results. Technical Report O-96163, Norwegian Institute for Water Research, Oslo, Norway, 1997. 36 pp.

Hoaglin D., Mosteller F. and Tukey J. *Understanding Robust and Exploratory Data Analysis*. John Wiley & Sons, Inc., New York, 2nd edition, 2000. 472 pp.

Hocking R.R. The analysis and selection of variables in linear regression. *Biometrics*, 32:1–49, 1976.

Hoerl A.E. and Kennard R. Ridge regression: biased estimation for nonorthogonal problems. *Technometrics*, 12:55–67, 1970.

Hollander M. and Wolfe D.A. *Nonparametric Statistical Inference*. John Wiley & Sons, Inc., New York, 1973. 548 pp.

Hotelling H. Analysis of a complex of statistical variables into principal components. *Journal of Educational Psychology*, 24:417–441, 1933.

Howarth R.J. and Garrett R.G. The role of computing in applied geochemistry. In I. Thornton and R.J. Howarth, editors, *Applied Geochemistry in the 1980s*, pages 163–184. Graham and Trotman, London, 1986.

Huber P.J. Robust estimation of a location parameter. *The Annals of Mathematical Statistics*, 36:1753–1758, 1964.

Huber P.J. *Robust Statistics.* John Wiley & Sons, Inc., New York, 1981. 308 pp.

Hubert L. and Arabie P. Comparing partitions. *Journal of Classification*, 2:193–218, 1985.

Hubert M. and Van Driessen K. Fast and robust discriminant analysis. *Computational Statistics and Data Analysis*, 45:301–320, 2004.

Huberty C.J. *Applied Discriminant Analysis.* John Wiley & Sons, Inc., New York, 1994. 466 pp.

Ihaka R. and Gentleman R. R: A language for data analysis and graphics. *Journal of Computational and Graphical Statistics*, 5:299–314, 1996.

Inselberg A. The plane with parallel coordinates. *Visual Computer*, 1:69–91, 1985.

Jackson J.E. *A User's Guide to Principal Components.* John Wiley & Sons, Inc., New York, 2003. 592 pp.

Janousek V., Farrow C.M. and Ebran V. Interpretation of whole-rock geochemical data in igneous geochemistry: Introducing Geochemical Data Toolkit (*GCDkit*). *Journal of Petrology*, 47:1255–1259, 2006.

Jianan T. *The Atlas of Endemic Diseases and Their Environments in the People's Republic of China.* Science Press, Beijing, China, 1989. 194 pp.

Johnson R.A. and Wichern D.W. *Applied Multivariate Statistical Analysis.* Prentice Hall, Upper Saddle River, New Jersey, USA, 5th edition, 2002. 767 pp.

Jolliffe I.T. *Principal Component Analysis.* Springer, New York, 2nd edition, 2002. 487 pp.

Kadunas V., Budavicius R., Gregorauskiene V., Katinas V., Kliaugiene E., Radzevicius A. and Taraskevicius R. *Lietuvos Geocheminis Atlasas. Geochemical Atlas of Lithuania.* Geological Institute, Vilnius, Lithuania, 1999. 91 pp. and 162 maps.

Kaiser H.F. The Varimax criterion for analytic rotation in factor analysis. *Psychometrika*, 23:187–200, 1958.

Kanji G.K. *100 Statistical Tests.* Sage Publications, London, 1999. 215 pp.

Kashulina G., de Caritat P., Reimann C., Räisänen M.-L., Chekushin V. and Bogatyrev I.V. Acidity status and mobility of Al in podzols near SO_2 emission sources on the Kola Peninsula, NW Russia. *Applied Geochemistry*, 13:391–402, 1998a.

Kashulina G., Reimann C., Finne T.E., de Caritat P. and Niskavaara H. Factors influencing NO_3 concentrations in rain, stream water, ground water and podzol profiles of eight small catchments in the European Arctic. *Environmental Pollution*, 102:559–568, 1998b.

Kashulina G., Reimann C., Finne T.E., Halleraker J.H., Äyräs M. and Chekushin V.A. The state of the ecosystems in the central Barents Region:scale, factors and mechanism of disturbance. *The Science of the Total Environment*, 206:203–225, 1997.

Kaufman L. and Rousseeuw P.J. *Finding Groups in Data.* John Wiley & Sons, Inc., New York, 1990. 342 pp.

Kaufman L. and Rousseeuw P.J. *Finding Groups in Data.* John Wiley & Sons, Inc., New York, 2005. 368 pp.

Kendal M. A new measure of rank correlation. *Biometrika*, 30:81–89, 1938.

Kleinbaum D.G. and Klein M. *Logistic Regression.* Springer, New York, 2nd edition, 2002. 513 pp.

Kleiner B. and Hartigan J.A. Representing points in many dimensions by trees and castles. *Journal of the American Statistical Association*, 76:260–269, 1981.

Koljonen T., editor. *The Geochemical Atlas of Finland, Part 2: Till.* Geological Survey of Finland, Espoo, Finland, 1992. 218 pp.

Krumbein W.C. and Graybill F.A. *An Introduction to Statistical Models in Geology.* McGraw-Hill Inc., New York, 1965. 475 pp.

Kruskal J.B. Multidimensional scaling by optimizing goodness of fit to a nonmetric hypothesis. *Pyschometrika*, 29:1–27, 1964.

Kruskal W.H. and Wallis W.A. Use of ranks on one-criterion variance analysis. *Journal of the American Statistical Association*, 47:583–621, 1952.

Kshirsagar A.M. *Multivariate Analysis.* Marcel Dekker, New York, 1972. 355 pp.

Kürzl H. Exploratory data analysis: recent advances for the interpretation of geochemical data. *Journal of Geochemical Exploration*, 30:309–322, 1988.

Lahermo P., Ilmasti M., Juntunen R. and Taka M. *The Geochemical Atlas of Finland Part 1. The Hydro-geochemical Mapping of Finnish Groundwater*. Geological Survey of Finland, Espoo, Finland, 1990. 66 pp.

Lahermo P., Väänänen P., Tarvainen T. and Salminen R. *Geochemical Atlas of Finland, Part 3: Environmental Geochemistry – Stream Waters and Sediments*. Geological Survey of Finland, Espoo, Finland, 1996. (In Finnish with English summary) 149 pp.

Lance G.N. and Williams W.T. Computer programs for hierachical polythetic classification (similarity analyses). *Computer Journal*, 9:60–64, 1966.

Lawley D.N. and Maxwell A.E. *Factor Analysis as a Statistical Method*. Butterworths, London, 1963. 117 pp.

Le Maitre R.W. *Numerical Petrology*. Elsevier Scientific Publishing Company, Amsterdam, The Netherlands, 1982. 281 pp.

Lehmann E.L. *Testing Statistical Hypothesis*. John Wiley & Sons, Inc., New York, 1959. 369 pp.

Leisch F. *Ensemble Methods for Neural Clustering and Classification*. PhD thesis, Institut für Statistik, Wahrscheinlichkeitstheorie und Versicherungsmathematik, Technische Universität Wien, Vienna, Austria, 1998. 130 pp.

Leisch F. Bagged clustering. Working Paper 51, SFB "Adaptive Information Systems and Modeling in Economics and Management Science". Technical report, Wirtschaftsuniversität Wien, Vienna, Austria, 1999. 11 pp.

Leisch F. A toolbox for k-centroids cluster analysis. *Computational Statistics and Data Analysis*, 51:526–544, 2006.

Levene H. Robust tests for equality of variances. In I. Olkin, editor, *Contributions to Probability and Statistics*, pages 278–292. Stanford University Press, Palo Alto, CA, USA, 1960.

Levinson A.A. *Introduction to Exploration Geochemistry*. Applied Publishing, Wilmette, 1974. 614 pp.

Li J. and Wu G. *Atlas of the Ecological Environmental Geochemistry of China*. Geological Publishing House, Beijing, China, 1999. 209 pp.

Lis J. and Pasieczna A. *Geochemical Atlas of Poland, 1:2,500,000*. Polish Geological Institute, Warsaw, Poland, 1995. 72 pp.

Loska K., Cebula J., Pelczar J., Wiechula D. and Kwapilinski J. Use of enrichment and contamination factors together with geoaccumulation indexes to evaluate the content of Cd, Cu, and Ni in the Rybnik water reservoir in Poland. *Water, Air, and Soil Pollution*, 93:347–365, 1997.

MacQueen J. Some methods for classification and analysis of multivariate observations. In L.M. Le Cam and J. Neyman, editors, *Proceedings of the Fifth Berkeley Symposium on Mathematical Statistics and Probability*, pages 281–297. University of California Press, Berkeley, California, 1967.

Mahalanobis P.C. On the generalised distance in statistics. *Proceedings of the National Institute of Science of India*, A2:49–55, 1936.

Mandelbrot B.B. *The Fractal Geometry of Nature*. W.H. Freeman and Company, New York, 1982. 480 pp.

Mankovská B. *Geochemical Atlas of Slovakia: Forest Biomass*. Ministry of the Environment of the Slovak Republic & Geological Survey of Slovakia, Bratislava, Slovakia, 1996. 87 pp.

Mann H.B. and Whitney D.R. On a test of whether one of two random variables is stochastically larger than the other. *Annals of Mathematical Statistics*, 18:50–60, 1947.

Mardia K.V., Kent J.T. and Bibby J.M. *Multivariate Analysis*. Academic Press, London, 6th edition, 1997. 518 pp.

Maronna R., Martin D. and Yohai V. *Robust Statistics: Theory and Methods*. John Wiley & Sons Canada Ltd, Toronto, ON, 2006. 436 pp.

Massart D.L., Vandeginste B.G.M., Deming S.N., Michotte Y. and Kaufman L. *Chemometrics: A Textbook. Data Handling in Science and Technology – Volume 2*. Elsevier, Amsterdam, The Netherlands, 1988. 488 pp.

Matschullat J., Ottenstein R. and Reimann C. Geochemical background – can we calculate it? *Environmental Geology*, 39:990–1000, 2000.

McGrath S.P. and Loveland P.J. *The Soil Geochemical Atlas of England and Wales.* Blackie Academic & Professional, London, 1992. 101 pp.

McMartin I., Henderson P.J. and Nielsen E. Impact of a base metal smelter on the geochemistry of soils of the Flin Flon region, Manitoba and Saskatchewan. *Canadian Journal of Earth Sciences*, 36:141–160, 1999.

Miesch A.T. Estimation of geochemical threshold and its statistical significance. *Journal of Geochemical Exploration*, 16:49–76, 1981.

Millard S.P. and Neerchal N.K. *Environmental Statistics with S-Plus.* CRC Press, Boca Raton, USA, 2001. 830 pp.

Miller R.L. and Goldberg E.D. The normal distribution in geochemistry. *Geochimica et Cosmochimica Acta*, 8:53–62, 1955.

Milligan G.W. and Cooper M.C. An examination of procedures for determining the number of clusters. *Psychometrika*, 50:159–179, 1985.

Monmonier M. *How to Lie with Maps.* The University of Chicago Press, Chicago, Illinois, 2nd edition, 1996. 207 pp.

Murrell P. *R Graphics.* Chapman & Hall/CRC, New York, 2006. 301 pp.

National Environmental Protection Agency of the People's Republic of China. *The Atlas of the Soil Environmental Background Value in the People's Republic of China.* China Environmental Science Press, Beijing, China, 1994. 196 pp.

Neinaweie H. and Pirkl H. Bewertung von Schwermetallverteilungen in Böden und Flußsedimenten mit Hilfe angewandt mineralogischer und geostatistischer Werkzeuge. Berichte der Geologischen Bundesanstalt, 1996. (In German), 67 pp.

Neykov N., Filzmoser P., Dimova R. and Neytchev P. Robust fitting of mixtures using the Trimmed Likelihood Estimator. *Computational Statistics & Data Analysis*, 52:299–308, 2007.

Niemelä J., Ekman I. and Lukashov A., editors. *Quaternary Deposits of Finland and Northwestern Part of Russian Federation and Their Resources 1:1,000,000.* Geological Survey of Finland, Espoo, Finland, 1993.

Niskavaara H., Reimann C. and Chekushin V. Distribution and pathways of heavy metals and sulphur in the vicinity of the copper-nickel smelters in Nikel and Zapoljarnij, Kola Peninsula, Russia, as revealed by different sample media. *Applied Geochemistry*, 11:25–34, 1996.

Niskavaara H., Reimann C., Chekushin V. and Kashulina G. Seasonal variability of total and easily leachable element contents in topsoils (0–5cm) from eight catchments in the European Arctic (Finland, Norway and Russia). *Environmental Pollution*, 96:261–274, 1997.

Niskavaara H., Kontas E. and Reimann C. Au, Pd and Pt in moss and the O-, B- and C-horizon of podzol samples from a 188,000 km² area in the European Arctic. *Geochemistry, Exploration, Environment, Analysis*, 4:143–159, 2004.

O'Connor P.J. and Reimann C. Multielement regional geochemical reconnaissance as an aid to target selection in Irish Caledonian terrains. *Journal of Geochemical Exploration*, 47:63–89, 1993.

O'Connor P.J., Reimann C. and Kürzl H. An application of exploratory data analysis techniques to stream sediment surveys for gold and associated elements in County Donegal, Ireland. In D.R. MacDonald and K.A. Mills, editors, *Prospecting in Areas of Glaciated Terrain 1988*, pages 449–467. The Canadian Institute of Mining and Metallurgy, 1988.

Ontario Ministry of Environment and Energy. *Ontario Typical Range of Chemical Parameters in Soil, Vegetation, Moss Bags and Snow.* Ontario Ministry of Environment and Energy, Toronto, Ontario, 1993. 263 pp.

Ottesen R.T., Bogen J., Bølviken B., Volden T. and Haugland T. *Geokjemisk Atlas for Norge. Del 1: Kjemisk samsettning av flomsedimenter. (Geochemical Atlas of Norway. Part 1: Chemical Composition of Overbank Sediments.)* Geological Survey of Norway special publication, Trondheim, Norway, 2000.(In Norwegian with English summary) 140 pp.

Pearce T.H. A contribution to the theory of variation diagrams. *Contributions to Mineralogy and Petrology*, 19:142–157, 1968.

Pison G., Rousseeuw P.J., Filzmoser P. and Croux C. Robust factor analysis. *Journal of Multivariate Analysis*, 84:145–172, 2003.

Räisänen M.L., Kashulina G. and Bogatyrev I. Mobility and retention of heavy metals, arsenic and sulphur in podzols at eight locations in northern Finland and Norway and the western half of the Kola Peninsula. *Journal of Geochemical Exploration*, 59:175–195, 1997.

Ramsey M.H. Measurement uncertainty arising from sampling: implications for the objectives of geo-analysis. *Analyst*, 122:1255–1260, 1997.

Ramsey M.H. Appropriate rather than representative sampling, based on acceptable levels of uncertainty. *Accreditation and Quality Assurance: Journal for Quality, Comparability and Reliability in Chemical Measurement*, 7:274–280, 2002.

Ramsey M.H. Sampling the environment: 12 key questions that need answers. *Geostandards and Geoanalytical Research*, 28:251–261, 2004a.

Ramsey M.H. When is sampling part of the measurement process? *Accreditation and Quality Assurance: Journal for Quality, Comparability and Reliability in Chemical Measurement*, 9:727–728, 2004b.

Rank G., Kardel K., Pälchen W. and Weidensdörfer H. *Bodenatlas des Freistaates Sachsen, Teil 3, Bodenmessprogramm*. Freistaat Sachsen, Landesamt für Umwelt und Geologie, Freiberg, Germany, 1999. 119 pp.

Rapant S., Vrana K. and Bodiš D. *Geochemical Atlas of Slovakia: Groundwater*. Ministry of Environment of the Slovak Republic and Geological Survey of Slovakia, Bratislava, Slovakia, 1996. 127 pp.

Reimann C. Untersuchungen zur regionalen Schwermetallbelastung in einem Waldgebiet der Steiermark. In *Umweltwissenschaftliche Fachtage – Informationsverarbeitung für den Umweltschutz*, pages 39–50. Forschungsgesellschaft Joanneum, Graz, Austria, 1989.

Reimann C. Geochemical mapping – technique or art? *Geochemistry, Exploration, Environment, Analysis*, 5:359–370, 2005.

Reimann C. and de Caritat P. *Chemical Elements in the Environment: Factsheets for the Geochemist and Environmental Scientist*. Springer Verlag, Berlin, Germany, 1998. 398 pp.

Reimann C. and de Caritat P. Intrinsic flaws of element enrichment factors (EFs) in environmental geochemistry. *Environmental Science & Technology*, 34:5084–5091, 2000.

Reimann C. and de Caritat P. Distinguishing between natural and anthropogenic sources for elements in the environment: regional geochemical surveys versus enrichment factors. *The Science of the Total Environment*, 337:91–107, 2005.

Reimann C. and Filzmoser P. Normal and lognormal data distribution in geochemistry: death of a myth. Consequences for the statistical treatment of geochemical and environmental data. *Environmental Geology*, 39:1001–1014, 2000.

Reimann C. and Garrett R.G. Geochemical background – concept and reality. *The Science of the Total Environment*, 350:12–27, 2005.

Reimann C. and Melezhik V. Metallogenic provinces, geochemical provinces and regional geology – what causes large-scale patterns in low-density geochemical maps of the C-horizon of podzols in Arctic Europe? *Applied Geochemistry*, 16:963–984, 2001.

Reimann C., Niskavaara H., de Caritat P., Finne T.E., Äyräs M. and Chekushin V. Regional variation of snowpack chemistry from the surrounding of Nikel and Zapoljarnj, Russia, northern Finland and Norway. *The Science of the Total Environment*, 182:147–158, 1996.

Reimann C., Boyd R., de Caritat P., Halleraker J.H., Kashulina G., Niskavaara H. and Bogatyrev I. Topsoil (0–5 cm) composition in eight arctic catchments in northern Europe (Finland, Norway and Russia). *Environmental Pollution*, 95:45–56, 1997a.

Reimann C., de Caritat P., Halleraker J.H., Finne T.E., Kashulina G., Bogatyrev I., Chekushin V., Pavlov V., Äyräs M. and Niskavaara H. Regional atmospheric deposition patterns of Ag, As, Bi, Cd, Hg, Mo, Sb and Tl in a 188,000 km² area in the European Arctic as displayed by terrestrial moss samples – long range atmospheric transport versus local impact. *Atmospheric Environment*, 31:3887–3901, 1997b.

Reimann C., de Caritat P., Halleraker J.H., Volden T., Äyräs M., Niskavaara H., Chekushin V.A. and Pavlov V.A. Rainwater composition in eight arctic catchments of northern Europe (Finland, Norway and Russia). *Atmospheric Environment*, 31:159–170, 1997c.

Reimann C., Äyräs M., Chekushin V., Bogatyrev I., Boyd R., de Caritat P., Dutter R., Finne T.E., Halleraker J.H., Jæger Ø., Kashulina G., Lehto O., Niskavaara H., Pavlov V., Räisänen M.L., Strand T.T. and Volden T. *Environmental Geochemical Atlas of the Central Barents Region*. NGU-GTK-CKE Special Publication, Geological Survey of Norway, Trondheim, Norway, 1998. 745 pp.

Reimann C., de Caritat P., Niskavaara H., Finne T.E., Kashulina G. and Pavlov V.A. Comparison of elemental contents in O- and C-horizon soils from the surroundings of Nikel, Kola Peninsula, using different grain size fractions and extractions. *Geoderma*, 84:65–87, 1998b.

Reimann C., Banks D., Bogatyrev I., de Caritat P., Kashulina G. and Niskavaara H. Lake water geochemistry on the western Kola Peninsula, north-west Russia. *Applied Geochemistry*, 14:787–805, 1999a.

Reimann C., Halleraker J.H., Kashulina G. and Bogatyrev I. Comparison of plant and precipitation chemistry in catchments with different levels of pollution on the Kola Peninsula, Russia. *The Science of the Total Environment*, 243-244:169–191, 1999b.

Reimann C., Banks D. and de Caritat P. Impacts of airborne contamination on regional soil and water quality: the Kola Peninsula, Russia. *Environmental Science and Technology*, 34:2727–2732, 2000a.

Reimann C., Banks D. and Kashulina G. Processes influencing the chemical composition of the O-horizon of podzols along a 500 km north-south profile from the coast of the Barents Sea to the Arctic Circle. *Geoderma*, 95:113–139, 2000b.

Reimann C., Kashulina G., de Caritat P. and Niskavaara H. Multi-element, multi-medium regional geochemistry in the European Arctic: element concentration, variation and correlation. *Applied Geochemistry*, 16:759–780, 2001a.

Reimann C., Niskavaara H., Kashulina G., Filzmoser P., Boyd R., Volden T., Tomilina O. and Bogatyrev I. Critical remarks on the use of terrestrial moss (*Hylocomium splendens and Pleurozium schreberi*) for monitoring of airborne pollution. *Environmental Pollution*, 113:41–57, 2001b.

Reimann C., Filzmoser P. and Garrett R.G. Factor analysis applied to regional geochemical data: problems and possibilities. *Applied Geochemistry*, 17:185–206, 2002.

Reimann C., Siewers U., Tarvainen T., Bityukova L., Eriksson J., Gilucis A., Gregorauskiene V., Lukashev V.K., Matinian N.N. and Pasieczna A. *Agricultural Soils in Northern Europe: A Geochemical Atlas*. Geologisches Jahrbuch, Sonderhefte, Reihe D, Heft SD 5. Schweizerbart'sche Verlagsbuchhandlung, Stuttgart, 2003. 279 pp.

Reimann C., Garrett R.G. and Filzmoser P. Background and threshold – critical comparison of methods of determination. *The Science of the Total Environment*, 346:1–16, 2005.

Reimann C., Arnoldussen A., Englmaier P., Filzmoser P., Finne T.E., Garrett R.G., Koller F. and Nordgulen Ø. Element concentrations and variations along a 120 km long transect in south Norway – anthropogenic vs. geogenic vs. biogenic element sources and cycles. *Applied Geochemistry*, 22:851–871, 2007.

Rock N.M.S. *Numerical Geology*. Lecture Notes in Earth Sciences 18. Springer Verlag, Berlin, 1988. 427 pp.

Rollinson H. *Using Geochemical Data: Evaluation, Presentation, Interpretation*. Longman Scientific and Technical, Essex, 1993. 352 pp.

Rose A.W., Hawkes H.E. and Webb J.S. *Geochemistry in Mineral Exploration*. Academic Press, London, 2nd edition, 1979. 657 pp.

Rousseeuw P.J. Least median of squares regression. *Journal of the American Statistical Association*, 79:871–880, 1984.

Rousseeuw P.J. and Leroy A.M. *Robust Regression and Outlier Detection*. John Wiley & Sons, Inc., New York, 1987. 329 pp.

Rousseeuw P.J. and Van Driessen K. A fast algorithm for the Minimum Covariance Determinant estimator. *Technometrics*, 41:212–223, 1999.

Rousseeuw P.J. and Van Driessen K. Computing LTS regression for large data sets. *Estadistica*, 54:163–190, 2002.

Rühling Å., editor. *Atmospheric Heavy Metal Deposition in Europe – Estimation Based on Moss Analysis*. Nord 1994:9, Nordic Council of Ministers, Copenhagen, 1994. 53 pp.

Rühling Å. and Steinnes E., editors. *Atmospheric Heavy Metal Deposition in Europe 1995–1996*. Nord 1998:15, Nordic Council of Ministers, Copenhagen, 1998. 66 pp.

Rühling Å. and Tyler G. An ecological approach to the lead problem. *Botaniske Notiser*, pages 121–321, 1968.

Rühling Å. and Tyler G. Heavy metal deposition in Scandinavia. *Water, Air and Soil Pollution*, 2:445–455, 1973.

Rühling Å., Steinnes E. and Berg T. *Atmospheric Heavy Metal Deposition in Europe 1995*. Nord 1996:37, Nordic Council of Ministers, Copenhagen, 1996. 46 pp.

Salminen R., Chekushin V., Tenhola M., Bogatyrev I., Glavatskikh S.P., Fedotova E., Gregorauskiene V., Kashulina G., Niskavaara H., Polischuok A., Rissanen K., Selenok L., Tomilina O. and Zhdanova L. *Geochemical Atlas of the Eastern Barents Region*. Elsevier, Amsterdam, The Netherlands, 2004. 548 pp.

Salminen R., Batista M.J., Bidovec M., Demetriades A., De Vivo B., De Vos W., Duris M., Gilucis A., Gregorauskiene V., Halamic J., Heitzmann P., Lima A., Jordan G., Klaver G., Klein P., Lis J., Locutura J., Marsina K., Mazreku A., O'Connor P.J., Olsson S.Å., Ottesen R.-T., Petersell V., Plant J.A., Reeder S., Salpeteur I., Sandström H., Siewers U., Steenfelt A. and Tarvainen T. *Geochemical Atlas of Europe. Part 1 – Background Information, Methodology and Maps*. Geological Survey of Finland, Espoo, Finland, 2005. 526 pp.

Sammon J.W. Jr. A non-linear mapping for data structure analysis. *Transactions IEEE*, C-18:401–409, 1969.

Scheffé H. *The Analysis of Variance*. John Wiley & Sons, Inc., New York, 1959. 477 pp.

Scheffé H. *The Analysis of Variance*. Wiley-Interscience, New York, 1999. 477 pp.

Scott D.W. *Multivariate Density Estimation. Theory, Practice and Visualization*. John Wiley & Sons, Inc., New York, 1992. 317 pp.

Seber G.A.F. *Multivariate Observations*. John Wiley & Sons, Inc., New York, 1984. 686 pp.

Seber G.A.F. and Wild C.J. *Nonlinear Regression*. John Wiley & Sons, Inc., New York, 2003. 792 pp.

Shacklette H.T. and Boerngen J.G. *Element Concentrations in Soils and Other Surficial Materials of the Conterminous United States*. Geological U.S. Survey Professional Paper 1270, 1984. 105 pp.

Shapiro S.S. and Wilk M.B. An analysis of variance test for normality. *Biometrika*, 52:591–611, 1965.

Siewers U. and Herpin U. *Schwermetalleinträge in Deutschland. Moos-Monitoring 1995/96*. Geologisches Jahrbuch, Sonderhefte, Reihe D, Heft SD 2. Schweizerbarth'sche Verlagsbuchhandlung, Stuttgart, 1998. 199 pp.

Siewers U., Herpin U. and Strassberg S. *Schwermetalleinträge in Deutschland. Moos-Monitoring 1995/96 – Teil 2*. Geologisches Jahrbuch, Sonderhefte, Reihe D, Heft SD 3. Schweizerbarth'sche Verlagsbuchhandlung, Stuttgart, 2000. 103 pp.

Sinclair A.J. Selection of threshold values in geochemical data using probability graphs. *Journal of Geochemical Exploration*, 3:129–149, 1974.

Sinclair A.J. *Applications of Probability Graphs in Mineral Exploration*. Special Volume 4. Association of Exploration Geochemists Special, Toronto, Ontario, 1976. 95 pp.

Skjelkvåle B.L., Andersen T., Fjeld E., Mannio J., Wilander A., Johansson K., Jensen J.P. and Moiseenko T. Heavy metal survey in nordic lakes; concentrations, geographical patterns and relation to critical limits. *Ambio*, 30:2–10, 2001a.

Skjelkvåle B.L., Henriksen A., Jónsson G.S., Mannio J., Wilander A., Jensen J.P., Fjeld E. and Lien L. *Chemistry of Lakes in the Nordic Region–Denmark, Finland with Åland, Iceland, Norway with Svalbard and Bear Island and Sweden*. Norwegian Institute of Water research (NIVA) Report, Acid Rain Research Report 5/2001, 2001b. 39 pp.

Slocum T.A. *Thematic Cartography and Geographic Visualization*. Prentice Hall, Upper Saddle River, New Jersey, USA, 1999. 224 pp.

Smirnov N.V. Table for estimating the goodness of fit of empirical distributions. *Annals of Mathematical Statistics*, 19:279–281, 1948.

Spearman C.E. "General Intelligence" objectively determined and measured. *American Journal of Psychology*, 15:201–293, 1904.

Stanley C.R. NUGGET; PC software to calculate parameters for samples and elements affected by the nugget effect. *Exploration and Mining Geology*, 7:139–147, 1998.

Student (Gosset W.S.). The probable error of a mean. *Biometrika*, 6:1–25, 1908.

Taylor S.R. and McClennan S.M. *The Continental Crust: Its Composition and Evolution*. Blackwell, Oxford, UK, 1985. 312 pp.

Taylor S.R. and McLennan S.M. The geochemical evolution of the continental crust. *Reviews of Geophysics*, 33:241–265, 1995.

Templ M., Filzmoser P. and Reimann C. Cluster analysis applied to regional geochemical data: problems and possibilities. Technical Report CS-2006-5, Vienna University of Technology, Vienna, Austria, 2006. 40 pp.

Tenenhaus M. *La Regression PLS. Theorie et Practique*. Editions Technip, Paris, 1998. (In French), 254 pp.

Tennant C.B. and White M.L. Study of the distribution of some geochemical data. *Economic Geology*, 54:1281–1290, 1959.

Thalmann F., Schermann O., Schroll E. and Hausberger G. *Geochemischer Atlas der Republik Österreich 1:1,000,000*. Geologische Bundesanstalt Wien, Vienna, Austria, 1989. Maps and 141 pp.

Thompson M. and Howarth S.R. A new approach to the estimation of analytical precision. *Journal of Geochemical Exploration*, 9:23–30, 1978.

Tukey J.W. *Exploratory Data Analysis*. Addison-Wesley, Reading, Massachussetts, USA, 1977. 506 pp.

Tyler G. Moss analysis–a method for surveying heavy metal deposition. In H.M. Englund and W.T. Berry, editors, *Proceedings of the Second International Clean Air Congress*, pages 129–132. Academic Press, New York, 1970.

van Helvoort P.J., Filzmoser P. and van Gaans P.F.M. Sequential factor analysis as a new approach to multivariate analysis of heterogeneous geochemical datasets: an application to a bulk chemical characterization of fluvial deposits (Rhine-Meuse delta, The Netherlands). *Applied Geochemistry*, 20:2233–2251, 2005.

Van Huffel S. and Vandewalle J. *The Total Least Squares Problem: Computational Aspects and Analysis*. SIAM, Philadelphia, USA, 1991. 300 pp.

Vandervieren E. and Hubert M. An adjusted boxplot for skewed distributions. In J. Antoch, editor, *Proceedings in Computational Statistics*, pages 1933–1940. Springer-Verlag, Heidelberg, 2004.

Velleman P.F. and Hoaglin D.C. *Applications, Basics and Computing of Exploratory Data Analysis*. Duxbury Press, Boston, 1981. 354 pp.

Venables W.N. and Ripley B. *Modern Applied Statistics with S*. Springer, New York, 4th edition, 2002. 495 pp.

Vistelius A.B. The skew frequency distributions and the fundamental law of the geochemical processes. *The Journal of Geology*, 68:1–22, 1960.

Volden T., Reimann C., Pavlov V.A., de Caritat P. and Äyräs M. Overbank sediments from the surroundings of the Russian nickel mining and smelting industry on the Kola Peninsula. *Environmental Geology*, 32:175–185, 1997.

Weaver T.A., Broxton D.E., Bolivar S.L. and Freeman S.H. *The Geochemical Atlas of Alaska*. GJBX-32(83). Geochemical Group, Earth and Space Sciences Division, Los Alamos National Laboratory, US Department of Energy, 1983. 55 pp.

Webb J.S. and Howarth R.J. Regional geochemical mapping. *Philosophical Transactions of the Royal Society of London. Series B, Biological Sciences*, 288:81–93, 1979.

Webb J.S., Fortescue J.A.C., Nichol I. and Tooms J.S. *Regional Geochemical Reconnaissance in the Namwala Concession Area, Zambia*. Technical Communication 47. Imperial College of Science and Technology, London, Geochemical Prospecting Research Centre, 1964. 42 pp.

Webb J.S., Thornton I., Thompson M., Howarth R.J. and Lowenstein P.L. *The Wolfson Geochemical Atlas of England and Wales*. Oxford University Press, Oxford, UK, 1978. 74 pp.

Wedepohl K.H. The composition of the continental crust. *Geochimica et Cosmochimica Acta*, 59:1217–1232, 1995.

Weissberg S. *Applied Linear Regression*. John Wiley & Sons, Inc., New York, 1980. 283 pp.

Wilcoxon F. Individual comparisons by ranking methods. *Biometrics*, 1:80–83, 1945.

Wold H. Soft modeling by latent variables: the non-linear iterative partial least squares approach. In J. Gani, editor, *Perspectives in Probability and Statistics. Papers in Honour of Bartlett M.S.*, pages 117–142. Academic Press, London, 1975.

Woronow A. and Butler J.C. Complete subcompositional independence testing of closed arrays. *Computers & Geosciences*, 12:267–279, 1986.

Yeung K. and Ruzzo W. An empirical study on principal component analysis for clustering gene expression data. *Bioinformatics*, 17:763–774, 2001.

Zoller W.H., Gladney E.S. and Duce R.A. Atmospheric concentrations and sources of trace metals at the South Pole. *Science*, 183:199–201, 1974.

Index